Law, Economics, and Game Theory

Law, Economics, and Game Theory

John Cirace

LEXINGTON BOOKS
Lanham • Boulder • New York • London

Published by Lexington Books
An imprint of The Rowman & Littlefield Publishing Group, Inc.
4501 Forbes Boulevard, Suite 200, Lanham, Maryland 20706
www.rowman.com

Unit A, Whitacre Mews, 26-34 Stannary Street, London SE11 4AB

British Library Cataloguing in Publication Information Available

Library of Congress Cataloging-in-Publication Data Is Available

ISBN 978-1-4985-4908-0 (cloth: alk. paper)
ISBN 978-1-4985-4909-7 (electronic)

∞™ The paper used in this publication meets the minimum requirements of American National Standard for Information Sciences—Permanence of Paper for Printed Library Materials, ANSI/NISO Z39.48-1992.

Printed in the United States of America

Contents

Acknowledgments

I wish to thank the many students in my Law and Economics classes at Brooklyn Law School. They provided a great sounding board for both the substance and pedagogical presentation of the ideas in my book. By the time these second- and third-year law students enrolled in my class they had been imbued with lawyers' distrust for theory unmoored from and unrelated to facts; they constantly made me clarify and justify the usefulness of the theory I was presenting.

I also wish to thank an anonymous reviewer for correcting several errors and making many suggestions as to what material needed clarifying and the direction such changes should take. This is constructive criticism at its best. I owe a great debt to this reviewer.

My colleague, Jaspal Chatha, provided crucial input on chapters 1 and 10.

My wife, Evelyn Ackerman, gave me constant encouragement and moral support, tirelessly read and reread my manuscript, and was an invaluable proof reader.

Introduction

Two anecdotes give focus to this book. I gave a book on law and economics as a present to my late brother-in-law, a lawyer. Two weeks later, he returned the book, my present, saying, "Something is missing." This remark was typical of my brother-in-law: succinct and dismissive. However, after thinking about his response, I realized it was also perceptive. Second, a law professor once said to me, "The trouble with law and economics is that it's insightful but often not useful." What is missing in law and economics, and why is it insightful but often not useful to lawyers and judges when they must decide specific cases?

In order to fill in what is missing in law and economics and improve its usefulness by having theory hew closer to facts of specific cases, I propose to look at law and economics through lenses of legal logic and game theory as well as economic logic. As depicted in the figures below, law, economics, and game theory each has its own formal logic.

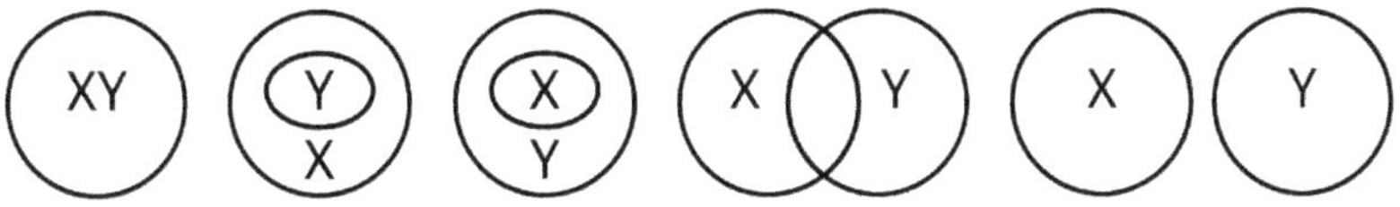

Figure I.1 **Law.** *Source*: Created by the author.

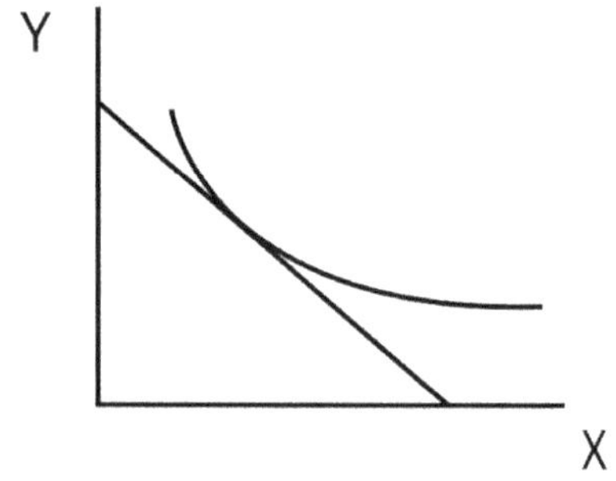

Figure I.2 Economics. *Source*: Created by the author.

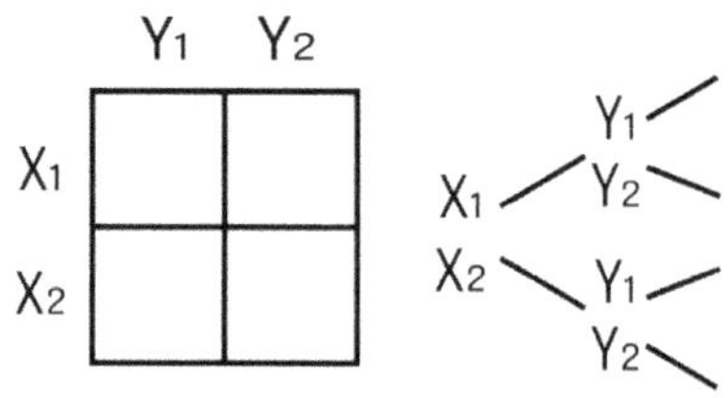

Figure I.3 Game Theory. *Source*: Created by the author.

These logics and their policy prescriptions are sometimes consistent but often conflict. When they conflict, the question is, which logic is more conducive to social welfare?

Consider three relationships: law and economics, economics and game theory, and game theory and law; what can each add to the interdisciplinary field of law, economics, and game theory?

LAW AND ECONOMICS

A major strength of economic analysis of law is that economists teach lawyers that economic principles cut across, explain, and integrate seemingly different legal subjects, such as contracts, torts, and property. A corresponding weakness is that economists seem to think that lawyers have nothing to teach economists. The views of lawyers, when considered at all, are often straw men. I believe lawyers have three lessons to teach economists.

First, legal rationality is a separate and distinct rational decision process that can be formalized by behavioral rules that are parallel to and comparable with the behavioral rules of economic rationality. For clarity of thought, explaining how economic rationality and the missing model of legal rationality relate to each other is important to economists, lawyers, and judges. Probably, few lawyers and judges are consciously aware that they employ two different rational decision processes: economic rationality and legal rationality. For example, when judges balance competing interests via marginal trade-offs, as in balancing civil rights versus national security, they are using economic reasoning. But, when judges base their reasoning on the recognition of similarity and difference (the ability to distinguish cases and categories), as in whether people in different legal categories are similarly situated with respect to equal protection of the laws, they are using legal reasoning. Sometimes economic rationality and legal rationality are consistent (*isomorphic*), sometimes inconsistent. When they conflict, which logic should trump the other? In Part I, "thinking like an economist" is compared with "thinking like a lawyer."

Second, for many economists, efficiency is usually the sole goal of economic analysis of law; however, law, in addition to efficiency, is also concerned with other goals, such as due process, equal protection of the laws, and distributional equity. Moreover, the goal of horizontal equity—like cases should be decided alike—is embedded in law's logic. Although economists theorize constantly about maximizing various things (e.g., utility or satisfaction) subject to constraints (e.g., budget constraints), when it comes to law and economics, many seem blind to the possibility and necessity of constraining efficiency by law's goals.[1]

For example, consider the rule that evidence must be excluded at trial if seized in violation of the Fourth Amendment's prohibition against unreasonable search and seizure. Is this rule merely "a judicially created remedy . . . grounded in deterrence," the applicability of which "must be resolved by weighing the costs and benefits of preventing the use" of such evidence,[2] or should a cost-benefit analysis be constrained by some minimum level of right of "the people to be secure in their persons, houses, papers, and effects"?[3] Should we decide whether or not to torture suspected terrorists solely by cost-benefit analysis?[4] If the Bill of Rights conflicts with economic efficiency, is it a nullity? Chapter 4 considers *equal protection* and *lexical* constraints on efficiency.

Third, economics and law have different, sometimes conflicting methodologies for determining the truth or falsity of economic theories, theorems, and empirical propositions. In economics, if a theory or empirical proposition explains, say, 80 percent of cases to which it is applicable, it is extremely insightful; it can also be said to be "true." If a particular example or case conflicts with a theory or empirical proposition, it is said to be an "outlier" or "anomaly," which can usually be ignored because it does not affect whether the theory or empirical proposition is correct in many, or even the overwhelming majority, of other cases.

In law, the truth or falsity of an economic theory is tested by whether it is consistent with and can explain decisions in hard cases. A *hard* case is a case that rises to the top of the appellate court pyramid because it involves conflicting principles or is interstitial with respect to those principles.[5] Hard cases rise to top appellate courts because they are difficult to decide; they are often outliers or anomalies. Hard cases are distinguished from easy cases, which are defined as cases that fall clearly within settled and uncontroversial rules of law. Easy cases are usually settled before trial, dismissed by judges without trial, or if tried, are rarely appealed by the losing party. Law professors impress on first year law students that even slight changes of facts in cases often result in opposite outcomes and that general rules of law must be tested by their ability to explain hard cases, actual and hypothetical. Had I not gone to law school, I never would have met a fact!

If an economic theory or theorem conflicts with the holding in a hard case that the legal community generally agrees to have been decided correctly, the theory cannot be used to decide (is not dispositive of) specific cases, unless there is a principled way to distinguish specific cases for which it is true, that is, an accurate guide to public policy, from cases for which it is false. If there is no principled way of distinguishing cases for which the theory is true from those for which it is false, the economic theory may be insightful in identifying a general tendency but it is not useful to judges in specific cases. In Part III, the seminal contributions of Rottenberg, Coase, Calabresi, and Posner are evaluated by both the general case methodology of economics and the hard case methodology of law.

ECONOMICS AND GAME THEORY

Because law and economics books tend to focus on one method—economic analysis of markets and bargaining—and one form of social control—the decisions of judges in common law cases—law and economics books have been mostly limited to economic analysis of contracts, torts, property, criminal law, and suit and settlement. The narrow focus on markets and judge-made laws creates a serious lacuna: there is usually no discussion of many areas of law that require cooperative action, such as is required to provide economic infrastructure, control public "bad" type externalities, and enact legislation. Consider recent US history: there is virulent disagreement about the role of government as regulator of the economy, for example, disagreement on whether regulations instituted after the financial meltdown of 2007–2008 are efficient and the role of government as provider of economic infrastructure and social welfare, intervener with respect to income inequality, controller of pollution, global warming and climate change, and arbiter of intergenerational equity. The economic analysis of common (judge-made) law paradigm, which ignores these divisive and important disputes, is much too narrow and limiting to support the interdisciplinary title "law and economics."

Game theory provides a bridge between markets and the missing discussion of cooperative action in law and economics. How? In the context of game theory, competitive markets are an example (subset) of game theory's Prisoners' Dilemma. Competitive markets are one of the few Prisoners' Dilemmas in which non-cooperative behavior makes a positive contribution to social welfare: although markets are good for buyers, who get the benefit of competitive prices, they are bad for sellers (Prisoners), who would be better off if they could collude (cooperate) or merge to form monopolies so as to charge higher prices.

The verbal paradigm of the Prisoners' Dilemma is based on an age-old interrogation technique. Two suspects, who are carrying concealed weapons, are arrested at the entrance to a bank before they draw their guns and are charged with two crimes: conspiring to rob the bank and carrying concealed weapons. The prisoners are interrogated by police in different rooms so that they cannot communicate with each other. If one confesses to conspiracy and agrees to testify against the other at trial and the other does not confess, the one who testifies for the prosecution goes free and the other, "the sucker," is sentenced to ten years in jail; if both confess and agree to testify against each other, both get reduced sentences of five years; and, if neither confesses, each is sentenced to one year for illegal possession of a weapon. Both prisoners are told that this is a one-time offer to confess and that the same offer with the same consequences is being made to the other prisoner.

No matter which strategy the other prisoner adopts, a *self-interested* prisoner is always better off confessing. Even though cooperation, not confessing, is in their *joint* self-interest, because their combined sentences (−1, −1) are lower than when they both confess (−5, −5), not confessing is unstable because one prisoner is always tempted to double cross the other, confess, and go free [(0, −10) or (−10, 0)], coupled with the fear that the prisoner who does not confess will get a ten-year sentence in jail. This is the Prisoners' Dilemma: a choice based on individual self-interest results in an outcome in which both are worse off than if they had adopted a cooperative strategy that is in their joint interest; but, adopting the cooperative strategy exposes a prisoner to a potential double cross by the other self-interested prisoner.

Game theory explains when the conflict between self-interested individuals and cooperation is or is not in the public interest. Why is it that when considering the benefits from competitive markets, the inability of self-interested sellers to cooperate, that is, the inability of sellers to form cartels and raise prices, is often cited as the main reason why markets produce benefits to consumers, but that in the political sphere, the difficulty self-interested legislators have in cooperating, so as to produce social infrastructure such as bridges, roads, medical care, and control negative externalities, such as pollution and global warming, as typified by the current "grid lock" in the US Congress, is decried? Game theory's Prisoners' Dilemma and Repeated Prisoners' Dilemma explain conflict between individual self-interested behavior and cooperation both in economic markets and in legislative bodies and demonstrate the need for social regulation in addition to free markets and judicial decisions in common law cases.

Littering, pollution, global warming, political interest group lobbying, providing social infrastructure, voting, and even self-sacrifice in battle can be modeled as Prisoners' Dilemmas, as can Adam Smith's "invisible hand" of competition. The fact that competitive markets are an example (subset) of the Prisoners' Dilemma exposes the fallacy of composition[6] behind the belief of those who argue that *if* self-interested behavior is socially beneficial in competitive markets, *then* self-interested behavior must be beneficial for society in all contexts, many of which are Prisoners' Dilemmas. In Part II, the Prisoners' Dilemma is used to explain the need for economic infrastructure and control of public "bad" type externalities such as pollution and global warming; it is also relevant to competitive markets and the economics of intergenerational equity.

In Prisoners' Dilemmas, short-term, individual self-interest is often antithetical to social welfare. Exclusive concern with one example of the Prisoners' Dilemma, competitive markets, results in a much exaggerated view of the efficacy and efficiency of unconstrained individual self-interest. As authors of a book on game theory said, the span in which individual self-interest is beneficial to society is extremely narrow.[7]

GAME THEORY AND LAW

In competitive markets, individual self-interest is controlled because all parties have alternative persons with whom they can deal. If a seller attempts to exploit a buyer, she can always say, "You're not the only seller of widgets." However, lawsuits usually involve one-on-one litigation or bargaining between two parties, not the myriad participants in markets.

Parties to litigation have locked horns; they cannot control an adversary's self-interest by threatening to deal with other, more tractable litigants. Analyzing one-on-one adversarial lawsuits from the perspective of competitive markets can be insightful but is often not useful because the assumption of self-interest constrained by markets does not carry over to legal disputes involving one-on-one bargaining in which the law gives one party superior bargaining power (Coase theorems, chapter 12). In order to import the efficiency properties of competitive markets into dispute resolution involving one-on-one bargaining where there is unequal bargaining power, cooperative rather than self-interested behavior must be implicitly assumed; but, assuming cooperative behavior is inconsistent with and contradicts the individual self-interest assumption of economics.[8]

Of the many law and economics writers, Judge Posner may be the one most aware of this problem. He uses the theory of bilateral monopoly (single buyer vs. single seller, for example, a small town in which wages are set by bargaining between the only employer and only union to which all workers belong) to show that where legal disputes involve one-on-one bargaining in which one party has the upper hand because legal institutions, rights, or rules favor it, there are many common law doctrines, such as mitigation of damages and the doctrine of waste, which are necessary to control self-interested behavior of the legally favored party (chapter 14).[9] These common law doctrines would be unnecessary if the legally favored party could be assumed to bargain cooperatively and fairly rather than self-interestedly. Competitive markets versus bilateral monopoly is also an example of the difference between the general case methodology of economics and the hard case methodology of law, which was discussed above.

Bilateral monopoly identifies the strategic problem that is inherent in one-on-one disputes and bargaining where one party has more bargaining power than the other. Game theory allows an analyst to add institutional and legal structure to bilateral disputes; models can be constructed that incorporate and predict the effect of different legal institutions, rights, and rules on the outcome of one-on-one bargaining.

Game theory has a natural framework, which is also used in this book. This natural framework is based upon two key variables which concern the temporal order of action and the state of players' information. As to action, do players act simultaneously or sequentially? As to information, do players have complete information or incomplete information? Since each of the two variables has two states, there are 2 × 2 or 4 possible combinations or games, which are game theory's natural framework: 1) simultaneous games of complete information, 2) sequential games of complete information, 3) simultaneous games of incomplete information, and 4) sequential games of incomplete information.[10]

In Part V, game theory's four-model framework is used in the analysis of the law and economics of civil obligation, which consists of torts (negligence), contracts, and unjust enrichment. In torts (negligence), where prevention of injury requires bilateral care (e.g., both automobiles and pedestrians must take care to prevent collisions), there are four negligence models: the classic negligence model, last clear chance model, defensive care model, and foreseeable use and misuse model. There are also four contracts and unjust enrichment models: the strict rules, competitive contract model; unilateral contract model; discretionary standards, imperfect competition model; and unjust enrichment model (promissory estoppel and quasi contract).[11]

Game theory's distinction between complete and incomplete information is also crucial in the two chapters on risk and insurance in Part IV. One chapter is concerned with risk and insurance under conditions of complete information, where the distribution of all possible risks is known. The other chapter is concerned with risk and insurance where information is incomplete and often asymmetric; questions of adverse selection, moral hazard, and principal-agent problem predominate.

Today, nearly two decades into the twenty-first century, when the view that unregulated markets and the decisions of common law judges are the only forms of social control necessary for both efficient and ethically "right" outcomes is being challenged, this book's expanded version of law, economics, and game theory is especially relevant (and possibly material).

NOTES

1. Douglas F. Greer, *Business, Government, and Society*, 3rd ed. (New York: Macmillan, 1993), 11.
2. *U.S. v. Leon*, 468 U.S. 897, 906–907 (1984).
3. 468 U.S. 943 (Justice Brennan dissenting).
4. Richard Posner, *Not A Suicide Pact: The Constitution in a Time of National Emergency* (New York: Oxford, 2006), 86–87.
5. Ronald Dworkin, *Law's Empire* (Cambridge, MA: Belknap Press, 1986), 106.
6. The error of assuming that what is true of a member of a group is true for the group as a whole. The Prisoners' Dilemma itself is an example of the fallacy of composition.
7. Avinash Dixit and Barry Nalebuff, *Thinking Strategically: The Competitive Edge in Business, Politics, and Every Day Life* (New York: W. W. Norton, 1991), 223.
8. Robert Cooter, "The Cost of Coase," *Journal of Legal Studies* 11 (1982): 17, 23.
9. Richard Posner, *Economic Analysis of Law*, 9th ed. (New York: Wolters Kluwer, 2014), 129–130.
10. Eric Rasmusen, *Games and Information: An Introduction to Game Theory*, 4th ed. (Cambridge, MA: Blackwell, 2006).
11. The unilateral contract and unjust enrichment models, which are suggested by the game theoretic framework, are not discussed.

Part I

WHEN ARE THINKING LIKE AN ECONOMIST AND THINKING LIKE A LAWYER CONSISTENT?

Part I consists of six chapters. Chapter 1 explains fundamental economic concepts and the two normative criteria, Pareto efficiency and Kaldor-Hicks criterion, which economists use to evaluate all economic decisions; these criteria have opposite strengths and weaknesses.

Chapter 2 demonstrates that rational economic behavior is defined by two simple behavioral rules and constraints on those rules. Economic rationality is completely ordered; that is, every economic choice is related to every other choice in a rigidly ordered way, which can be depicted in two or three special dimensions or mathematical vectors. Surprisingly, economic choices can even be rank ordered linearly.

Chapter 3 demonstrates that cases, both actual and hypothetical, in which judges balance the same marginal trade-offs can be totally ordered according to economic rationality.

Chapter 4 shows that courts often place restrictions on economic trade-offs. John Rawls's maximin principle, which limits or restricts economic inequality, and its relation to Pareto efficiency and the Kaldor-Hicks criterion is also explained.

In chapter 5, legal rationality (like economic rationality) is defined by two behavioral rules, which are constrained by the principle of precedent (stare decisis), which requires that like cases be decided alike. The process by which under-inclusive rules of law expand and over-inclusive rules of law shrink so that rules of law tend toward optimum level of generality is also explained.

Chapter 6 is concerned with the conditions under which economic rationality and legal rationality are consistent (isomorphic)? When they are inconsistent, the question is, which logic takes precedence or trumps the other?

Chapter 1

Pareto Efficiency and Kaldor-Hicks Criterion Compared

The purpose of this chapter is twofold: first, to explain and illustrate with examples, fundamental economic concepts—law of scarcity, opportunity cost, production possibilities frontier, law of increasing costs, and law of diminishing marginal returns; and second, to explain the relationship between these fundamental concepts and the two normative criteria that economists use to evaluate economic decisions. Pareto efficiency is the primary moral justification for a market economy; it is used to determine the circumstances under which voluntary actions by self-interested individuals in markets increase social welfare. Kaldor-Hicks criterion is used to determine when actions by courts or government, which explicitly or implicitly involve coercion, improve social welfare.

Pareto efficiency and Kaldor-Hicks criterion attempt to "maximize" a social good, such as income, wealth, or utility. Both criteria assume the income distribution of the status quo, whether fair or not, where the preferences of a rich person count more than the preferences of a poor person because one's preferences only count if one has the both the willingness and the *ability* to pay for goods and services. The two criteria differ in the way in which they account for direct and indirect benefits and costs of market transactions and government actions. In terms of social welfare, Pareto efficiency and Kaldor-Hicks criterion have opposite strengths and weaknesses, which are discussed in this chapter and summarized in the Conclusion.

LAW OF SCARCITY AND OPPORTUNITY COST

The fact that individuals must choose among competing goods or ends because resources are scarce relative to their wants is called the *law of scarcity*. If an individual's income is taken as given, he or she cannot spend more on clothes without either reducing expenditure on food, other products, services, or by reducing saving; that is, there are necessary trade-offs between *goods*. The law of scarcity is relevant even to an enormously wealthy individual who presumably can afford anything she wants because time is a scarce resource to her; she must allocate her scarce time among competing pleasures. With time taken into account, scarcity is a fact in every one's life.

The law of scarcity is also relevant to the choices an entire economy faces in the sense that society has less resources, products, and services than it desires because the supply of resources of land, labor, capital, entrepreneurial ability, and time are strictly limited; therefore, society cannot fulfill its apparently unlimited demand for products and services. For example, the United States is just beginning to deal with the necessity of choice due to scarcity in resources devoted to health care.

Because resources are scarce in relation to wants, choices must be made among a limited set of possibilities; a decision to have more of one thing means that we must give up some of another thing. In terms of the choices that society as a whole confronts (if we assume full employment), we cannot have more refrigerators unless we have fewer air conditioners, dishwashers, or other products which are alternative uses for the steel that could be used in refrigerator production. Given full employment, a society cannot increase military goods, unless it increases taxes or reduces the amount private goods or social programs such as Social Security, Medicare, welfare, etc. Trade-offs due to scarcity are implicit in every social decision.

The opportunity cost of any choice or decision is the foregone alternative that is not chosen, or "the road not taken." Choosing to become a lawyer usually means that one foregoes the opportunity to become a business person. Buying a luxury automobile often means saving and investing less of one's income. Choosing to have another child often has a cost in terms of personal career advancement and vacations.

A well paid football "quarterback" once said that if he had not played professional football he would have been "pumping gas" in his home town. In this example, in which the foregone alternative is a low paying job, economists would say that he has a low opportunity cost and that the difference between his low opportunity cost and his high professional football salary is *rent* due the scarce abilities of throwing a football and making the right decision while facing physical duress from opposing players, which abilities are highly valued by the market for athletes. If his talents were easily replicable by other players, that is, his talents were not scarce, his wage would be competed down to his opportunity cost. Conversely, when a highly paid business person chooses to enter public service at a substantial pay cut, the opportunity cost of that decision is very costly to her or him.

When one considers a multi-product economy rather than a simple choice between two alternatives, there is always more than one alternative to any particular decision; therefore, the *opportunity cost* of a decision to use resources in a given way is defined as the *next best alternative use* of the resources. The real opportunity cost of a product or service is not money but the alternative products or services that the resources could have produced. In the context of opportunity cost, money can be thought of as a claim that allows an individual to direct resource use rather than something of intrinsic worth. The opportunity cost concept means that choices are implicit in every economic decision.

Explicit and Implicit Opportunity Costs

Opportunity costs are often implicit or hidden as well as explicit. If an individual decides to attend a private college or law school full-time, out-of-pocket costs for tuition, books, and incidental costs are, say, $50,000 a year. These out-of-pocket costs represent explicit opportunity cost because that money could be spent on other products, services, or savings. However, there is also a hidden or implicit opportunity cost of attending school full-time. If a student did not attend college or business school, she could have worked full-time and earned a salary of at least $40,000 per year, which is an implicit opportunity cost. She has a choice between either paying $50,000 to attend school full-time or working full-time for a year and earning $40,000. Thus, her full opportunity cost of going to school for a year is the difference between a –$50,000 and a +$40,000, which equals a total opportunity cost of $90,000 per year.

Consider another example of latent or implicit opportunity cost: assume an individual, who has $100,000 to invest, starts his own business; assume further that at the end of one year in business he subtracts all costs of doing business from total sales revenue and determines that he now has a total of $140,000, or an accounting profit of $40,000. Has he been successful? Although he has an accounting profit that must be reported to the Internal Revenue Service as income, economists would say that he has suffered an economic loss. Why? Consider his opportunity cost or alternatives to opening his own business. Assume he could have invested the $100,000 in a safe, relatively liquid investment for a year at 5% interest rate and earned $5,000 in interest; in addition, he could have worked for someone else at $40,000 per year. His opportunity cost for opening his own business is $40,000 + $5,000 = $45,000. Thus, an economist would say that this businessman suffered an economic loss of $45,000 – $40,000 = –$5,000. The opportunity cost concept means that a normal return on one's capital, or *normal profit*, plus the return on one's labor are opportunity costs of doing business. In economic theory, true "profit" is the return over and above the opportunity cost of capital and labor.

Special Interest Groups and Opportunity Cost

All special interest groups, usually have at least one thing in common, they abhor the concept of opportunity cost. Whether it be those who advocate a strong national defense "to protect our way of life," the subsidization of court costs because "every person is entitled to her or his day in court," increased support for AIDS research in order to rid the world of this modern plague, increased spending on education so as to ensure equality of opportunity and to prepare this nation for the global competition in the twenty-first century, or increased spending on entitlements such as Social Security, Aid to Dependent Children, disability benefits, and health care, both as a "entitlement" and to ensure that children, the old, and infirm are protected by a social safety net, they all think of their interest as unique. Each special interest group attempts to portray its interest as being of overriding and special importance. For example, senior citizens don't want to hear it said that Social Security cost of living increases must be weighed against more spending on prenatal care for expectant mothers or spending on K–12 education. Advocates of more spending on AIDS, cancer, and heart research argue that there should not be a competition for funds among these groups; however, there is competition for medical research funds as well as competition with every other object of government spending. The opportunity cost concept means that all governmental expenditures compete with all other possible government expenditures as well as with tax cuts, since the opportunity cost for government expenditure is the taxation that is raised from individuals and corporations who pay for such expenditure.

Every social decision has an implicit opportunity cost. In the face of political groups who claim that their interest is unique and incomparable with other interests, economists employ the maxim "there is no such thing as a free lunch"[1] to remind society that honesty and consistency (rationality) require explicit consideration of opportunity cost of every decision in terms of all alternative resource uses. One of the implications of economic logic in general and in the opportunity cost concept in particular is that in economics *everything* has its price in terms of everything else that must be forgone. That is why it is sometimes said that "economists know the price of everything but the value of nothing."

However, there are many legal issues, for example, cases involving fundamental rights, about which many in the legal community are not willing to balance interests via opportunity cost trade-offs. For these individuals, fundamental rights "trump" other interests (see chapters 4–6).

Opportunity Cost Defined as Marginal Trade-Offs

Consider a simple example that relates the law of scarcity and opportunity cost to marginal trade-offs and choices. Suppose a survey is conducted and you are asked the following question: "Are you in favor of a strong national defense?" This question is put in terms of an all-or-nothing choice; either you prefer a strong national defense or prefer a weak or no national defense. Almost everyone will answer yes to this question because a person who answered no to this question could be considered unpatriotic. This is clearly a biased or "loaded" question.

To an economist, all-or-nothing questions usually result in inefficient choices and decisions. In the context of spending on national defense, the question should be couched in marginal or incremental choices: given the current amount of spending on national defense and other programs, for example, social programs, would you be in favor of shifting 1 billion dollars currently being spent on social programs to national defense? If the answer is yes, then the next question is, would you be in favor of shifting a second billion dollars from social programs to national defense? These questions are couched in terms of opportunity cost and marginal trade-offs. If the answer is yes again, then the person is asked about shifting a third billion dollars, and so on. As resources are shifted from social programs to national defense, a point will be reached at which the budget for social programs will have shrunk so much that it will be difficult for social programs to accomplish their goals (resources for social programs will be very scarce), while military spending will have expanded so much that the increased expenditure will be used to fund less than necessary or even wasteful expenditure. Not everyone will have the same stopping place. In economics, rational people can differ, just as in law, where reasonable people can differ. Determining how much to spend on national defense in the context of opportunity cost, the law of scarcity, and marginal trade-offs is part of the process of allocating government resources efficiently.

Sometimes society is presented with all-or-nothing choices: as examples, a court may have to decide whether or not to impose an injunction that will require a cattle feedlot to cease operation, as in *Carpenter v. Double R Cattle Co.* discussed below; or, the US Congress must decide whether or not to ratify international trade treaties (discussed below). But, the point of this section is that small or marginal choices are usually more efficient than all-or-nothing choices.[2]

PRODUCTION POSSIBILITIES FRONTIER

The marginal trade-offs that an entire economy faces are illustrated by the *production possibilities frontier*. In Figure 1.1, the two products that are used to illustrate the production possibility frontier, automobiles and grain, depict the marginal opportunity cost trade-offs that society confronts when it must choose between food and other goods and services. If all of society's resources are devoted to the production of grain (food products), there will be a maximum amount of grain that can be produced per year; exactly how much grain will be produced is determined by the resources available to the economy in question (land, labor, and capital) and technological efficiency with which they are used. If all resources are used to produce grain, so that no resources are used to produce automobiles, we assume that 5 billion bushels of grain will be produced and 0 automobiles; this combination of 5 million bushels of grain and 0 automobiles is labeled point F = (5, 0). At the other extreme, society could employ all its resources to produce automobiles and no grain (in which case individuals would have to live off of previous stores of grain). Again, because resources are scarce only some maximum number of automobiles can be produced, say 15 million automobiles. The combination is labeled point A = (0, 15).

Points A and F represent the extreme possibilities. In between the extremes, there are myriad intermediate possibilities. If we are willing to trade-off some grain, we can have more automobiles; if we are willing to trade-off still more grain, we can have

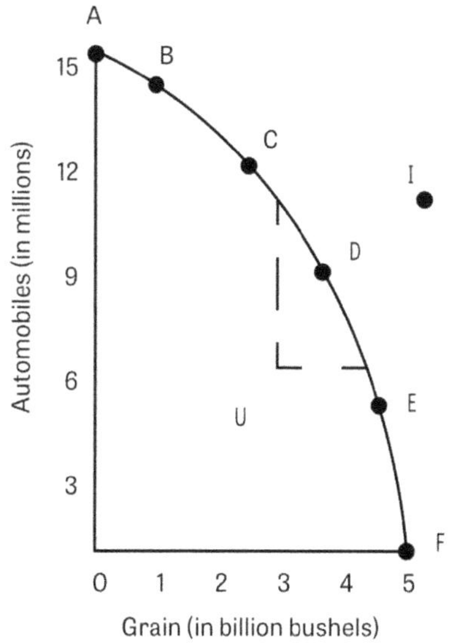

Figure 1.1 Production Possibilities Frontier. *Source*: Created by the author.

still more automobiles. Four of the possible intermediate combinations where there is some grain and some automobiles are represented by points B = (1, 14), C = (2, 12), D = (3, 9), and E = (4, 5). For example, at point C, society is producing the combination of 2 billion bushels of grain and 12 million automobiles. If we move from point C to point B, where society is producing the combination of 1 billion bushels of grain and 14 million automobiles, society has chosen to produce more automobiles and less grain; grain is transformed into automobiles, not physically, but by diverting resources from one use to the other use; in order to produce more automobiles and less grain, workers and resources leave farms and enter automobile factories. In a multi-product economy rather than the simple two-product case, there is always more than one alternative to any particular decision, so *the* opportunity cost of a decision is the next best alternative use for resources that is not chosen.

Given the current level of resources, point I, which is outside the production possibility curve, represents an unobtainable choice; I stands for an Impossible combination of products. In the future, point I may be obtainable if the production possibilities frontier shifts outward due to technological change or increases in land, labor, or capital. Point U, which is inside the production possibility frontier, represents circumstances in which resources are not fully employed; U stands for Unemployed resources. Clearly, the combination of products at point U does not represent full employment because it is possible to move toward the production possibilities frontier and produce more automobiles, more grain, or more of both; all points inside the production possibilities frontier represent an economy that is under-employed. Any point in the triangle-like area, which has U as its vertex, is preferred to U; that is, any point in this area has more grain and at least as many guns, more guns and at least as much grain, or more of both. Given full employment of all society's resources, all possible combinations of automobiles and grain that society can produce are represented by the continuous curve that is called the production possibilities frontier; it depicts society's menu of choices.

If an economy is producing a combination of the two goods, automobiles and grain, and is on its production possibilities frontier, it is "productively efficient." However, if consumers would rather have less grain and more automobiles than is currently being produced, the theory of supply and demand predicts that the price of grain will fall due to an oversupply of grain and the price of automobiles will rise due to a shortage of automobiles, which price changes will cause business persons to move resources from grain production to automobile production.

Although the former combination was productively efficient, it was not allocatively efficient because resources were not being employed in a way that was consistent with consumer preferences as expressed in markets. *Allocative efficiency* means that an economy is both productively efficient, that is, operating on its production possibilities frontier, and that the specific combination of goods and services that it produces on the frontier are dictated by consumer preferences as expressed through supply and demand. The theory of supply and demand is explained in chapter 9.

The production possibilities frontier demonstrates that the concept of economic efficiency has two components: first, the technological component requires that all productive resources must be fully employed; second, the values component requires that the particular mix of goods that is produced (where society is on the production possibilities frontier) is what consumers desire as indicated by their preferences as expressed in markets. In short, the concept of efficiency is not a purely technological, non-normative, concept. (An allocatively efficient economy is synonymous with a Pareto efficient economy, as defined below.)

Law of Increasing Costs

The shape of an economy's production possibilities frontier is assumed to be concave to the origin (bowed out) because of the law of increasing costs, which is based on the fact that resources tend to be specialized, at least somewhat, so that some of their productivity is lost when they are transferred from doing what they are relatively good at doing to what they are relatively bad at doing. If society tries to produce more grain and fewer automobiles, it has to produce each additional billion bushels with workers, capital, and land that are less and less suitable for farming, so that the marginal trade-off (opportunity cost) of automobiles for each successive billion bushels of grain increases.

For example, in Figure 1.1, if society chooses to produce an additional 1 billion bushels of grain by moving from point C to point D (from 2 billion bushels of grain at point C to 3 billion bushels of grain at point D), the marginal trade-off as shown in Figure 1.1 is 1 billion bushels of grain for 3 million automobiles (from 12 million automobiles at point C to 9 million automobiles at point D). If society moves from point D to point E (from 3 billion bushels of grain at point D to 4 billion bushels of grain at point E), the marginal trade-off is now 1 billion bushels of grain for 4 million automobiles (from 9 million automobiles at point D to 5 million automobiles at point E). Conversely, as society tries to produce more automobiles and less grain, the marginal trade-off of automobiles for grain increases; that is, increasing amounts of grain must be foregone for equal increases in automobiles.

If resources were not specialized, the marginal trade-offs would be constant (every increase of 1 million bushels of grain would cost the same amount in terms of thousands of automobiles forgone and vice versa) and the production possibility frontier would be a straight line.

Law of Diminishing Returns

The law of diminishing returns is another reason why an economy's production possibilities curve is generally assumed to be concave to the origin. The law of *eventually* diminishing marginal (extra) returns states that if any input into the production process, say land on a farm, is held *fixed* or constant, and we add successive units of other *variable* production inputs, such as fertilizer, labor, capital (machines), and seeds to the fixed input, then, even if not immediately, at some point, as the variable inputs are continually increased, total output will continue to rise but by smaller and smaller amounts; that is, the extra or *marginal* output will eventually get smaller.

The law of diminishing returns tells us that if society attempts to produce more grain by expanding production on *existing* farms of a fixed size, the marginal cost of additional tons of grain in terms of the trade-off in automobiles will rise. However, the law of diminishing returns does not apply if additional land is converted to farming because no resource is fixed in amount.

In sum, the production possibilities frontier is concave with respect to the origin because of the law of increasing costs and the law of diminishing returns.

PARETO EFFICIENCY

This section begins with a definition and explanation of the social welfare implications of Pareto[3] efficiency, which allows economists to demonstrate that voluntary transactions by individuals acting in competitive markets best serve consumers' interests. Since Pareto efficient markets rely entirely on voluntary transactions, it cannot be used to evaluate government action requiring coercive transfers. The Kaldor-Hicks criterion, which justifies coercive court or government action, is discussed below. The two efficiency criteria are similar in that the preferences of a rich person count more than the preferences of a poor person because one's preferences only count if one has both the willingness and *ability* to pay for goods and services. As discussed below, the criteria differ in the manner in which they account for the direct and indirect costs and benefits of market transactions and government actions.

Pareto Improvement or Pareto Superior Choice

The paradigm that is often used to explain Pareto efficient markets is a single bilateral transaction between two people entered into *voluntarily*; for example, one buys a loaf of bread, a computer, or swaps clothes with someone. A single, voluntary transaction is the basis of the concept of a *Pareto improvement*, which is used to determine when a change or move from one state to another improves social welfare (the resulting state is said to be *Pareto superior* with respect to the former state). Stated as a principle, "change which harms no one and which makes some people better off (in their own estimation) must be considered an improvement."[4]

One should not confuse being better off with being totally satisfied. When a buyer agrees to buy a used car from a seller for $3,000, they may both be less than totally satisfied because the buyer wanted a lower price, $2,000, and the seller wanted a higher price, $4,000. Nevertheless, they both must be better off for having made the transaction or they would not have agreed voluntarily. Settlement of a lawsuit is another example of a Pareto improvement or superior choice. Again, both sides may not be totally satisfied with the settlement, but they must be better off after settlement in their own estimation or they would have litigated the dispute rather than having settled it.

A distinction is sometimes made between a weak Pareto improvement and a strong Pareto improvement. A *weak* Pareto improvement is a change that makes at least one person better off from the change and no one worse off from it. A *strong* Pareto improvement is a change that makes everyone better off. By making extreme assumptions with respect to continuity and divisibility of benefits from a transaction, the weak and strong Pareto superior choice versions are equivalent: whenever a single individual's situation is improved, we can conceive of a situation in which some of the extra benefit is given away by her or him and divided up as many times as necessary so that everyone gets at least a little bit more.

The requirement that no one is harmed from a Pareto improvement also means that there are no indirect negative effects or "externalities" that harm third parties. This requirement is violated where a bilateral transaction between a manufacturer of steel and a buyer results in smoke being released into the atmosphere, which smoke pollution contributes to respiratory diseases and global warming. Pollution affects the economy indirectly by imposing costs that harm third parties because these costs are external to the market or transaction that generated these costs. These indirect costs distort prices; that is, the prices of goods or services that cause pollution as negative side effects are inefficiently low because they do not account for the costs inflicted on third parties. Inefficient prices distort incentives and misallocate resources.

To count as a voluntary transaction that does not harm others, there must be either no significant negative externalities, which are indirect effects on third parties, or potential externalities must be "internalized" either by regulation that removes the externality (for example, mandating smoke stack "scrubbers" for factories) or the addition of effluent taxes to the product's price equal to the cost of indirect damage to others caused by the externality (see chapter 8). In short, the existence of significant indirect (external) negative effects on third parties unequivocally impairs or destroys Pareto efficiency. Admittedly, the concept of a negative externality or third party effect is an elastic concept, which runs the gamut from significant, such as pollution that causes global warming, to *de minimis*, such as displeasure of those who do not like the color of a neighbor's newly painted house.

Second, when economists say that third parties are not harmed by voluntary transactions, they do not mean that markets do not produce losers as well as winners. For example, consider a person who has lost income due to technological change. Assume this person produces and sells compact discs (CDs) containing music, but that due to technological innovation, music is now sold digitally over the internet, so that few people buy CDs; that is, few buyers now want to make voluntary transactions with the seller of CDs. Seller may feel personally "harmed" by this innovation, but refusal of individuals to make voluntary Pareto transactions with that person is not harm recognized by the Pareto criterion (no transaction means no harm, no foul!).

This process, by which new products and processes destroy the old, is called "creative destruction";[5] it results in winners and losers and is inherent in allocation of resources by markets. The loss that sellers of CDs suffer because individuals no longer want to buy their products goes directly through competitive markets via supply and demand; this loss is required by efficient pricing and is not an indirect, third party effect or externality like pollution. The CD example illustrates the interdependence and interrelatedness of all markets, which constantly revalue prices of products and services through supply and demand, producing both winners and losers. The losers in the process of creative destruction may or may not be reemployed in new or different industries; if reemployed, they may have higher, lower, or the same wages.

In sum, a single, bilateral transaction between two individuals has two important characteristics: one, it is voluntary and both parties to the transaction benefit from it in their own estimation; and, two, a transaction must harm no one, including having no significant negative side effects on third persons. Until recently, the social or moral justification for markets and Pareto efficiency seemed self-evident due to the over half-century contest between the former Soviet Union's command economy and the Western world's market system, which the latter won decisively. It was assumed that markets produce as many or more winners than losers, so that the Pareto criterion's failure to consider the losses of losers was not significant. However, as income inequality has increased over the last forty years, along with allegations that most gains from markets go to the top 1 percent of the income distribution, middle class jobs have tended to disappear, and wages have grown slowly or stagnated, the fact that Pareto criterion ignores the losses of losers makes it more problematic as a criterion of social welfare.

Pareto Efficient Economy

A single transaction is just one of the many billions of Pareto improvements occurring in a market economy. Every sale of a pair of shoes, haircut, car, or accounting service is a Pareto improvement. What limits the number of Pareto improvements that can occur in a competitive economy? The maximum number of Pareto improvements is achieved when it is not possible to find two additional parties who are willing and *able* to make another transaction. (Being able to make a transaction means an individual must have income in order for her or his demand for a good or service to be effective.) No additional transaction can occur because the only individuals who want to make additional transactions are either sellers, who want to sell at a price that exceeds the price potential buyers are willing to pay, or buyers, who want to buy at a price that is less than the price at which potential sellers are willing to sell. Stated as a principle, a Pareto efficient state is one in which it is not possible to make a change from the current state without making at least one person worse off.

In terms of the production possibilities frontier depicted in Figure 1.1, any move within the triangle-like area starting from point U, which describes a state in which there are some unemployed resources, toward the production possibilities frontier is a Pareto improvement in the sense that it is possible to have more grain, more automobiles, or more of both goods; that is, it is possible to make a change which benefits some through added employment and output without anyone being harmed by these transactions. Once the production possibilities frontier is reached, all possible Pareto improvements have been made and a Pareto efficient state is reached.

Any point on the frontier (productive efficiency) represents a *potential* Pareto efficient state. A movement along the frontier indicates a movement from one potential Pareto efficient state to another. However, since consumers' income and tastes are expressed through supply and demand in competitive markets, only one point on the production possibilities frontier is actually Pareto efficient, which is synonymous with being allocatively efficient. A movement from one Pareto efficient state to another occurs whenever consumers' income, tastes, or prices change, so that consumers desire more of some goods than are being produced currently and less of others. Changes in consumers' desires are transmitted through the economy via changes in supply and demand. However, as shown in the next subsection, Pareto efficiency, which mandates that no one can be harmed by a change, cannot be used to justify taxation or government coercion of any kind.

There are two reasons for the importance of Pareto efficiency. One, it is *the* major philosophical justification for organizing an economy on the basis of markets and for reducing the role of government. Pareto efficiency produces the maximum possible output that consumers desire. Two, in the context of Pareto efficiency, negative externalities or third party side effects from transactions are inefficient by definition: the prices of goods or services that cause pollution as negative side effects are inefficiently low because they do not account for the costs inflicted on third parties, and prices of products negatively affected by third party effects are inefficiently high. Inefficient prices distort incentives and misallocate resources.

Pareto Criteria as a Justification for Individual Choice over Government Regulation

The following examples use Pareto efficiency to demonstrate that governmental restrictions on individual choice often reduce consumer welfare. Assume a legislature passes a rent control law, which operates prospectively, so that the maximum rental price on any new apartments that are built will be limited to $1,700 per month. Potential investors, who can either invest in housing or in other profit-making activities, may determine that the legislative restrictions on rent will cause new apartment housing to earn a lower rate of profit than other activities in which they can invest their funds with the result that they choose other investment activities and no new rental housing will be built. Assume further that if investor-landlords were allowed to charge a rental price of $2,000 per month they would be willing to invest in new apartment houses and tenants would be willing to rent these apartments. The rent control legislation results in a Pareto inefficient state because it prevents landlords and tenants from making voluntary transactions that result in both being better off in their own estimation and which do not harm anyone else (What if voluntary transactions result in a shortage of affordable housing?).

As a second example of how the Pareto efficiency can be used as a justification for individual choice over government control, consider the question of whether the poor should be provided aid in the form of money or "food stamps." Should a poor person be given a stipend of say $300 per month or should that person be given coupons that can only be used to purchase $300 worth of food? Those who favor giving the poor food stamps argue that one of the reasons people are poor is that they do not spend their money on the "right" things. They fear that the poor may spend this money stipend on alcohol, drugs, and gambling rather than nutritious food for themselves and their families. However, according to Pareto efficiency, individuals are the best judges of what is best for themselves; government restrictions on how people use financial aid interfere with individual choice, are paternalistic, reduce social welfare, and are inconsistent with the absence of restrictions on others who can freely choose how to allocate their financial resources. According to Pareto efficiency, any financial aid to the poor should be given without "strings." That is not to say that choosing to give aid in the form of food stamps is illegitimate or wrong; there are good arguments supporting such restrictions. It is only that such restrictions are inconsistent with Pareto criteria.

Pareto Criteria Do Not Permit Involuntary Transfers

Consider the following hypothetical example, which Ackerman constructed in order to demonstrate the restrictive character of Pareto efficiency with respect to government action. Assume that the government in a two-person society must determine how to distribute the single source of sustenance, manna. Under Policy A, one citizen, Glutton, will receive 100 grains of manna and the other citizen, Starvation, will receive 0. Under Policy B, however, both Glutton and Starvation will each receive 99 grains of manna.[6] Given the definition of a Pareto improvement, one cannot say that a change from Policy A to Policy B is preferable. Even though Policy B would greatly improve Starvation's position, Glutton's position would slightly deteriorate. Since a Pareto improvement is one that harms no one and makes some people better off in their own estimation, Policy B would not be a Pareto superior to Policy A, unless Policy B permitted Glutton to retain 100 grains of manna while giving some manna to Starvation.

Moreover, assuming that the government can only choose between Policies A and B, whichever policy the government adopts first would be Pareto efficient: if the government adopts Policy A, it is not possible to change to Policy B without harming Glutton; the same problem exists if the government initially adopts Policy B and wants to adopt Policy A, except that Starvation would be harmed. Thus, Pareto efficiency is a very restrictive test with which to evaluate government activity because social preferences or collective action, which economists say require "interpersonal utility" comparisons, is not allowed.

This hypothetical example, which deliberately emphasizes the restrictive character of Pareto efficiency as a guide to government policy is, however, misleading because it makes Pareto efficiency appear trivial. Pareto efficiency is best thought of as the moral foundation or justification for a market economy. Its enduring appeal is due to two characteristics: One, it forbids coercion of any kind. Given an equitable status quo (discussed below), it is hard to think of a reason to reject a change which makes some people better off and no one worse off. Second, negative externalities are by definition Pareto inefficient because they distort prices, incentives, and resource allocation. On the other hand, as discussed in the next section, Pareto efficiency is totally indifferent to relative income inequality.

INCOME INEQUALITY AND WELFARE ECONOMICS OF REDISTRIBUTION

If the income of the rich remains the same and the income of the poor rises so that the differential in income between the two groups is reduced, the change would be a Pareto improvement because some individuals are better off and no one is harmed; however, if the income of the poor stays the same and the income of the rich rises so that the differential in income between the two groups is increased, this change would also be a Pareto improvement. The Pareto criteria are indifferent to relative changes in income distribution.

Income Inequality in the United States

The past thirty-five years have seen a large shift in the income distribution in the United States. The rich have become richer while middle income and blue collar wages have stagnated. Since 1980, 80% of net income gains went to the top 1% of the income distribution, boosting their share of total income to levels unseen since before the Great Depression. The share of income of the top 1% increased from 10% in 1980 to 21% in 2010 an increase of 11 percentage points; the share of income of the top 10% increased from 30–35% in 1980 to 45–50% in 2010, an increase of 15 percentage points.[7] The income share of what Piketty calls the "middle class," those with incomes between the top 10%–50% of the income distribution, is about 40%, and the share of income of the bottom 50% is less than 20%.[8]

According to Piketty, the shift in favor of the rich in the United States and Europe in the late twentieth century and early twenty-first century is a return to the income distribution of the nineteenth and early twentieth centuries. He argues that the period from 1950 to 1980 when job creation, technological change, and real wages all seemed to rise together was an aberration, resulting from the major shocks of two World Wars, the Great Depression that was sandwiched between them, and government fiscal policies (high marginal tax rates, up to 90 percent, and a great deal of government spending) in response to these major shocks. In that era, it was generally assumed that those who lost jobs through technological change would find employment in new industries so that the returns to labor would keep pace with the return to capital. Additionally, the Western market system clearly bested the former Soviet Union's planned economy. As memory of that era receded, government fiscal policies that reduced income inequality were reversed.[9]

Gini Coefficient of Income Inequality

A widely used measure of income inequality is the Gini coefficient, named for Italian statistician Corrado Gini, who constructed the concept. The coefficient measures income distribution on a scale from 0 to 1, where 0 represents perfect income equality and 1 represents perfect income inequality (one person has all the income). Figure 1.2 depicts how the Gini coefficient is constructed. The vertical axis depicts the cumulative share of income from 0% to 100% in quintiles; the horizontal axis also depicts the cumulative share of income in quintiles, but ranks income starting from the poorest to the richest; that is, the poorest 20%, the poorest 40%, poorest 60%, poorest 80%, and 100%.

If there were complete income equality, so that the poorest 20% had 20% of the income, the poorest 40% had 40% of the income, and so on, the income distribution would be represented by the 45° line. As the income becomes more unequal, the actual income distribution curve bows out from the 45° line. For example, point C in Figure 1.2 represents an income distribution in which the poorest 40% of income earners have 20% of total income. The Gini coefficient is calculated by dividing area A, which is the area between the 45° line and the bowed income distribution line, by the total area of the triangle (A + B); that is, the Gini coefficient = A/(A+B). If income of everyone is equal, area A = 0, so the Gini coefficient is 0/(0 + B) = 0. If one person has all the income, the curve bows out to the maximum possible, occupying the entire triangle; area B = 0, so the Gini coefficient is 1/(1 + 0) = 1.

The Gini coefficient is also useful because it introduces the concept of a *normalized* variable on scale from 0 to 1, which can be used to rank qualitative as well as quantitative variables like income.

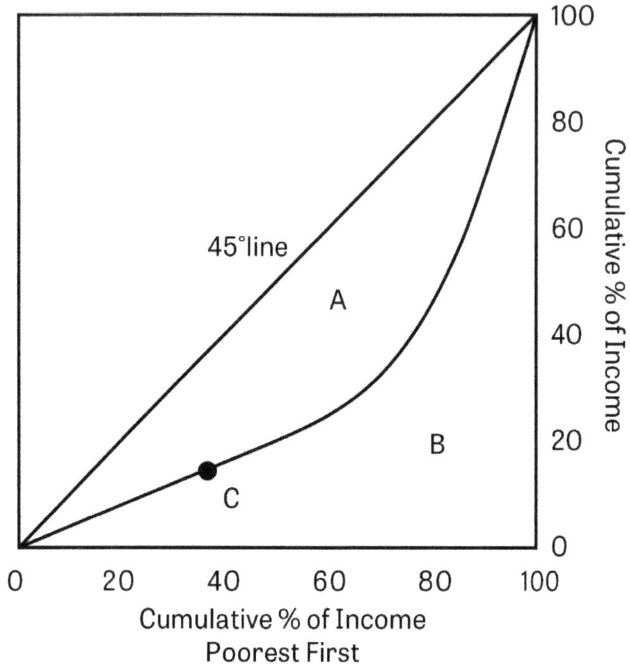

Figure 1.2 Gini Index of Income Inequality. *Source*: Created by the author.

In the United States, the Gini coefficient declined from .37 in 1945 to around .35 in the early 1970s, after which it began a steady rise to .44 in 2006 (some estimates range as high as .47). As Piketty has emphasized, from the vantage point of the twenty-first century, the era when the Gini coefficient declined, from 1930 to 1970, is an historical anomaly due to political shocks and high marginal income tax rates. In terms of international comparisons, Sweden's Gini index is .250, Japan .332, United Kingdom .360, Russia .399, Mexico .461, and Brazil .570.

Welfare Economics and Income Redistribution

In common English usage, the term "welfare" usually refers to income supplements from government to those who are indigent, unemployed, dependent, or disabled. In economic theory, however, *welfare economics* refers to the branch of economics that involves normative evaluation of economic action. What economists call the Second Welfare Theorem of Economics states that in principle, government can tax individuals and redistribute endowments of goods so as to alter how much wealth individuals have, after which competitive markets will determine a new Pareto efficient allocation. In theory, a government's tax and redistribution policy could achieve every Pareto efficient allocation on the production possibilities frontier in Figure 1.1.[10]

> The Second Welfare Theorem implies that the problems of distribution and efficiency can be separated. . . . The market mechanism is distributional neutral; whatever your criteria for a good or a just distribution of welfare, you can use competitive markets to achieve it.[11]

Although efficiency and distribution are separable in theory, in practice they are often very difficult to separate. Attempts to alter the endowment of goods through taxes will usually distort incentives, both of those who are taxed (they may work less) and those who receive the transfers (they may also work less), such that taxes will result in a distorted allocation of resources (not Pareto efficient). Economists refer to these potential disincentive effects with the metaphor that redistribution of income by government is like transporting water with a "leaky bucket"; some water (income) is lost in the transfer.

Theoretically, only a *lump-sum tax*, which is a transfer of purchasing power from some individuals to the government that does not affect anyone's behavior, will be non-distortionary and satisfy the Second Welfare Theorem's requirement that redistribution not impair efficiency. For example, assume each consumer is required to transfer to the government money that would be earned from twelve hours of her labor each week. This tax is independent of how many hours a person actually works or sells.[12] More realistically, since most people's labor supply, which includes entrepreneurial effort, is relatively insensitive to variations in tax rates or remuneration (during the mid-twentieth century, when the United States was arguably at its economic peak, marginal tax rates were 90 percent), the efficiency loss from taxing labor and capital at less than confiscatory rates may not be too large.

> [T]he message of the Second Welfare Theorem is important. Prices should be used to reflect scarcity. Lump-sum transfers of wealth should be used to adjust for distributional goals. To a large degree, these two policy decisions can be separated.[13]

KALDOR-HICKS CRITERION

A second normative principle, Kaldor-Hicks criterion, which is also called the potential compensation principle, potential Pareto improvement, or wealth maximization, is used to determine when action by courts or government, which explicitly or implicitly involves coercion, improves social welfare. Stated as a principle: "A change is an improvement if those who gain evaluate their gains at a higher figure than the value which the losers set upon their losses."[14] Gains and losses are measured by individual's potential willingness and ability to pay in dollars. Kaldor-Hicks justifies government or court action that results in substantial gains for some individuals and substantial losses for others. Actual compensation to losers is not required. In addition, the usual formulation of the Kaldor-Hicks criterion is ambiguous as to whether one is comparing total gains to total losses or marginal gains to marginal losses as shown in the following example. (A third normative criterion, marginal cost-benefit analysis, for example, the Judge Hand test which is used throughout this book, justifies an action if marginal benefits or gains are greater than marginal costs or losses without regard to who gains and who loses.)

Kaldor-Hicks Can Justify Inaction against Significant Negative Externalities

According to Pareto efficiency, negative externalities are unequivocally inefficient because they distort prices and misallocate resources. However, Kaldor-Hicks can justify inaction against negative externalities if the total utility of an activity outweighs its total negative effects. Consider the case of *Carpenter v. Double R Cattle Co.*,[15] in which homeowners who claimed that an expanded cattle feedlot created a nuisance due to noxious odors, pests, air and water pollution, sought an injunction and damages. The jury was told to weigh the alleged injury to the homeowners against the "social value" of the feedlot and to consider "the interests of the community as a whole" in determining whether a nuisance existed. In other words, the jury was told to weigh *total gains* from the feedlot against *total losses* to the homeowners (Kaldor-Hicks criterion). The jury found no nuisance. The Court of Appeals reversed the jury verdict on the ground that the jury received improper instructions.

> [T]he test for existence of a nuisance no longer depends solely upon the balance between the total gravity of harm versus the total utility of the conduct. . . . [A]n invasion may be regarded as unreasonable even though the utility of the conduct is great and the amount of harm is relatively small. . . . [U]tility of the activity alleged to be a nuisance is a proper factor to consider in the context of injunctive relief; but that damages may be awarded regardless of utility. Evidence of utility does not constitute a defense against recovery of damages where the harm is serious and compensation is feasible. Were the law otherwise, a large enterprise, important to the local economy, would have a lesser duty to compensate its neighbors for invasion of their rights than would a smaller business deemed less essential to the community.

> [O]ur view is not based simply upon general notions of fairness; it is also grounded in economics. . . . Where an enterprise externalizes some burdens upon its neighbors, without compensation, our market system does not reflect the true cost of products or services provided by that enterprise. Externalities distort the price signals essential to the proper functioning of the market.[16]

The Supreme Court of Idaho reversed the Court of Appeals and reinstated the jury verdict. It held: "The State of Idaho is sparsely populated and its economy depends largely upon the benefits of agriculture, lumber, mining and industrial development. To eliminate the utility of the conduct . . . in determining whether a nuisance exists . . . would place an unreasonable burden upon these industries."[17] In short, the Supreme Court of Idaho reinstated the Kaldor-Hicks criterion based upon a comparison of total gains versus total losses as the test for nuisance.

To repeat, in either-or, that is, all-or-nothing, contexts, Kaldor-Hicks criterion can provide a rationale for inaction against negative externalities if total gains from the activity outweigh total losses due to its negative third party effects. This use of Kaldor-Hicks also violates the economic principle discussed above that decisions based upon comparing totals are less efficient than marginal decisions.

International Trade Agreements as All-or-Nothing Choices

The North American Free Trade Act (NAFTA) which removed tariff and quota barriers from trade among Canada, Mexico, and the United States is an all-or-nothing decision made by comparing total benefits to total costs. In the context of Kaldor-Hicks,

US government removal of trade restrictions is justified if total gains from trade to US consumers in the form of lower prices for goods and services, which occur when each country specializes in producing what it does most efficiently, outweigh total losses of income to those who are displaced from jobs in the United States. For example, when an automobile factory is built in Mexico, where automobiles can be produced with cheaper labor than in the United States, some auto workers in Detroit lose their jobs.

However, free trade among Canada, Mexico, and United States reduces prices for goods and services so that consumers' real income, that is, their spending power, increases even if their nominal income stays constant. Some of the gains to consumers could be taxed by government in order to provide unemployment and retraining benefits for the losers from international trade agreements; in this way gainers could potentially compensate losers. However, according to Kaldor-Hicks, a change is justified whether or not gainers actually compensate losers.

Next, consider NAFTA in the context of Pareto efficiency: To the extent that NAFTA is interpreted as merely the removal of government restrictions on voluntary transactions between individuals in different countries, restriction removal can be said to result in Pareto improvements from more efficient production of products and services. In the context of Pareto efficiency, the losses to factory workers in Detroit and displaced farmers in Mexico, which result from bilateral transactions that no longer occur, do not count as harms that negate a Pareto improvement. As discussed above, the market process of creative destruction creates both winners and losers; losses of losers in markets, that is, individuals who used to be able to make transactions but are no longer able to find willing buyers, do not count in Pareto efficiency; it only considers benefits from markets transactions; the losses of losers in the process of creative destruction can be ignored.

Is Pollution Abatement All-or-Nothing or Marginal Decision?

The two previous examples, the decision of courts in *Double R Cattle Co.* whether to grant an injunction and the decision of the federal government to join NAFTA, are truly either-or, all-or-nothing decisions. Now I ask, is the question of pollution elimination or abatement an all-or-nothing or a marginal decision? Should pollution be unregulated if total gains to individuals from economic development outweigh total losses to other individuals? What would a society with totally unregulated pollution be like? Or, should the amount of pollution abatement be decided by weighing marginal gains and marginal losses? It is almost certain that starting from a position of no controls on pollution, the marginal gains from beginning pollution abatement (e.g., smoke stack "scrubbers" which trap smoke) in terms of cleaner air would be large and the marginal losses in terms of economic growth would be small. Marginal efficiency decisions require that pollution abatement be increased as long as the marginal gains of abatement exceed or equal the marginal losses in economic growth. In the context of marginal analysis, pollution abatement can be significant without having a drastic impact on growth. The Kaldor-Hicks either-or, all-or-nothing approach completely distorts incentives and can justify total inaction against global warming.

Possible Extreme Redistributive Consequences of Kaldor-Hicks

Kaldor-Hicks, which is based on comparing gains of gainers to the losses of losers, can be used to support arguments for either an increase or a decrease in income inequality due to problems in making a complete accounting of all the direct and indirect gains and losses. At one extreme, Kaldor-Hicks can support an argument for significant redistribution of income toward equality (if most gains from markets go to the top 1 percent); at the other extreme, it can support an argument for increased inequality ("trickle down" or "supply side" theories).

The following hypothetical example was constructed by Polinsky to demonstrate that Kaldor-Hicks can permit extreme involuntary transfers that can exacerbate income inequality. He supposes that the government must decide whether to build a dam and that the dean and a professor of Stanford Law School are the only two individuals affected by this government decision.[18] Without the dam, the dean's income is $65 and the professor's income is $35, so that their combined total income is $100. The dam would cost $30 to build, consisting of $30 worth of the professor's labor but none of the dean's. The dam would create benefits worth $40, which would go only to the dean because the dam is going to be built on his property. If the dam is built, the dean would have an income of $105 and the professor would have $5, so their combined total income would be $110. According to Kaldor-Hicks, the government should build the dam because it creates benefits worth $40 to the dean and costs of only $30 to the professor, resulting in a net social gain of $10. However, Polinsky says: "If promoting equity is very important, it might be desirable to sacrifice some efficiency for more equity by not building the dam (in other words, 'damn' the Dean)."[19]

Probably no aspect of law and economics has caused as much controversy among judges and lawyers as the possible distributional implications of Kaldor-Hicks based upon a comparison of total gains versus total losses rather than marginal gains versus marginal losses.[20] Although Polinsky asks whether the government should adopt Kaldor-Hicks and build the dam in view of the conflict between efficiency and distributional equity, he does not ask whether government could force the professor to comply with its directive to build the dam or whether he or she would leave Stanford for Berkeley or points east. Moreover, since the dean-professor example is a thinly disguised master-slave relationship, it violates the prohibition against slavery or involuntary servitude and equal protection of the laws clauses of the Thirteenth and Fourteenth Amendments to the US Constitution.

In the context of the antebellum South, a Kaldor-Hicks analysis of the total gains to gainers versus total losses to losers from slavery could theoretically support slavery. At that time, when it was illegal to educate slaves, the opportunity cost to them of slavery was low because their wages in unskilled occupations would be low. On the other hand, the gains to plantation owners from slave labor was high. Therefore, it is quite possible to argue that total gains from slavery in the antebellum South outweighed its total losses. Polinsky's example and historical conditions under which slavery existed demonstrate that Kaldor-Hicks could possibly support action that has extremely negative implications for income distribution fairness.

It is often emphasized that government tax and expenditure policies, which attempt to reduce income inequality by transfers of income from rich to poor, run grave risks of creating severe negative effects on the incentive to work. The disincentive effects influence both those who are taxed and those who receive the transferred income. However, the United States is just beginning to debate whether government policy that substantially increases income inequality may also have severe disincentive effects on effort and efficiency. Are the significant number of men dropping out of the labor force and the opioid crisis due to the poor prospects of making a living wage?

CONCLUSION

Pareto efficiency and Kaldor-Hicks criterion are the two normative criteria that economists use to evaluate economic decisions. Both criteria assume the income distribution of the status quo, whether fair or not, in which the preferences of a rich person count more than the preferences of a poor person because one's preferences only count if one has both the willingness and the ability to pay for goods and services.

With respect to income distribution, Pareto efficiency is indifferent to relative increases or decreases in income inequality; for example, if the rich get richer and the poor's income stays the same, this change counts as a Pareto improvement and vice versa. However, the Second Theorem of Welfare Economics states that lump-sum taxes can be used to redistribute endowments of goods toward greater equality without impairing Pareto efficiency.

Kaldor-Hicks criterion can be used to support arguments either for an increase or a decrease in absolute or relative income inequality due to the difficulty in making a complete accounting of all direct and indirect gains and losses from income redistribution; Kaldor-Hicks can support either an argument for significant redistribution of income toward equality or an argument for increased inequality.

In terms of social welfare, Pareto efficiency and Kaldor-Hicks criterion have opposite strengths and weaknesses. Pareto efficiency is the primary moral justification for a market economy; it has two characteristics that are undeniably good for society. One, all transactions in markets are entered into voluntarily and make both parties to the transaction better off in their own estimation. Two, negative externalities from transactions, such as pollution, global warming, and dumping of toxic chemicals, are unequivocally bad or inefficient because they distort incentives and prices and misallocate resources; that is, prices of goods or services that cause pollution as negative side effects (e.g., electricity generated by coal that results in greenhouse gases) are inefficiently low because they do not account for the costs inflicted on third parties.

On the negative side, Pareto efficiency considers only gains to gainers (buyers and sellers) from transactions in markets; losses to losers, those who lose their jobs due to changes in technology or global competition, do not count; that is, losses of those who used to be able to make voluntary transactions but no longer can because there are now no willing buyers for their goods or services are not counted. (No transaction, "no harm, no foul.") Second, because Pareto efficiency requires that no one be harmed from an action, it cannot be used to justify non-voluntary or coercive action by courts or government.

The Kaldor-Hicks criterion has opposite positive and negative attributes. On the positive side, Kaldor-Hicks justifies coercive (non-voluntary) action by courts or government. Second, it considers both gains and losses. Under Kaldor-Hicks, court or government action is justified if gains to gainers outweigh losses to losers. Gainers could potentially compensate losers, but need not do so; actual compensation to losers is not required.

One of Kaldor-Hicks's most socially negative aspects is that in the context of either-or, that is, all-or-nothing choices based upon comparing total gains to total losses, it can support arguments for total inaction against negative externalities like pollution or public nuisances as in *Double R Cattle Co.,* discussed above. With respect to negative externalities, marginal cost-benefit efficiency, which is based upon comparing marginal gains to marginal losses, should be used to determine how much to tax or reduce pollution rather than whether or not to reduce pollution, yes or no, based upon weighing total gains versus total losses.

NOTES

1. The maxim apparently alludes to the practice in early-twentieth-century saloons of providing a so-called free lunch for patrons who bought alcoholic drinks. Who paid for the "free lunch?"

2. As discussed below, the Kaldor-Hicks criterion, which justifies government or court coercive action if the gains to the gainers outweigh the losses to the losers, is ambiguous as to whether the comparison is between marginal gains and marginal losses or total gains and total losses. The usual rule in economics is that the marginal comparisons are efficient whereas total comparisons are not.

3. Vilfredo Pareto (1848–1923) was an Italian economist who first formulated the normative concept of economics.

4. William Baumol, *Economic Theory and Operations Analysis*, 3rd ed. (Englewood Cliffs, NJ: Prentice Hall, 1972), 400.

5. Joseph Schumpeter, *Capitalism, Socialism, and Democracy* (New York: Harper, 1942), 83.

6. Bruce Ackerman, *Economic Foundations of Property Law* (Boston: Little, Brown, 1975), xi–xii.

7. Thomas Piketty, *Capital in the Twenty-First Century* (Cambridge, MA: Belknap Press, 2014), 292–294.

8. *Ibid.*, Table 7.1, 247.

9. *Ibid.*, 291–303.

10. Hal Varian, *Intermediate Microeconomics: A Modern Approach*, 7th ed. (New York: W. W. Norton, 2006), 586. The First Welfare Theorem of Economics states that *if* all markets are competitive, *then* all possible gains from trade will be achieved and such markets will be Pareto efficient. The First theorem says nothing about whether the income distribution in a Pareto efficient economy is fair or just. *Ibid.*, 580.

11. *Ibid.*, 586.

12. *Ibid.*, 587.

13. *Ibid.*, 587–588.

14. Baumol, note 4, 402 (Kaldor test). The Hicks test states that a change is justified if maximum amount the losers would pay the gainers to forgo the change is less than the minimum amount the gainers would accept to agree. Both criteria do not require actual compensation, only that it potentially can be made.

15. 105 Idaho 320 (1983) (Judge Burnett).

16. 105 Idaho 329–330.
17. 108 Idaho 602 (1985).
18. A. Mitchell Polinsky, *An Introduction to Law and Economics*, 4th ed. (New York: Wolters Kluwer, 2011), 8–9.
19. *Ibid.*
20. Ronald Dworkin, “Is Wealth a Value?” in *A Matter of Principle* (Cambridge, MA: Harvard University Press, 1985), ch. 12.

Chapter 2

Rational Economic Behavior and Logic Defined

The goal of this chapter is to explain what it means to *think like an economist*. Three specific characteristics of economic logic are emphasized:

1. Rational economic behavior and logic are defined by two simple behavioral rules and constraints on such behavior. Chapter 5 will show that legal logic is also defined by two simple, but different, behavioral rules and constraints.
2. The consistency criterion economic behavior, which is called *transitivity* in economics is explained. Consistency in economic logic requires that every economic decision must be consistent with every other decision. An individual who is "penny wise and pound foolish" violates the consistency requirement and is not economically rational; for example, an individual, who spends half an hour driving around searching for gasoline that is two cents cheaper per gallon and then goes into a retail store and makes an impulse purchase of a $1,000 computer, without checking whether the computer is selling for less at other stores or on the internet, is acting inconsistently and is not economically rational. So too, a person who exercises a lot and smokes violates the consistency requirement. To repeat, in economic rationality, every choice must be consistent with every other choice. Chapter 5 demonstrates that legal logic has a similar but different concept of consistency or transitivity.
3. It is shown that economic logic is completely or totally ordered; that is, every choice is related to every other choice in a rigidly ordered way. With respect to every choice being related to every other choice, recall the discussion in chapter 1 of marginal opportunity cost, which was used to show that, for example, all public expenditure on schools, police, fire, and so on are related to each other because they compete as alternative uses for the same resources (tax dollars). In addition, when economically rational individuals make choices involving two competing goods, all possible choices can be ordered and depicted in strict relation to each other in two-dimensional space. Three competing interests can be ordered in three-dimensional space; four or more interests can be ordered on mathematical vectors. Moreover, economic rationality is also linearly ordered in the sense that all choices

> can be further rank ordered on a single line (some points on the line may have more than one choice). The rigid total and linear ordering characteristics of economic logic which are demonstrated below are quite remarkable!

The reason the complete ordering of economics is important in law is that when judges employ economic logic in cases, say they balance the same competing interests in several cases, these case decisions must be completely (spatially) ordered in relation to each other (chapter 3). Violation of this requirement means such decisions are inconsistent and not economically rational.

Readers are not expected to be able to reproduce or to have a working knowledge of the economic theory in this chapter. It is enough if readers take away an understanding of the three specific characteristics of economic logic, two behavioral rules and constraints, transitivity, and total or linear ordering. I hope readers will find interesting the discussion of choices individuals make between income and leisure, which is used to explain the characteristics of economic logic, and will enjoy making a list of Binary Choice Preferences that reveal her or his preferences between work and leisure.

The first section of this chapter contains a statement of the three essential requirements that define any rational decision-making process: purpose, consistency, and constraints. Next, a "methodological" discussion shows why the assumption of rational behavior is useful in economics even if individuals do not always act rationally: economic rationality can be used as 1) a criterion of consistency by which to determine whether an individual's actual choices are consistent with each other; 2) a normative (values) standard, that is, to determine what decisions an individual "ought" to make (e.g., what is an economically efficient tax structure); 3) a technique for predicting the future, called positive or predictive economics; or, 4) a methodological principle, which is employed because it has proved to be useful.

Next, economic rationality and logic are defined in terms of two simple behavioral rules in the context of an extended example involving choices between income and leisure. In this context, the "transitivity" and "total ordering" properties of economic rationality are explained. The requirement that any rational process is subject to constraints, which in economics are budget, income, or resource constraints, is used to complete the model of optimal choice among competing goods. Finally, if judges balance competing interests according to economic rationality, their decisions must be completely ordered in relation to each other; if not, such decisions are inconsistent and not economically rational.

THE THREE CRITERIA OF A RATIONAL DECISION PROCESS

A rational decision process is defined by three criteria: purposefulness, consistency, and constraints. Philosopher Robert Nozick said that rationality is a goal-directed or instrumental process;[1] that is, rational decision-making is purposeful. Three different rational decision processes are described in this book: economic rationality, legal rationality, and game theory.

In the textbook definition of economics, individuals and firms are explicitly assumed to act purposefully: microeconomics is the study of how individuals allocate scarce resources among competing goals so as to maximize (optimize) their satisfaction or utility. Business firms are assumed to allocate resources so as to maximize profits.

Correspondingly, the decision process in law, the legal adjudication of disputes, is purposeful because judges and lawyers are constantly refining and clarifying the goals and issues that are to be decided in cases. One cannot determine what facts of a case are material (significant) unless one knows the purpose for which the facts are probative. For example, in negligence cases, if the goal is to deter careless behavior, one set of facts, concerning whether the parties took reasonable precautions to avoid injury, is material; however, if the goal is risk distribution or risk spreading so as to distribute financial losses over many persons rather than have one person bear a crushing loss, another set of facts, concerning who is the cheapest insurer, is material.[2]

When appealing an adverse decision by a trial court, lawyers begin appellate court briefs (challenges) with a question that frames the legal dispute in a way that focuses an appellate court's attention on facts most favorable to its side of the case. An appellate brief begins by posing a question which suggests that the dispute should be resolved so as to achieve a goal consistent with that party's interest. For example, in a dispute concerning government restrictions on free speech, the plaintiff, who objects to government restrictions on speech, will emphasize that the goal of protecting free speech is enshrined in the First Amendment; the defendant, government, will emphasize that reasonable restrictions on free speech are necessary to the goal of ordered liberty. ("No one has the right to shout 'fire' in a crowded theater.")

Nozick also wrote, "Rationality is a matter of reliability."[3] The reliability element of rationality accounts for the emphasis that both economists and lawyers place on consistency in their explanations of economic and legal decision-making.[4] The different purposefulness and consistency criteria of economics and of law result in different permissible relationships between cases in the two disciplines.[5]

With respect to the third element, constraints, a rational decision process must take into consideration constraints that hinder an actor from achieving her or his goals. In economic rationality, various scarcities in the form of budget, cost, income, and resource constraints place limits on achievement of one's purpose. The principal institutional constraint on overbroad judicial decisions in law is stare decisis, or the principle of precedent, which mandates that judges must decide like cases alike (chapter 5).

ECONOMIC RATIONALITY AS CRITERION OF CONSISTENCY, NORMATIVE STANDARD, PREDICTIVE TECHNIQUE, OR METHODOLOGICAL PRINCIPLE

The assumption of economics that individuals act rationally is not universally accepted. Even so, this section contains a discussion of why the assumption of rational behavior is useful in economics even if individuals do not always act rationally. I begin by stating the obvious criticisms of economic rationality and explain why these criticisms do not negate the use of economic rationality as a criterion of consistency, normative standard, predictive technique, or methodological principle.

Sociologists, Freudians, behavioral economists, and others criticize the assumption of economic rationality because they believe that social conditioning, biases, and irrational desires play a large and sometimes controlling influence on the motivation of individuals. This criticism seems to be borne out by daily stories in the news media that report violence and inexplicable or "irrational" cruelty inflicted by human beings on each other. Criticism by behavioral economists has stressed the inability of individuals to make the complex calculations or do the consistent reasoning required by rational economic behavior. In the behaviorists' view, individuals may try to choose the best feasible option, but they often make significant mistakes.[6]

On the other hand, the effect of computers, which increase individual rational decision-making capacity manyfold, and global competition, which ineluctably grinds down local economic rules and preferences that are not cost justified (not economically rational), are very powerful forces toward increased economic rationality that push national economies toward Pareto efficiency, like it or not.

However, there are four uses of the rationality principle in economics that do not require that individuals always act rationally; that is, rationality can be used as a criterion of consistency, normative standard, predictive technique, or methodological principle.

First, assume that we wish to evaluate whether an individual or business does in fact make decisions that are rational in the sense of whether this person's decisions are directed to a purpose and are internally consistent. For example, assume that a government claims that its tax and transfer policy is based on "one's ability to pay." Given this stated purpose, is the government acting consistently (rationally) with respect to that purpose if it requires a young married couple with two children and an annual income of $50,000 to pay taxes in order to sustain Social Security ("transfer") payments to a retired couple who have an annual income of $90,000?[7] Consider another example: if society professes to value human life over animal life, is it consistent (rational) to spend hundreds of thousands of dollars to save a whale trapped in Arctic ice flows but not to spend $500 on a woman for prenatal care in order to prevent birth defects in her fetus? In these two examples, where there are apparent inconsistencies between avowed purposes and actual choices, the method of rationality is used as a criterion of consistency with which to test or criticize behavior or decisions.

Second, if the assumption of rationality is used to suggest how individuals *ought* to behave, what *values* they ought to hold, whether or not individuals behave this way in fact, rationality is used as normative standard. For example, when economists attempt to determine an "optimal" or "efficient" tax structure, they are practicing normative economics.

Third, normative economics is distinguished from positive or predictive economics where economists use the rationality assumption to predict consequences of choices without advocating a position. As an example of positive economics, the rationality principle can be used as a method to predict the consequences of an increase in income tax rates on government revenue. Separating positive or predictive economics from normative economics is difficult or maybe impossible because the evaluative criteria (Pareto efficiency or Kaldor-Hicks criterion) of economics are based upon utilitarian principles, in which aggregate welfare is preferred to individual welfare.

According to the definition of rationality as goal-directed behavior that is consistent and predictable, even a frog or other animals act rationally in their consistent and predictable pursuit of food or sex. Assuming for the moment that individuals are rational in the sense of engaging in goal-directed predictable behavior, the concept of rationality does not require that the purpose or goal that guides an individual's choices be socially acceptable. A Nazi, racist, sexist, or murderer could be rational. We may abhor their purposes and preferences, but as long as such preferences are followed consistently, such individuals are acting rationally according to the definition of rationality. However, to say that such individuals are rational does not imply that they are reasonable in the sense of the law's reasonably prudent person. A reasonably prudent person is a rational person whose preferences conform to those of society at large. Thus, a Nazi or a racist may be rational but, in our society, he is not a reasonable person. All reasonable persons are rational, but not all rational persons are reasonable; that is, reasonable persons are a subset of rational persons. On the other hand, a person's motive or purpose, say a person who goes on a rampage and kills dozens of people, may be so antithetical to humanity that even if he has a well thought out, consistent plan to accomplish his heinous purpose, we say that such purpose in itself may suggest he is irrational or crazy. Yet, in mass murder cases, juries very rarely accept the insanity defense.

Fourth, the logical or methodological status of the rationality principle in economics was explained by Bruce Caldwell. He says that the rationality principle, which stipulates that individuals act "adequately or appropriately . . . in accordance with the situation" can be interpreted in three ways. One interpretation is that the rationality principle merely postulates that individuals are "purposeful" and "goal-directed," without specifying specific goals or how they are achieved. This interpretation of rationality is untestable. A "second way of interpreting the rationality principle is as the equivalent of a universal law within the social sciences." If the rationality principle is interpreted as a law regarding how individuals behave in different circumstances (e.g., consistently), it is testable and "actually false, though a good approximation to the truth." Third, the rationality principle "need not be considered a statement at all, but rather a methodological principle that is retained because it has shown to be particularly fruitful in the past."[8]

Whether the rationality principle is accepted as a statement of purpose, good approximation to the truth, or pure methodological principle, this assumption defines economics in the sense that although economists are usually willing to alter other assumptions where appropriate, economists, despite the criticisms of behavioral economists,[9] rarely jettison the rationality assumption. In sum, whether or not individuals are strictly rational, the principle of rationality is the sine qua non of economics.

THE TWO BEHAVIORAL RULES OF ECONOMIC RATIONALITY

Recall that a rational decision process was defined by three criteria: purposefulness, consistency, and constraints. Economic rationality and logic will be defined in terms of two simple behavioral rules, which satisfy the purpose and consistency requirements of rational decision-making. Next, two characteristics or properties of economic rationality, transitivity and total ordering, are derived and explained. Then, the third criterion, constraints on economic rationality, such as budget, income, or resource constraints which complete the model, will be explained.

Defining economic rationality and its logic serves two purposes: one, the definition of economic rationality is necessary so that it can be compared to the parallel definition of legal rationality in chapters 5 and 6; and two, the discussion supports the conclusion that whenever judges balance the same competing interests in cases according to economic rationality, all cases can be consistently and totally ordered in relation to each other. In the next chapter, three examples will show how legal cases involving two or three competing interests can be totally ordered on graphs or ordered linearly.

The first behavioral principle that defines economic rationality states the purpose or goal of rational individuals and compels the ordering of some choices; the second behavioral principle completes total ordering of all other choices.

More of a Good Is Preferred to Less

The purpose or goal of rational economic individuals is to maximize or optimize satisfaction or utility. (Business firms are assumed to maximize profits.) Practically or operationally, an individual's goal is stated as a simple behavioral rule; it is then explained:

> Rule E1: More of a good is strictly preferred to less.[10] (Business firms are assumed to prefer more profits to less.)

A *good* represents a product, service, interest, or end. Goods (plural) represent products, interests, or ends. Examples of specific goods are: automobiles, computers, lawyers' services, and haircuts. Although an individual can become sated with any one good, economic choices individuals really care about concern choices they must make among goods in circumstances where, due to budget or income constraints, they have less than they want of most goods.[11]

Goods are distinguished from economic "bads" like crab grass and pollution. Rational individuals want less or fewer bads. Goods also refer to interests, such as civil rights and public security. The concept of a good also refers to abstract concepts like liberty and freedom. In the context of Rule E1, it makes sense to say that a rational individual prefers greater individual freedom in the United States to lesser freedom in China, that is, a rational individual can prefer more freedom to less and can rank order choices among states of freedom.

For simplicity, consider rational decisions in the context of two competing goods: income and leisure. When an individual chooses between different amounts of income and leisure, choices can be represented in two-dimensional space as depicted by coordinates in Figure 2.1 below. Income is depicted on the vertical axis and hours/week of work is depicted on the horizontal axis. The vertical axis has the usual mathematical interpretation that more income is preferred to less. However, since fewer hours/week of work are preferred to more hours, that is, more leisure is preferred to less, the usual representation on the horizontal axis from low numbers to high numbers is reversed; a low number of hours/week, which represents more leisure, is preferred to a higher number of hours/week.

The choices among two goods are not independent of each other, but rather are chosen together in "bundles," which are collections of goods, such that each bundle has two dimensions, one dimension for income and one dimension for hours/week of

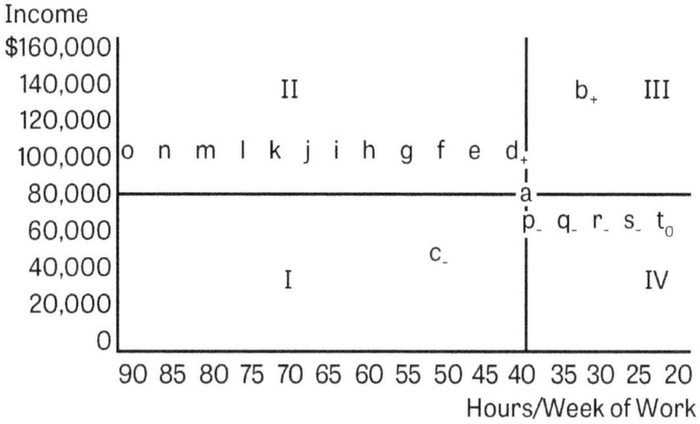

Figure 2.1 Income and Leisure Trade-Off. *Source*: Created by the author.

work. For example, an individual may be asked to choose between a bundle consisting of a job that pays $80,000 and requires 40 hours/week of work (40, $80,000) and a bundle consisting of a job that pays $100,000 and requires 60 hours/week of work (60, $100,000). (I follow the usual mathematical convention of putting the number on the horizontal axis first, the number on vertical axis second.)

In order to demonstrate that a rational economic individual's choices are transitive, totally ordered in two-dimensional space, and capable of being linearly ranked, we begin by choosing an arbitrary bundle; any bundle is just as good as any other bundle as a starting point. We begin by choosing bundle *a*, which has an income of $80,000 and 40 hours/week of work (40, $80,000). Next, we draw two lines through the arbitrarily chosen bundle *a*, one perpendicular to the vertical income axis, one perpendicular to the horizontal work axis. I repeat that any bundle in the two-dimensional space could have been chosen as bundle *a*. In Figure 2.1, these two perpendicular lines through bundle *a* divide two-dimensional space into four quadrants, counter clockwise: I, II, III, and IV.

First, we consider bundle a in relation to bundles in Quadrants I and III. In each of the four following comparisons, Rule E1, more is strictly preferred to less, logically requires that one bundle must be preferred to another; that is, all economically rational individuals must decide choices among these bundles the same way.

Consider any bundle in Quadrant III, say for example, bundle b (35, $140,000), in relation to bundle a (40, $80,000). According to Rule E1, a rational economic person must strictly prefer bundle b (or any bundle in Quadrant III) to bundle a because more is preferred to less: bundle b has both more income and more leisure (less work) than bundle a. In Figure 2.1, subscript + is put next to b (b_+), indicating that it is preferred to a.

Next, consider any bundle in Quadrant I, say for example, bundle c (50, $40,000), in relation to bundle a (40, $80.000). According to Rule E1, a rational economic person must strictly prefer bundle a to bundle c (or any bundle in Quadrant I) because more is preferred to less: bundle a has both more income and more leisure (less work) than bundle c. In Figure 2.1, subscript – is put next to c (c_-), indicating it is inferior to a.

The consistency criterion of economics, called "transitivity," is defined as follows: for any three bundles, if x is preferred to y, and y is preferred to z, then x must be preferred to z. Bundles b, a, and c are transitive. Bundle b is preferred to bundle a, and bundle a is preferred to bundle c, therefore bundle b must be preferred to bundle c. The relations among bundles b, a, and c are logically required by Rule E1 and the principle of transitivity. All rational individuals must rank bundles b, a, and c in that order.

Now consider bundle d (40, $100,000), which is on the line separating Quadrant II from III; this line also runs through bundle a (40, $80,000). According to Rule E1, a rational economic person must prefer bundle d to bundle a because bundle d has as much work-leisure as bundle a (40), but has more income ($100,000 to $80,000). According to Rule E1, any bundle that has as much of all competing goods as another bundle and more of one good must be strictly preferred. In Figure 2.1, subscript + is put next to bundle d (d_+), indicating that it is preferred to a.

PREFERENCES BETWEEN BUNDLES THAT HAVE MORE OF ONE GOOD AND LESS OF ANOTHER GOOD

As shown above, the relationships between the arbitrarily chosen bundle a and bundles b, c, and d are required by Rule E1. However, the relationships between bundle a and bundles in Quadrants II and IV are not logically determined by Rule E1. For example, consider the relationship between bundle a (40, $80,000) and bundle e (41, $100,000). Rule E1 does not allow a rational economic individual to determine whether he or she prefers bundle a to bundle e because each bundle has more of one good and less of another. Bundle e has more income than bundle a ($100,000 is more than $80,000), but bundle a has more leisure-less work than bundle e (40 hours of work gives an individual more leisure than 41 hours of work does).

Rational people prefer more income to less; rational people also prefer fewer hours/week of work. And, rational people prefer a bundle of more income and fewer hours of work to a bundle with less income and more hours of work. However, rational people differ with respect to trade-offs between income and hours worked. For example, Americans seem to be willing to work (trade-off) additional hours of work for higher incomes while Europeans seem to be willing to give up (trade-off) some income in order to work fewer hours and have more leisure.

Thus, not all rational people make the same trade-offs between income and work. A second behavioral rule or principle of economic rationality is required to enable total ordering of choices among bundles where one bundle has more of one good and the other bundle has more of another good.

When choices are made between two bundles that consist of competing goods where one bundle contains more of one good and the other bundle contains more of another, Rule E2 states that a rational individual must be able to make choices that are totally ordered as shown below.

> Rule E2: For any two bundles of goods, an individual can determine whether he or she prefers one bundle to the other or is indifferent between them.[12]

How Rule E2 requires a rational economic individual to make transitive and totally ordered decisions between bundle a and any bundle in Quadrants II and IV is demonstrated as follows: Assume an individual is asked whether she prefers bundle a (40, $80,000) to bundle e (41, $100,000), or prefers bundle e to bundle a, or is indifferent between these two bundles. Someone who greatly prefers leisure to work may prefer bundle a to bundle e, but most individuals would probably be willing to work an additional 1 hour per week for an additional $20,000 a year. Most rational individuals will prefer bundle e to bundle a, but not all. Rational individuals can have different preferences; however, their preferences must be transitive and totally ordered as explained below.

The reader is asked to do the following thought experiment below. Compare bundle a with each of the 14 bundles from bundle b to bundle o listed serially in Binary Choice Preferences below and decide in each case whether you prefer bundle a or the alternative bundle. If you prefer the alternative bundle, put a subscript + next to the alternative bundle's symbol (on a separate sheet of paper on which the Binary Choice Preferences as listed); if you prefer bundle a to the alternative bundle, put a subscript – next to alternative bundle's symbol; and, if you are indifferent between the two bundles put a subscript 0 next to the alternative bundle's symbol. In the Binary Choice Preferences, all rational economic individuals must answer the first three choices, 1), 2), and 3) as required by Rule E1, as explained above (these choices are also depicted in Figure 2.1 above). The answers for these three choices, which are b_+, c_-, and d_+, are filled in Binary Choice Preferences below.

Rule E1 is irrelevant to choices 4) through 14) because each of these choices has more of one good and less of the other good. But, Rule E2 mandates that a rational individual must be able to answer all questions 4) through 14) with a +, –, or 0; agnosticism is not allowed by Rule E2, which says that there are only three possibilities and a rational individual must choose one of them. Although different rational individuals may have different preferences and different answers, transitivity requires that a rational individual's choices must be consistent with each other. This thought experiment also demonstrates that rational economic choices are not all-or-nothing choices but are an infinite series of marginal choices between two alternative bundles, which consist of two or more goods or interests.

In Binary Choice Preferences 4) through 14) the annual income in each of these bundles is held constant at $100,000, but the number of hours/week of work is larger in each successive bundle. Please replicate the Binary Choice Preferences on a piece of paper and decide questions 4) through 14) before reading on.

Answer

1. a ($80,000 and 40 hrs./week) vs. b ($100,000 and 30 hrs./week) b+
2. a ($80,000 and 40 hrs./week) vs. c ($60,000 and 50 hrs./week) c–
3. a ($80,000 and 40 hrs./week) vs. d ($100,000 and 40 hrs./week) d+
4. a ($80,000 and 40 hrs./week) vs. e ($100,000 and 41 hrs./week) e
5. a ($80,000 and 40 hrs./week) vs. f ($100,000 and 45 hrs./week) f
6. a ($80,000 and 40 hrs./week) vs. g ($100,000 and 50 hrs./week) g
7. a ($80,000 and 40 hrs./week) vs. h ($100,000 and 55 hrs./week) h
8. a ($80,000 and 40 hrs./week) vs. i ($100,000 and 60 hrs./week) i

9. a ($80,000 and 40 hrs./week) vs. j ($100,000 and 65 hrs./week) j
10. a ($80,000 and 40 hrs./week) vs. k ($100,000 and 70 hrs./week) k
11. a ($80,000 and 40 hrs./week) vs. l ($100,000 and 75 hrs./week) l
12. a ($80,000 and 40 hrs./week) vs. m ($100,000 and 80 hrs./week) m
13. a ($80,000 and 40 hrs./week) vs. n ($100,000 and 85 hrs./week) n
14. a ($80,000 and 40 hrs./week) vs. o ($100,000 and 90 hrs./week) o

BINARY CHOICE PREFERENCES

In Figure 2.1 above, bundles e through o are depicted from right to left in Quadrant II on an imaginary line that is perpendicular to the horizontal axis at $100,000. Consider the two extreme bundles, bundle e on the right (41, $100,000) and bundle o on the left (90, $100,000): most people will have attached + to bundle e because they will want to trade-off 1 additional hour /week of work for annual increase in income of $20,000; and, most people will have attached – to bundle o because they will not want to trade-off 50 additional hours/week of work (for a total of 40 + 50 = 90 hours/week) for an increase in annual income of only $20,000. (Persons with extreme preferences in favor of leisure may have attached – to bundle e and those with extreme preferences for income could have attached + to o).

Now look at the bundles between e_+ and o_-. Starting at e_+ and going to the left toward o_-, at some point, either + become – or there is some bundle with the label of "0" in between + and –. Not everyone will have the same number of + and –, and some may have a 0 in the list, some may not. Persons with extreme preferences might have all + or all –. As examples, consider List 1) a person's preferences from o to e might be – – – + + + + + + + + and List 2) a person's preferences might be – – – – – – – – 0 + + +. List 1) preferences depict a high preference for income over leisure; he or she may be a workaholic; List 2) shows a high preference for leisure over income. These two lists depict different but transitive (consistent) preferences. List 3) preferences might be – – + – – + + + + + +, which are inconsistent (intransitive). In order for a person's preferences to be internally consistent, that is, transitive, every – must be to the left of every +; and + and – cannot be interspersed. In other words, any preference to the left of – must be –, and any preference to the right of + must be +. List 3) of preferences are intransitive and inconsistent because the switch from – back to + states that he or she prefers working more hours/week, 80, to working fewer hours/week, 75, for the same income, $100,000, which violates Rule E1, the assumption that more leisure is preferred to less.

The 0 in List 2) preferences means that this person is indifferent between bundle a (40, $80,000) and bundle h (55, $100,000). In List 1) preferences, the point of indifference, which is interstitial, is not shown; it is between bundles where the + change to –. A finer gradation of choices than intervals of 5 hours/week would be necessary in order to reveal the zero point of indifference. In theory, all goods are infinitely divisible, so a 0 point can in principle always be found.

Indifference Curves and Transitivity

Now assume another thought experiment is conducted, which is also depicted in Figure 2.1, where bundle a (40, \$80,000) is compared to bundle p (40, \$60,000), which is on the vertical line between Quadrants I and IV. Bundle a must be preferred to bundle p. Why? Answer: Rule E1. Next, bundle a is compared to successive bundles in which income is held constant at \$60,000, but the number of hours/week of work in each successive bundle is less: bundle q (35, \$60,000), bundle r (30, \$60,000), bundle s (25, \$60,000), bundle t (20, \$60,000) and so on. In the previous list, most economically rational individuals started with + on the right ended with – on the left; in this list, individuals start with – on the left because bundle a is preferred to bundle p, and end up with + somewhere on the right.

Comparisons of bundle a to bundles in Quadrant IV along the \$60,000 income level with different numbers of hours/week of work reveal another point of indifference between bundle a and some other bundle, say bundle t (20, \$60,000). This person is indifferent between bundle a (40, \$80,000) and bundle t (20, \$60,000). See Figure 2.1, where bundles p through s have – attached to them and bundle t has a 0 attached to it.

An infinite number of experiments are possible comparing bundle a to bundles at different levels of income and work in Quadrants II and IV. Each experiment would yield a point where the person is indifferent between bundle a and some other bundle at another level of income. When all the points of indifference are connected, they form an *indifference curve*. In Figure 2.2 below, assume that a particular rational individual is *indifferent* between bundle a (40, \$80,000) and bundle h (55, \$100,000); assume this person is also indifferent between bundle a (40, \$80,000) and bundle t (20, \$60,000); this means that bundles h, a, and t represent three bundles on the same indifference curve. The individual, whose preferences are represented by this curve, is indifferent among these three bundles even though they contain different amounts of income and work-leisure. Bundles on indifference curves are another example of transitivity in economic rationality: if h = a and a = t, then h = t.

As this individual moves from bundles t to a to h (right to left) along the indifference curve, the slope (ΔIncome/Δwork) of the indifference curve rises in absolute value (where Δ means "change in"), indicating that for each additional increment of \$20,000, this individual is willing to give up less leisure. As shown in Figure 2.2, in going from bundle t to bundle a, for an additional \$20,000 this individual is willing to work an additional 20 hours/week, which is a marginal trade-off of (ΔIncome)/(ΔWork) of \$20,000/20 = \$1,000/1; however, in going from bundle a to bundle h, for an additional \$20,000, this individual is only willing to work an additional 15 hours/week, which is a marginal trade-off of (ΔIncome)/(ΔWork) of \$20,000/15 = \$1,333/1. Thus, along the indifference curve from right to left, the slope of the curve rises because more income is required for each additional hour/week of work; that is, there is an *increasing marginal trade-off* of income for work.

This analysis is consistent with one's intuition, which is that as one good gets more plentiful (income) and another good gets more scarce (leisure), in order to be indifferent among choices, an individual is willing to give up less and less of the scarce good for equal additional amounts of the relatively plentiful good. Correspondingly, as we

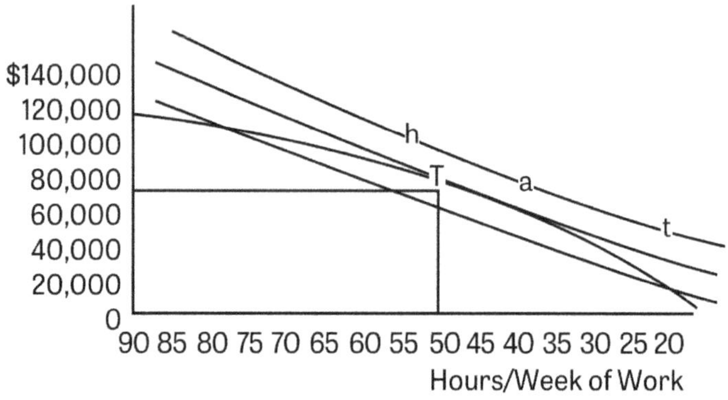

Figure 2.2 Optimal Choice. *Source*: Created by the author.

move from left to right in the direction of decreasing income and increasing leisure, the slope must decrease because income is becoming scarcer relative to leisure.

In Figure 2.2, the indifference curve, on which the bundles h, a, and t are, can also be used to illustrate the transitivity (consistency) property of economics. According to Rule E1, any bundle *above* this indifference curve is preferred to any bundle *on* the indifference curve, and any bundle on this indifference curve is preferred to any bundle *below* this indifference curve; therefore, any bundle above the indifference curves must be preferred to any bundle below the indifference curve.

Economic Rationality Is Totally and Linearly Ordered

Recall that bundle a was an arbitrarily chosen bundle. By choosing any other arbitrary bundle as the reference bundle, then drawing lines through it which are perpendicular to the horizontal and vertical axes and making a similar infinite series of bilateral comparisons, additional indifference curves for this person can be generated. In Figure 2.2 above, three such indifference curves are depicted. No two indifference curves of an individual can intersect due to the transitivity of consistency principle. (However, indifference curves of different individuals, who have different preferences, can

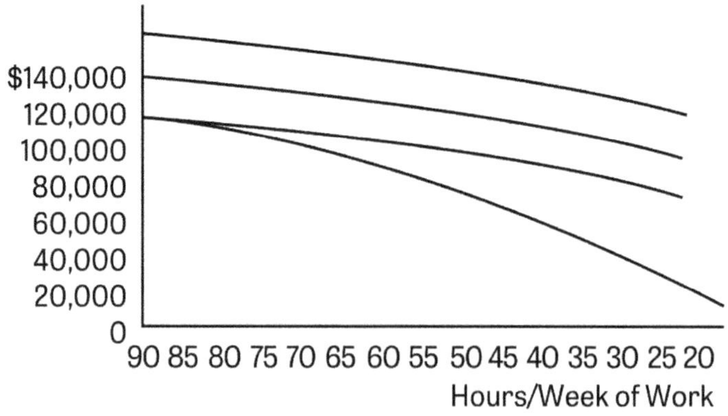

Figure 2.3 Extreme Indifference Curves (Corner Solution). *Source*: Created by the author.

intersect: those who prefer income to leisure have indifference curves with flatter slopes; those who prefer leisure to income have indifference curves with steeper slopes. Why?)

Indifference curves are said to be *everywhere dense*, which means that every possible bundle of work and income must be on some indifference curve. In other words, in economic rationality, all bundles are totally ordered. Moreover, since according to Rule E1, a higher indifference curve is preferred to a lower indifference curve, these indifference curves can also be rank ordered along a line from least preferred indifference curve to most preferred indifference curve. Since one indifference curve contains many bundles among which an individual is indifferent, many bundles can be stacked at the same point on the line on which indifference curves are rank ordered.

ECONOMIC RATIONALITY MAXIMIZES (OPTIMIZES) GOALS SUBJECT TO CONSTRAINTS

Thus far, we have discussed a rational individual's preferences with respect to income and leisure, which are depicted in the Figure 2.2 by indifference curves. However, an individual's preferences (wishes) on income and leisure are constrained by job prospects in the labor market. Despite her or his wishes, not everyone can get a job as an executive of a large corporation or become a professional athlete.

A depiction of an individual's prospects in the labor market is given by the job possibilities curve in Figure 2.2, which is shaped like the production possibilities frontier discussed in chapter 1 for similar reasons (the law of diminishing returns and law of increasing costs). In addition to being limited by one's skills, an individual is probably most efficient working from 35 to 60 hours/week. More hours/week than this range usually leads to fatigue and poorer performance (diminishing returns); fewer hours/week than this range usually limits an individual's opportunities to lower paying jobs. The concave (bowed out) to the origin job possibilities curve is also consistent with the assumption that individuals choose among bundles of work and income rather than income and hours/week of work as independent variables.

Given the market constraint on job choices, a rational economic individual chooses the job on the highest attainable indifference curve that is just tangent to one's job possibilities curve; this is called the *tangency solution*. In Figure 2.2, the tangency solution is at point T; this individual's optimal job requires work of 50 hours/week and pays $85,000 (50, $85,000). Indifference curves, such as the one with the bundles h, a, and t on it, which are above the job possibilities curve, are unattainable; indifference curves that intersect the job possibilities curve or are below it provide less than optimal satisfaction or utility.

As discussed above, rational decision-making is purposeful, has a criterion of consistency, and is subject to constraints. Rules E1 and E2 provide purpose and consistency; the constraints on an individual's choices are provided by the job possibilities curve, over which an individual does not have control.

WELL-BEHAVED (CONVEX) VERSUS EXTREME (CONCAVE) PREFERENCES

The next chapter will make use of the following, seemingly arcane, but important distinction between convex and concave indifference curves. As explained in the next several paragraphs, in Figure 2.2, indifference curves that are convex to the origin (bowed in) reveal preferences that are *well behaved*; whereas, in Figure 2.3, indifference curves that are concave to the origin (bowed out) reveal preferences that are extreme, which in the context of socially divisive issues like abortion rights, gay rights, gun control, and, religious disputes (chapters 3 and 4) are preferences that are absolute and brook neither compromise nor marginal trade-offs.

Rules E1 and E2 dictate that indifference curves must have a negative slope (indifference curves must descend from left to right). However, these two rules do not require that indifference curves be convex to the origin; they could also be straight lines or be concave to the origin and still satisfy the two behavioral rules. Economists assume indifference curves are convex and rule out indifference curves that are concave because of their different behavioral implications. When real-world individuals are asked to choose among varying amounts of any two goods, like income and leisure, their behavior indicates that they generally prefer some of each good rather than all of one good and none of the other. Thus, specialization in consumption is usually ruled out by economists because it is contradicted by casual observation of the way individuals behave.

Convex indifference curves, which imply that individuals want some of many goods, are said to be well behaved. In Figure 2.2, convex indifference curves result in a tangency at (50, $85,000), which is an *interior solution*, because the point of tangency, T, is between but not on either the horizontal work axis or the vertical income axis.

Concave indifference curves imply specialization in consumption and reveal extreme preferences. Depending upon the slope of concave indifference curves, an individual will choose either a bundle with as much income as possible, regardless of hours/week of work, or a bundle with as much leisure as possible, least hours/week of work, regardless of income. Concave indifference curves result in tangencies on the horizontal axis or the vertical axis and are called *corner solutions*. In Figure 2.3, the individual's concave indifference curves result in corner solution at (90, $125,000) on the vertical axis.

When individuals, judges, or legislatures make decisions in commercial and common law contexts, the assumption that judges' preferences between two goods, such as property owners' rights versus rights of those with security interests in property, are convex or well behaved is reasonable and for the most part is factually accurate (government and court decisions reveal marginal trade-offs between these two goods). However, when individuals, judges, and legislatures make decisions concerning divisive social issues, such as abortion rights, gay rights, gun control, and religious disputes, absolute or extreme preferences, which are concave rather than convex, cannot be ruled out and are all too common. The discussion of abortion rights in chapters 3 and 4 will describe some preferences that are well behaved (convex) and some preferences that are extreme (concave) and result in corner tangency solutions.

ECONOMIC RATIONALITY, TRANSITIVITY, AND TOTAL (LINEAR) ORDERING IN CASE LAW

If judges who decide common law and commercial cases have shared values and decide cases according to economic rationality, we should be able to take the next logical step, which is that these case decisions must exhibit the two crucial properties of economic rationality, transitivity and total ordering. When judges balance, say two competing interests in cases, these cases can be totally ordered in relation to each other on two-dimensional graphs and can be rank ordered on a line. Violation of these requirements means such decisions are not economically rational. In the next chapter, three examples will be given of cases in which judges balance two or three competing interests, demonstrating that these decisions can be totally ordered according to economic rationality on graphs or rank ordered linearly.

NOTES

1. Robert Nozick, *The Nature of Rationality* (Princeton, NJ: Princeton University Press, 1993), 63–64.
2. See *O'Brien v. Muskin Corp.*, 463 A.2d 298, 305–06 (N.J. 1983).
3. Nozick, note 1, 64.
4. Joseph Stiglitz, *Economics* (New York: Norton, 1997), 28–29; Frank Esterbrook, "Ways of Criticizing the Court," *Harvard Law Review* 95 (1982): 802, 812.
5. See chapters 5 and 6. "The strong claim that legal reasoning is not different from reasoning in other fields is misleading in a significant respect. The characteristics of a field's data and language, as well as the laws of logic, determine the transformations, or 'moves', that can or cannot be made within a field." Steven Burton, *An Introduction to Legal Reasoning* (New York: Aspen, 1985), xv.
6. David Laibson and John List, "Principles of (Behavioral) Economics," *American Economic Review: Papers and Proceedings* 105 (2015): 385–390.
7. US Senator Warren Rudman, "60 Minutes," CBS Television Network, program broadcast (Sunday, December 6, 1993, 7–8 p.m.).
8. Bruce J. Caldwell, "Clarifying Popper," *Journal of Economic Literature* 29 (1991): 14, 17–20.
9. See Colin F. Camerer, *Behavioral Game Theory* (Princeton, NJ: Princeton University Press, 2003). Decisions individuals make concerning the future when such decisions are interdependent with the decisions of others (e.g., stock markets) will be discussed in chapter 7.
10. In chapter 6, the two rules are defined as binary relations; their mathematical properties are also discussed.
11. Hal Varian, *Intermediate Microeconomics: A Modern Approach*, 7th ed. (New York: Norton, 2006), 44.
12. More precisely, Rule E2 states that an individual can determine whether one bundle is "weakly preferred" to another bundle or he or she is indifferent between them, where "weakly preferred" means one bundle is "at least as good as" another bundle. See chapter 6, where the two rules are defined as binary relations and their mathematical properties are stated.

Chapter 3

Judicial Use of Economic Rationality (Efficiency)

Chapter 2 demonstrated that all possible bundles of competing economic goods are totally and linearly ordered and can be depicted in spatial dimensions: for example, all possible bundles consisting of two goods can be depicted in two-dimensional space, bundles of three goods can be depicted in three-dimensional space, and four or more goods are ordered on mathematical vectors.

This chapter demonstrates that cases, both actual and hypothetical, in which judges use economic rationality as the normative criterion can be totally *ordered* according to economic rationality, either in spatial dimensions or linearly. Three examples are considered: the first is from the law of torts; it concerns products liability cases in which the issue is whether automobiles are "crashworthy;" second, Lon Fuller's[1] well-known, hypothetical cases, which balance property rights of original owners of goods against rights of good faith buyers after theft or fraud, will be ordered according to economic rationality.[2] The third example concerns the three interests that are balanced via trade-offs in *Roe v. Wade*.

CRASHWORTHY AUTOMOBILE CASES

At one time, a dispute existed as to whether automobiles ought to be designed so as to reduce injuries from a collision. In *Evans v. General Motors Corp.*,[3] a new, X-frame automobile design, which increased overall stability but which turned out to be less safe than a conventional rectangular frame when an automobile was hit from the side, was the subject of a design defect, products liability suit. The Seventh Circuit Court of Appeals held that the manufacturer was under no duty to design an accident-proof or crash-proof car because the intended use of automobiles did not include participation in collisions. However, in *Larsen v. General Motors Corp.*,[4] the Eighth Circuit disagreed with the Seventh Circuit, holding that an automobile manufacturer is under a duty to use reasonable care in designing its vehicle so as to avoid subjecting the user to an unreasonable risk of injury in the event of a collision. Furthermore, the *Larsen* court held that the manufacturer is liable for that portion of the damage or injury over and above the damage or injury that would have occurred as a result of the impact from the collision absent the defective design. A long line of cases supported each side.[5] The *Larsen* view ultimately prevailed and even the Seventh Circuit changed its opinion, repudiating *Evans*.[6] Courts now generally agree that automobile manufacturers owe a duty to occupants of automobiles to make them crashworthy, that breach of this duty subjects a manufacturer to liability based on a defective design theory, and that the cases concern the extent of such duty.

The conflict between the *Evans* and *Larsen* lines of cases suggests that all of these cases cannot be ordered consistently. However, economic theory does not postulate that all cases in which judges use economic rationality must be capable of a consistent ordering, but only that if a consistent ordering is not possible a mistake must have been made. In cases involving economic issues, such as the crashworthy cases, the assumption that judges' preferences are both compatible and consistent with economic efficiency is reasonable. As indicated above, the Seventh Circuit eventually conceded that a mistake had been made in *Evans*. Thus, the following discussion, which shows how crashworthy cases are totally ordered according to economic rationality, will be confined to cases that are compatible with the *Larsen* view; that is, those cases in which a total consistent ordering is possible.

Judges essentially decide crashworthiness cases as ordinary negligence cases using cost-benefit (risk-utility) efficiency in which the cost of reducing injury to individuals from an accident through the provision of additional impact protection is balanced (traded-off) against the benefit from such added impact protection, using the Judge Hand negligence test, $B < PL$, where P is the *probability* that injury will occur if precaution against injury is not taken; L is the *loss* in dollars if injury occurs (PL is expected loss or the loss discounted by its probability of occurring); and B is the *burden* or cost in dollars of taking precaution to prevent injury from occurring. If $B < PL$, an actor is negligent for not taking precaution because the burden of taking care is less than the expected damage that would result if care is not taken; if $B \geq PL$, an actor need not take care because the cost of eliminating the possibility of injury equals or exceeds the expected damage from injury should it occur.

There is an ambiguity in the Judge Hand test, similar to the ambiguity in the Kaldor-Hicks criterion (chapter 1), with respect to whether the variables B and L concern total costs and benefits or marginal costs and benefits. In order to apply the Judge Hand test correctly, the comparison between B and PL is not made on the basis of the relation of total burden, B, to total expected damage, PL, but at the *margin* for a small change, by comparing $\Delta B < \Delta PL$ (where the symbol "Δ" means "change in").[7] If a collision occurs, the benefit from adding a particular protection against injury (say, adding padded steering wheels or a driver's air bag) is the decrease in the marginal expected damage (ΔPL), where P is the probability of an accident occurring, multiplied by ΔL, which is the marginal loss in dollars from that portion of the damage or injury over and above the damage or injury that would have occurred from the collision impact with added protection.

The increase in marginal cost of impact protection measured in dollars is ΔB, which is the burden of increased precaution against accident impacts; the marginal burden is equal to the cost of redesign per automobile. Posner said: "Fortunately the common law method facilitates a marginal approach, simply because it will usually be difficult for courts to get information on other than small changes in the safety precautions taken by, or should have been taken by, the injurer."[8] The cost of additional increments of impact protection increases because it is assumed that manufacturers increase impact protection by employing the least expensive method first, the next least expensive method second, and so on. For example, impact protection can be increased least expensively by providing seat belts, then by adding increased structural support to cars, then by developing air bag technology, and so on.

The only difference between the Judge Hand test version of cost-benefit efficiency and risk-utility analysis is that, on its face, the Judge Hand test does not consider the utility of a product, such as an individual's "taste" for driving in a convertible car with the top down in spite of the slightly higher risk of injury from an automobile roll-over. However, utility is easily accommodated in the Judge Hand test through an increase in the burden factor, B; that is, the more useful a product, the greater the burden on society to do without it. Under this formulation of the Judge Hand test, both the cost of increased impact protection and the disutility of being encumbered by bulky impact protection devices are considered in the burden factor, B.

In determining the expected damage from a design defect, the trier of fact must consider not only the expected damage from the accident that occurred, but also the expected damage from any other similar, reasonably foreseeable accidents that the defective design might cause. For example, when a railroad engineer fails to blow the train's horn as a warning that it is about to move, which cost of care, B, is small in dollars, there is not only a probability, P_1, that hearing the horn would save the life of "an inspector who had crawled under a car," ΔL_1, but also a probability, P_2, that hearing the horn would save "an inspector leaning on a car," ΔL_2, and probability, P_3, that hearing the horn would save "a railroad employee doing repairs on top of a car," ΔL_3, or "a brakeman straddling two cars," ΔP_4L_4, "and anyone else who might have business in or on (as well as under) a car," ΔP_5L_5.[9] To be rigorously accurate, the right side of the Hand test should be written as $\sum\Delta PL = \Delta P_1L_1 + \Delta P_2L_2 + \Delta P_3L_3 + \Delta P_4L_4 + \Delta P_5L_5$ (where $\sum$ stands for summation) Thus, to be accurate, the Hand test for liability should be written as $\Delta B < \sum\Delta PL$. (For simplicity in subsequent discussion, the symbols for summation, $\sum$, and marginal change, Δ, will be omitted.)

In Figure 3.1, six crashworthiness cases have been ordered by the Judge Hand test in the context of economic rationality. Marginal Impact Damage, ΔPL, is measured in dollars on the vertical axis; Marginal Cost of Impact Protection, ΔB, is measured in dollars on the horizontal axis. All cases in which there is liability for design defects according to the Judge Hand test are where $\Delta B < \Delta PL$. All cases in which there is no liability for design defects are where $\Delta B \geq \Delta P$; consumers bear the risk of collision damage in these cases. Optimal design safety occurs where $\Delta B = \Delta PL$.

First, consider two automobile accident cases in which persons were thrown against sharp objects in the passenger compartment as a result of a collision. In both cases, the marginal burden per automobile, B, of redesigning the passenger compartment to remove sharp objects is quite low; moreover, such hazards have no countervailing utility. In *Ellithorpe v. Ford Motor Co.*,[10] the plaintiff suffered severe facial injuries when her face struck a raised plastic emblem on the steering wheel as a result of an accident. Since the probability is extremely high that a driver will come in contact with sharp objects protruding from the steering wheel as a result of almost any

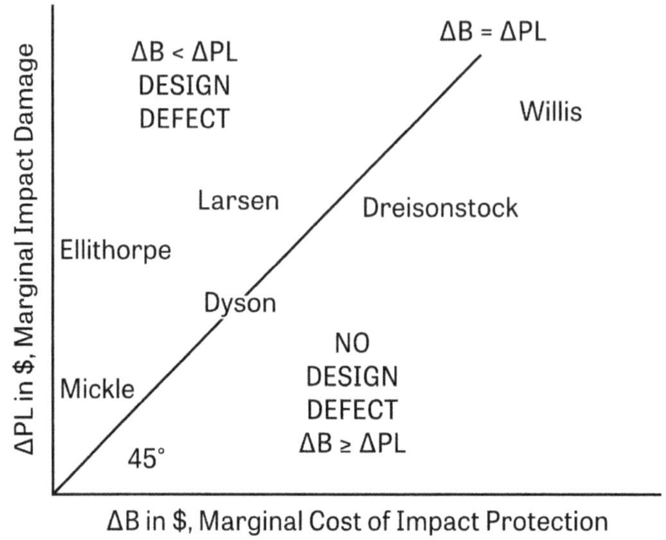

Figure 3.1 Cost of Damage versus Cost of Protection. *Source*: Created by the author.

significant front-end collision, and the marginal injury resulting from such impacts will often be substantial, PL is greater than B, added cost of a padded steering wheel. In Figure 3.1, *Ellithorpe* belongs at the intersection of a relatively low value for B and a relatively high value for PL.

In *Mickle v. Blackmon*,[11] a passenger was thrown against the gear shift knob that protruded two inches beyond the rim of the steering wheel and was of a color that caused it to become brittle in sunlight; due to the impact, the knob broke and she was impaled upon the gear shift lever, resulting in paralysis. Although the marginal injury in this particular accident was severe, a very high L, the probability of such accidents occurring, P, is very much smaller than that of injuries from being thrown against a sharp object on the steering wheel; *Mickle* involves a freak accident. On the PL axis in Figure 3.1, PL in *Mickle* is substantially less than PL in *Ellithorpe,* so *Mickle* is placed at a lower level than is *Ellithorpe*. However, since the burden of redesigning the gear shift knob so as not to become brittle due to sunlight is very small, B < PL, the defendant is also liable in *Mickle*.

Next consider three cases in which the marginal burden, B, is significant due to substantial redesign costs and because the initial designs were chosen because they had at least some positive utility, so that either doing without these designs or providing added protection involves a significant increase in the burden, B. In *Larsen*, the plaintiff, who was driving a car that had been designed with a rear-mounted engine in order to compete with a popular imported car that had a rear engine, was injured seriously in a head-on collision after the steering shaft was thrust rearward. Even though it is obvious that drivers of rear-engine cars are more vulnerable to injury from head-on collisions, this car was particularly vulnerable to the type of accident that occurred because the steering column protruded beyond the front tires; that is, PL is large.[12] The marginal B for redesigning this particular vehicle depends on the cost of providing protection from rearward displacement of the steering column, which was substantial.

In comparing B to PL, it can be argued that since head-on collisions are both likely and often result in severe injuries, the PL is probably quite large and in excess of the cost of increased impact protection, so that it should have been provided.[13] In Figure 3.1, *Larsen* is depicted well within the area or liability.

In *Dyson v. General Motors Corp.*,[14] a two-door (hardtop) car, which was designed without either center column supports for the roof or full-frame doors in order to give it a sporty appearance, left the road and rolled onto its roof. The right side of the roof collapsed under the weight of the car, and the plaintiff was injured severely. The burden B is not only the cost of providing support for the roof, such as a roll bar or side posts, but also the reduction in utility due to the less attractive styling. However, if consumers are unaware that the sporty appearance of a hardtop has a significant cost in terms of reduced safety, the reduction in utility from the redesign will be overstated. Since the cost of redesign plus the reduction in utility is additive, the B in *Dyson* is arguably as large as the B in *Larsen.* The court denied the defendant's motion for judgment on the pleadings and rejected the contention that a verdict for plaintiff in this case would be equivalent to declaring that all soft-top automobiles without roll-bars are unreasonably dangerous per se. The court said that a manufacturer was not under an obligation to provide a hardtop model which would be as resistant to roll-over damages as a four-door sedan but was required to provide a hardtop automobile which was a reasonably safe version of such model, and which was not substantially less safe than other hardtop models.

Because few accidents will result in roll-overs, P is small, but L, the impact damage to occupants of a vehicle without protection from a roll-over should one occur, is large. Since the expected damage, PL, to the occupant of a vehicle without protection from a roll-over is less than that to the occupant of a vehicle without protection from head-on collisions, *Dyson* belongs lower on the PL axis than does *Larsen.* As indicated above, the marginal burden, B, of redesign in *Dyson,* is arguably as large as the B in *Larsen*; therefore, on the B axis, *Dyson* is placed at the same level as *Larsen* on the B axis. Since the B in *Dyson* is substantially greater than in cases involving sharp objects in the passenger compartment, Dyson belongs substantially to the right of *Ellithorpe* and *Mickle* on the B axis. Finally, because reasonable persons may differ on whether $B < PL$ or $B \geq PL$ in *Dyson*, it is represented as straddling the 45° line.

In *Dreisonstock v. Volkswagenwerk, A.G.*,[15] the plaintiff sued a car manufacturer for enhanced leg injuries she sustained as a passenger in a microbus, which crashed into a telephone pole as it traveled downhill at more than 40 miles per hour. The plaintiff claimed that the vehicle had insufficient crush space in front of the seats. The court stated that this van-type, multipurpose vehicle was uniquely developed in order to provide the maximum amount of cargo or passenger space in an inexpensive vehicle of such dimensions as to make possible easy maneuverability. The driver's seat was brought forward in close proximity to the front of the vehicle to add to the cargo or passenger space. In addition, the utility of the design was attested to both by the popularity of the vehicle and its imitation by other manufacturers. The court said that there was no evidence that the crashability of the design could be improved consistent with the specific purposes of its design. In terms of the Judge Hand test, the great utility of the vehicle and the difficulty of improving the crashworthiness of its design (given the state of art for automobile safety in the 1970s) without interfering with its utility means that B is large and exceeds PL even though PL, which represents potential injury from a front-end collision and is therefore arguably equal to that of *Larsen,* is substantial. Thus, in Figure 3.1, *Dreisonstock* is depicted on a level with *Larsen* in terms of PL, but is to the right in terms of B and therefore is below the 45° line.

Finally, in *Willis v. Chrysler Corp.*,[16] a policeman was killed in a collision with another car. At the time of impact, the police car was traveling between 65 and 70 miles per hour when it veered into the lane of oncoming traffic. Although there was no estimate of the speed of the other car, it left 205 feet of skid marks and continued in its direction of travel for 81 feet after impact. The police car broke into two sections just behind the front seat, both parts coming to rest 100 feet apart. All occupants of both vehicles were killed. The plaintiffs alleged a design defect because the car separated into two pieces on impact. Although the court granted a summary judgment based on *Evans*, it used language that would be equally applicable to a decision following *Larsen*: an automobile manufacturer is under no duty "to design an automobile that could withstand a high speed collision and maintain its structural integrity." Although the probability of a high-speed collision is less than that of a low speed collision, the expected damage including loss of life in a high-speed collision is much higher than from a low speed collision. Thus, the marginal expected impact damage, PL, which could be prevented if vehicles were designed to withstand a high-speed collision, is greater than for a low speed collision. Hence, *Willis* is depicted at a higher level on the PL axis than *Larsen* or *Dreisonstock.* In terms of the marginal burden, B, of redesigning cars to withstand a high-speed collision, *Willis* is depicted far to the right and below the 45° line because of the immense burden that would be placed on manufacturers if they had a duty to design automobiles to protect occupants from injury and death from a high-speed collision. Thus, there is no liability in *Willis* because $B > PL$.

When the Judge Hand test is used to determine liability, the complete ordering property of economic rationality dictates that every case must be related to every other case. Although rational people can differ as to the placement of these cases, the relatedness of every case to every other case is beyond dispute.

Economic Optimization, Line Drawing, and Stare Decisis

Interest balancing through marginal trade-offs often results in a line of demarcation that separates cases that should be decided one way from those that should be decided the other way. This line of demarcation suggests a relation between economic rationality and the principle of stare decisis, which postulates that like cases should be decided alike. When a court engages in interest balancing through marginal trade-offs and line drawing, all cases on one side of the line of demarcation are alike and therefore belong in one legal category; cases on the other side of the line are alike and belong in another legal category.

For example, in the crashworthy automobile cases of Figure 3.1, all cases in which $B < PL$ are alike, that is, all cases above the line of demarcation, $B = PL$, are within the legal category of design defects or negligence; all cases on the line and below it, $B \geq PL$, are within the legal category of due care. This relationship between interest balancing through marginal trade-offs, line drawing, and stare decisis can be expressed by a mathematical analogy: all cases on one side of the line of demarcation can be mapped onto one legal category; all cases on the other side can be mapped onto another legal category. In the context of the Judge Hand test, the line of demarcation is B = PL. In chapter 6, the conditions under which economic rationality and legal rationality are consistent or "isomorphic" will be discussed.

LON FULLER'S HYPOTHETICAL CONTRACT CASES

Hypothetical cases have a very important role in both legal and economic reasoning for three reasons: (1) hypothetical cases can be constructed so as to fill in gaps between actual cases, so that the number of cases can be as dense as the analyst chooses to make them; (2) the greater the number of cases, the more defined becomes the line that demarcates cases that should be decided one way from cases that should be decided the other way; and (3) hypothetical cases can be constructed so as to isolate only those goals or interests that the analyst wishes to balance.

Lon Fuller was the author of five hypothetical contract cases,[17] in which two competing interests are balanced: the interests of owners of personal property, whose interest is protected so that property will be transferred voluntarily through markets (contracts) rather than coercively via theft or deceitfully via fraud, are *balanced* (traded-off) against the interests of buyers of personal property, whose reasonable expectations that a seller can transfer good title are protected in order to encourage commerce and exchange.[18] Initially, these cases will be discussed as a court would in terms of legal categories and rules of law; then, they will be analyzed as problem in economic rationality.

Fuller's Five Hypothetical Cases

Fuller constructed the five cases in order to demonstrate to a philosopher with whom he had a disagreement that the legal decision process is rational.[19] He discussed the cases in the order in which they might be decided in a common law jurisdiction. He said, "In the early cases the courts did not have imagination enough to foresee the situations of fact that did later arise, but when those situations arose they contained nothing that was not accessible to the minds of the judges from the beginning."[20] The cases are decided according to the personal property law as modified by cases in which courts realize that property rights must be balanced against the reasonable expectations of those engaged in commerce. The parties in each case are different persons, but to facilitate comparisons Fuller used symbols to indicate the role played by a particular party: thus, O designates the original owner of personal property (a horse), T stands for thief, and G stands for "good faith purchaser," which is defined as one who 1) acts honestly, 2) pays valuable consideration, and 3) is without notice of theft or defect in title; that is, a good faith purchaser has neither knowledge nor reason to know that a seller of personal property does not have good title.

Case 1 (good faith purchaser after theft): T, a thief, steals a horse from O, the owner, and sells it to a good faith purchaser, G_T, where the subscript T means good faith purchaser after theft (line of transfer: O – T – G_T). In the suit of O versus G_T to recover the horse, O wins. A court might have held that "a thief has a 'void' title to goods and therefore cannot pass what he does not have"; title never left O, the owner.

Case 2 (fraud): O sells a horse to F who pays with a forged check. After delivering the horse to F, O discovers that he has been defrauded (line of transfer: O – F). In the suit of O versus F to recover the horse, O wins. The court, believing that the rule in Case 1 also covers Case 2, may restate the rule saying, "A person who acquires goods without acquiring valid title cannot pass a valid title."

Case 3 (good faith purchaser after fraud): O sells a horse to F who pays with a forged check; then, F sells the horse to a good faith purchaser G_F (line of transfer: O – F – G_F). This case is similar to Case 2 in that O sells a horse to F who pays with a forged check. This case is similar to Case 1 in that the third party, G_F, is a good faith purchaser, but different from Case 1 in that G_F bought the horse from a defrauder rather than from a thief. In the suit of O versus G_F to recover the horse, G_F wins. In Case 3, the balance of interests is in favor of the buyer because it would place an intolerable burden on commerce if purchasers of property were compelled to scrutinize the details of a transaction by which the former owner voluntarily delivered property into the hands of the person now offering it for sale. As between original owner and buyer, the former is blameworthy. "Fraud takes many and subtle forms; if the victim could not recognize it, it is unreasonable to ask a stranger to the transaction to determine whether it was present."[21]

In order to distinguish Cases 1 and 2 from Case 3 according to the property law paradigm, the court creates a new legal property concept or category called *voidable* title, which applies to transfers based upon fraud. The rule that is thought to cover all three cases is now: "a) a person who acquires goods with a *void* title (after theft) acquires no title and cannot transfer valid title; b) a person who acquires goods with a *voidable* title (after fraud) may transfer valid title to a good faith purchaser."

Case 4 (the "shelter rule"): This case is like Case 3 except that after buying the horse from F, the good faith purchaser, G_F, sold it to K_F, who paid value, was honest in fact, but had knowledge of the fraud before he bought the horse from G_F (line of transfer: O – F – G_F – K_F). In the suit of O versus K_F to recover the horse, K_F wins. O argues that K_F is not a good faith purchaser, since he had knowledge of F's fraud when he bought the horse. However, if O's argument were accepted, it would be possible for a person in the position of O to destroy the value of the title acquired by G_F simply by giving general publicity to the fact that F had induced the sale by fraud. Thus, in order to protect the good faith purchaser's right to sell property, K_F is sheltered by G_F.

The rule which is thought to cover all four cases is now same as the rule in Case 3 with the following amendment of part b) in italics: "b) a person who acquires goods with a voidable title may transfer valid title to a good faith purchaser *or to an assignee.*"

Case 5 ("laundering" and accomplice to fraud): This case is similar to Case 4 except that after buying the horse from F, G_F sold it to A_F, who not only knew of the fraud but was an accomplice who forged the name on the check used by F to pay O (line of transfer: O – F – G_F – A_F). In the suit O versus A_F to recover the horse, O wins. If those who acquire property by fraud could cut off the rights of the owner simply by transferring (laundering) property through a good faith purchaser and then back to an accomplice, fraud would be immeasurably easier to accomplish. To understand the implication of this laundering transaction, assume that G_F paid $100 to F for the horse, which $100 F gives to A_F who in turn buys the horse from G_F for the same $100 (plus a small profit so that G_F will have an incentive to transfer the property back to the conspirators), the net result of which is that F and A_F get the property after paying almost nothing for it.

The rule which is thought to cover all five cases is now same as the rule in Case 4 with the following amendment of part b) in italics: "b) a person who acquires goods with a voidable title has valid title if the person is a good faith purchaser or an assignee *who is not implicated in the voidable transaction.*"

It will be shown below that Fuller's five hypothetical cases, plus a sixth case, can be explained (described) by balancing of competing interests in the context of economic rationality's total ordering property, which dictates that every case be related to every other case.

Fuller's Cases Can Be Ordered in Economic Rationality

In order to encourage efficiency in a free market system, two competing interests, stability of private property and finality of voluntary transfers of property through markets, must be balanced. As is clear from Fuller's hypothetical cases, these two interests often compete or conflict. In Figure 3.2, Fuller's five cases (plus a sixth hypothetical case, discussed below), which courts decide by balancing the stability of property rights against finality of voluntary exchange, are ordered in relation to each other on a line on which Owners' Rights are at one end and Buyers' Rights are on the other. Owners' Rights are given greater legal protection to the left of the center dividing line (on which Buyers' Rights and Owners' Rights are in balance) because Owners' Rights outweigh Buyers' Rights; and, Buyers' Rights are given greater legal protection to the right of the center line because Buyers' Rights outweigh Owners' Rights. Recall from chapter 2 that economic rationality is not only completely ordered, it is linearly ordered.

The fact that the two competing interests, stability of property rights and finality of voluntary exchange, cannot be quantified is not a problem for economic rationality because the theory requires only ordinal preference ordering, not quantifiable (cardinal) preference ordering. (Recall indifference curves of chapter 2.) The theory of ordinal ordering was specifically constructed to divorce economic theory from the crude version of utilitarianism which assumed that quantification of individual preferences and desires was possible. The shift from cardinal to ordinal preference ordering in economic theory occurred in the late nineteenth century.

The placement of Cases 2 and 3 on Figure 3.2 is not difficult. In Case 2, the fraud case, the transfer of property (O – F) is not upheld. Fraudulent transfers, like those accomplished by theft, are not efficient because F got possession of the property without having exchanged anything of value. This implies the rights of fraudulent buyers should be given very little weight relative to Owners' Rights. Case 2 is placed close to the Owners' Rights end of the spectrum and far from Buyers' Rights end of the spectrum, indicating that Owners' Rights greatly outweigh Buyers' Rights in this case.

In Case 3, good faith purchaser after fraud, the transfer of property ($O - F - G_F$) is upheld. It is much easier for owners to prevent fraud than to prevent theft. Court

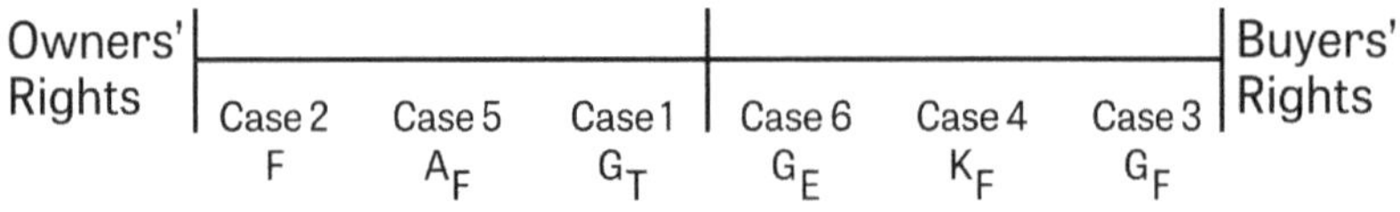

Figure 3.2 Owners' Rights versus Buyers' Rights. *Source*: Created by the author.

decisions in favor of buyers who are good faith purchasers after fraud merely reinforce the need for owners of goods to undertake reasonable investigations of the financial solvency, credit worthiness, and honesty of buyers before parting with goods. Case 3 is placed close to the Buyers' Rights end of the spectrum and far from the Owners' Rights end of the spectrum, indicating that Buyers' Rights greatly outweigh Owners' Rights in this case.

In Case 4, good faith purchaser shelters a buyer with knowledge, the transfer of property where (O – F – G_F – K_F) is upheld. In order to protect the right of owner, G_F, to sell property, K_F is sheltered. Although K_F wins because Buyers' Rights outweigh Owners' Rights in Case 4, the balance is not as overwhelmingly in favor of the Buyers' Rights as in Case 3, the simple fraud case; therefore, Case 4 is placed on the diagram slightly to the left of Case 3.

In Case 5, accomplice to fraud, the transfer of property (O – F – G_F – A_F) is not upheld. Although Owners' Rights greatly outweigh Buyers' Rights, the balance is not as overwhelmingly in favor of Buyers' Rights as in Case 2 because of additional market transactions. Therefore, Case 5 is placed on the spectrum slightly to right of Case 2.

In Case 1, good faith purchaser after theft, the transfer of property (O – T – G_T) is not upheld. However, the competing interests of property owner versus buyer (good faith purchaser) are approximately equal; both parties are without fault. Owners' Rights should be protected so that personal property will be transferred voluntarily through markets (contracts) rather than coercively via theft; and, Buyers' Rights should be protected in order to encourage voluntary exchange in markets. In Figure 3.2, Case 1 is placed slightly to the left of the center dividing line, on which Owners' Interests and Buyers' Interest are in balance.

The close balance of interests in Case 1 and difficulty of deciding it becomes clear when we consider that different societies may decide Case 1 differently. In the United States, the legal rule is that if there is theft in the chain of title, a thief does not have title to the property and therefore cannot pass title to subsequent purchasers, even if there is a long line of many subsequent good faith purchasers after the theft.[22] The rationale is that since personal property is easily and rapidly transported, protecting good faith purchasers of stolen goods may encourage hijacking and selling of stolen goods through third parties, called "fencing." However, in the past, a different rule, called the *market overt* rule, prevailed in much of Europe where after theft, a good faith purchaser acquired title to personal property which was bought from a reputable merchant in the ordinary course of business[23] (as opposed to a fly-by-night seller who is hawking goods from a suit case on a street corner). The market overt rule is very similar to but distinguishable from the entrustment provision of the Uniform Commercial Code, which is the subject of hypothetical example, Case 6, which is discussed below.

When judges, and lawyers who replicate the reasoning of judges, implicitly use economic rationality (balance interests through marginal trade-offs) and engage in line drawing, that is, the separation of cases that should be decided one way from cases that should be decided the other way, they are implicitly using the transitivity property of economic rationality, which was discussed in the chapter 2. As an example of transitivity, consider the relationship between Case 1 in Figure 3.2, which is decided in favor of owners, and Case 4, which is decided in favor of buyers. Any case, the facts of which require that it be ranked to the left of Case 1, should be decided in a similar fashion, that is, for owners; correspondingly, any case ranked to the right of Case 4 should be decided similar to Case 4, for buyers. Even if additional cases are found through subsequent research or new hypothetical cases are constructed that are ranked interstitially among the old cases, a case that was ranked to the left of Case 1 before new cases were added to the linear ranking will still be to the left of Case 1 and thus should be decided consistently with it; the same reasoning holds for cases to the right of Case 4.

Of course, it is possible to discover actual cases or to construct hypothetical cases that are ranked interstitially between cases that are decided one way and cases that are decided the opposite way, between Case 1 and Case 4; these interstitial cases may be difficult to decide because the side of the line of demarcation to which they belong is not clear. Such interstitial cases are the stock and trade of law professors, the bane of law students, and the subject of myriad appellate court opinions. In the context of Fuller's hypothetical contract cases, an example of an interstitial case between Case 1 and Case 4, is provided by the entrustment provision, § 2–403(2), of the Uniform Commercial Code (UCC).[24] This hypothetical case will be called Case 6.

Case 6 (entrustment of goods to a merchant who deals in goods of that kind): O entrusts a horse for the purpose of being shod to E who is a blacksmith and a horse trader. After entrustment, E sells the horse to a purchaser in the ordinary course of business, G_E, and absconds with the money from the sale (line of transfer: O – E – G_E). In the suit of O versus G_E to recover the horse, G_E wins. UCC §2–403(2) states that one who "entrusts" goods to a *merchant* who deals in goods of that kind gives the merchant the power to transfer all rights of the entruster to a buyer in the ordinary course of business. A person who buys in the "ordinary course of business" is a good faith purchaser who, in addition, buys goods from a reputable seller. In Case 6, the horse trader is a merchant who deals in goods of that kind; therefore this case is within the statute and should be decided in favor of G_E.

However, the competing interests of Owners' Rights and Buyers' Rights in Case 6 are closely balanced. Both parties are innocent; both the owner who entrusted the property and the buyer who bought the property dealt with someone who, to all outward appearances, is a reputable merchant. Case 6 is like Case 2, the fraud case, in that the entruster voluntarily parted with the property; but, Case 6 is also like Case 1, the good faith purchaser after theft case, in that the entruster did not intend to transfer title. In Figure 3.2, Case 6 is placed slightly to the right of the dividing line at which Owners' Rights and Buyers' Rights are in balance.

In the entrustment case, the competing interests are in such balance that if the issue were not definitively settled by the UCC, it is easy to conceive that different common law jurisdictions would be likely to decide such cases differently, that is, inconsistently.[25] The rationale for the UCC's entrustment provision is to protect consumers who buy an appliance from a retail store, which was selling the appliances on consignment that are still technically owned by a manufacturer or subject to a security interest. If the retailer goes bankrupt or absconds with the proceeds of appliance sales, absent the UCC's entrustment provision, the manufacturer could demand the return of the appliance from the innocent consumer who bought the appliance from what seemed to be a reputable retailer (market overt) and had neither reason to know nor notice of any defect in title to the appliance.

The legal rule, which covers Fuller's 5 hypothetical cases plus Case 6, is:

a. a person who acquires goods with a *void* title acquires no title and cannot transfer valid title;
b. a person who acquires goods with a *voidable* title has valid title if the person is a good faith purchaser or an assignee who is not implicated in the voidable transaction;
c. a buyer in the ordinary course of business, who bought goods that were entrusted to a merchant who deals in goods of that kind, has all rights of the entruster.

Query: Does the legal rule, a) to c), which covers the six hypothetical cases, elucidate or obfuscate the underlying balancing of economic interests between personal property owners versus buyers in commerce, which is described by economic rationality?

ROE V. WADE: TRADE-OFFS AMONG THREE COMPETING INTERESTS

Although *Roe v. Wade*[26] is no longer the governing precedent with respect to the right to abortion in the United States,[27] the decision is an example of balancing of three competing interests or goods: women have a fundamental right to privacy, which includes the right to have an abortion; states have a compelling interest in protecting the potential life of a fetus; and states also have a compelling interest in protecting the health of a pregnant woman. These three competing interests implicitly involve balancing three trade-offs:

One trade-off involves a woman's right to an abortion versus the state's interest in protecting a fetus; the greater the protection for a woman's right to an abortion, the less the protection given to a fetus, and vice versa.

A second trade-off involves the state's interest in protecting the health of a pregnant woman versus the state's interest in protecting a fetus; the greater the protection of a pregnant woman's health, the less the protection of a fetus, and vice versa.

A third trade-off involves a woman's right to an abortion versus the state's interest in protecting the health of the pregnant woman: in the first trimester of pregnancy, abortion performed under a doctor's care is as safe, if not safer, for a woman's health than is completion of the pregnancy; however, the longer the pregnancy, the higher the probability that abortion will be detrimental to the health of the woman.

According to the Supreme Court's decision in *Roe*, the right to an abortion differs in each of the trimesters: During the first trimester of the pregnancy, there can be no significant restriction on the right to abortion; that is, no trade-offs are allowed during the first trimester.

During the second trimester, only those regulations of abortion procedures that are reasonable and are narrowly tailored to protect the health of the woman are constitutional;[28] therefore, during the second trimester, only statutes restricting abortion that involve the third trade-off, a woman's right to abortion versus state's interest in protecting the health of the pregnant woman, are constitutionally permissible.

By the beginning of the third trimester or the end of the sixth month, the Supreme Court held that the state can demonstrate an important interest in the fetus' existence and may prohibit abortions except where necessary to protect the health or life of the woman.[29] The third trimester trade-offs are shown in Figure 3.3.

On the axis labeled Woman's Right To Abortion, 0 represents an absolute prohibition of abortion and A = 1 represents an absolute right to an abortion. On the axis labeled Protection of Fetus, 0 represents no protection of the fetus and F = 1 represents an absolute protection of the fetus, that is, no right to an abortion.

If a pregnant woman will die unless an abortion is performed, the probability that abortion will save her life is $P_s = 1$; that is equivalent to saying that the probability that an abortion is *not* necessary to protect a pregnant woman's life is zero, $(1 - P_s) = 0$; therefore, on the axis labeled Probability Abortion is *Not* Necessary To Protect a Pregnant Woman's Health or Life, 0 means that abortion is necessary to protect the life of a pregnant woman and 1 means that abortion is *not* necessary to save a pregnant woman's health or life.

Are Legislative Preferences "Well Behaved"?

Although it is not possible to observe the legislative preferences on abortion directly, a state statute concerning abortion reveals those preferences. In Figure 3.3, legislative

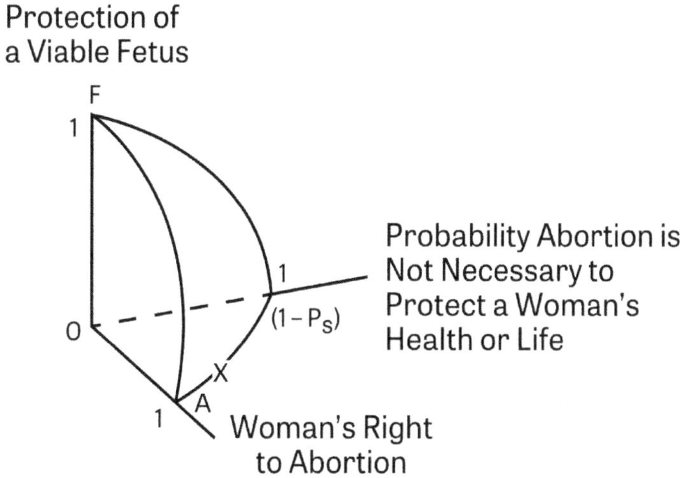

Figure 3.3 Three Trade-Offs in *Roe v. Wade*. *Source*: Created by the author.

preferences as *revealed* by the Texas statute in *Roe* are depicted in the context of economic theory. The Texas statute prohibited procuring or attempting the abortion of a human fetus except when necessary to save the life of the mother.[30] This statute is depicted by point A with coordinates (1, 0, 0), which indicates that the only circumstances under which a woman has a right to an abortion are when abortion is absolutely necessary to save her life; only in that event does the state of Texas express no interest in protecting a fetus. Extreme points like A, F, and $(1 - P_s) = 1$ are corner solutions as opposed to a point like X in Figure 3.3, which is an interior solution in the sense that it is on the convex sphere-like section depicted but not at any of the extreme points. For example, a state abortion statute that, depicted by point x on the curve segment AH, reveals second trimester, legislative preferences that give great weight to a woman's right to choose an abortion, some concern for the pregnant woman's health, and no protection to the fetus.

The preferences of the Texas legislature, A in Figure 3.3, are revealed as extreme because corner solutions can only result if indifference curves are concave to the origin (chapter 2). In *Roe*, the Supreme Court held that prohibition of abortion except where necessary to protect the life or health of the mother is constitutional only during the third trimester of pregnancy; during the first two trimesters, such legislative preferences are unconstitutional. The Texas statute is unconstitutional because it prohibits abortions in all three trimesters except when the mother's life is in danger.

In terms of judicial and legislative behavior, well-behaved (convex) preferences mean that judges and legislators are not absolutists or extremists. If judicial preferences are well behaved, as exemplified by the products liability and contract cases discussed above, judicial decisions will not be very sensitive to personnel changes on courts and will tend to be stable over time. Conversely, when judges make decisions involving controversial social issues, such as abortion or free speech, where preferences are more likely to be concave and extreme, law is more likely to become unstable over time due to personnel changes on courts.

NOTES

1. "The Forms and Limits of Adjudication," *Harvard Law Review* 92 (1978): 353, 375–377.
2. The analyses of the crashworthy automobile cases, Fuller's hypothetical cases, and *Roe v. Wade* first appeared in John Cirace, "When Judges Balance Interests Through Trade-Offs, They Implicitly Use Economic Theory to Order Cases," *University of Detroit Mercy Law Review* 74 (1996): 27–62.
3. 359 F.2d 822, 824–25 (7th Cir.), *cert. denied*, 385 U.S. 836 (1966). In other contexts, courts have had no difficulty in rejecting the narrow rationale of *Evans*. For example, in *Spruill v. Boyle-Midway, Inc.*, 308 F.2d (4th Cir. 1962), involving the death of a small child after swallowing furniture polish manufactured by the defendant, the court allowed a cause of action against the manufacturer even though, admittedly this was not the intended use of furniture polish.
4. 391 F.2d 495, 502, 503 (8th Cir. 1968).
5. See *Frericks v. General Motors Corp.*, 317 A.2d 494, 502–94 (Md. Ct. Spec. App. 1974).
6. See *Nada v. Ford Motor Co.*, 509 F.2d 213 (7th Cir. 1974).

7. John P. Brown, "Toward an Economic Theory of Liability," *Journal of Legal Studies* 2 (1973): 323, 331–335.

8. Richard Posner, *Economic Analysis of Law*, 9th ed. (New York: Wolters Kluwer, 2014), 192–193.

9. *Davis v. Consolidated Rail Corp.*, 788 F.2d 1260, 1264 (7th Cir. 1986) (Posner, J).

10. 503 S.W.2d 516 (Tenn. 1973).

11. 252 S.C. 202 (1969).

12. See, Ralph Nader, *Unsafe at Any Speed: The Designed-in Dangers of the American Automobile* (New York: New York Pocket, 1966). Chapter 1 described the alleged technological defects in the rear suspension of the Corvair model car that was involved in *Larsen*. This book is often credited with having created public support for passage of the National Traffic and Motor Vehicle Safety Act of 1966.

13. In *Larsen*, the 8th Circuit reversed a summary judgment in favor of the defendant. However, on remand, a jury decided for the defendant. Note, "Automobile Design Liability: Larsen v. General Motors and its Aftermath," *University of Pennsylvania Law Review* 118 (1969): 299, 312 note 67.

14. 298 F.Supp. 1064, 1073, 1073–1074 (1969). *Turner v. General Motors Corp.*, 514 S.W.2d 497 (Tex. 1974); 584 S.W.2d 844 (Tex. 1979).

15. 489 F.2d 1066, 1073–1074 (4th Cir. 1974).

16. 264 F.Supp. 1010, 1011–1012, 1012 (D.C. Tex. 1967). See *Dreisonstock*, 489 F.2d 1066, 1073, where the court said that although the *Willis* court relied on *Evans*, the holding is consistent with *Larsen*.

17. Fuller, note 1.

18. With respect to ownership transfers of real estate, automobiles, and miscellaneous other types of expensive personal property, there are systems for recording title transfers, so that only recorded transfers of title are protected by law.

19. Fuller, note 1, 381.

20. *Ibid.*

21. *Ibid.*, 376.

22. *Menzel v. List*, 24 N.Y.2d 91 (1969). In 1940, a painting by Marc Chagall was confiscated from its owner by Germans when they invaded Belgium in World War II. In 1962, when owner discovered the painting was in the United States in possession of defendant, owner sued and legally repossessed the painting.

23. Robert Cooter and Thomas Ulen, *Law and Economics*, 2nd ed. (New York: Addison-Wesley, 1997), 130.

24. "'Entrusting' includes any delivery and any acquiescence in retention of possession regardless of any condition expressed between the parties to the delivery or acquiescence and regardless of whether the procurement of the entrusting or the possessor's disposition of the goods have been such as to be larcenous under the criminal laws." UCC § 2–403(3).

25. In *Malewicz v. City of Amsterdam*, 362 F.Supp.2d 298 (2005), even though the State Department's approval was sought and received for the temporary loan of art work to two US museums, a district court held that although the museums were protected from seizure of the art work while in the United States under the Immunity from Seizure Act, the Act does not preclude suits for damages by the owners because the loan of art works is commercial in nature, not the type of sovereign act that was immune from suit.

26. 410 U.S. 113, 153, 162–163, 163, 163–164 (1973).

27. In *Planned Parenthood v. Casey*, 505 U.S. 833 (1992), a plurality of justices upheld the "essential holding" of *Roe*, that women have a constitutional right to an abortion but

overturned the *Roe* trimester framework in favor of a viability analysis. At and subsequent to viability, a state can regulate or possibly proscribe abortion "except where it is necessary, in appropriate medical judgment, for the preservation of the life or health of the mother" (879). Prior to viability, a state cannot pose an undue burden on a woman's fundamental right to abortion.

28. 410 U.S. 163–164.
29. 410 U.S. 163, 164–165.
30. 410 U.S. 164.

Chapter 4

Equal Protection and Lexical Constraints on Efficiency

In chapter 2, an example was given of an individual who chooses an optimum bundle of income and leisure subject to a market constraint given by her job possibilities curve. Economics is often about maximization subject to constraints, or constrained maximization. The purpose of this chapter is to demonstrate that courts and governments often place equal protection or lexical constraints on marginal trade-offs among competing goods.

The first example of constraints on trade-offs between competing goods is abortion rights in the context of equal protection of the laws. Then, the concept of lexical ordering is explained. A second example, the exclusionary rule for illegally seized evidence, is depicted as a lexical constraint on trade-offs between Fourth Amendment rights and public security. Then, John Rawls' difference or maximin principle,[1] which is an alternative to Pareto efficiency or Kaldor-Hicks criterion, is explained. Rawls believed that an unconstrained market system results in income inequality that cannot be socially justified. Under his maximin principle, social and income inequality is limited or constrained, unless such inequality improves the welfare of the worst-off individuals or group.

ABORTION AS AN EQUAL PROTECTION QUESTION

In chapter 3, the economic rationality approach to the abortion dispute, interest balancing via three marginal trade-offs in *Roe v. Wade* was discussed. The abortion dispute also may be thought of as an equal protection of the laws question.[2] Those who oppose any right to abortion, extreme pro-birth proponents, and those that favor an absolute right to abortion, extreme pro-choice proponents, tend to view abortion as an equal protection question. Extreme pro-birth proponents believe that laws with respect to abortion involve two classifications of individuals by government: pregnant women and fetuses. They view fetuses as human beings and argue that these two classifications by government, women and fetuses, are similar and are entitled to the same fundamental rights; therefore fetuses are entitled to equal protection of the laws.

If equal protection were extended to fetuses (the Supreme Court specifically refused to extend such protection in *Roe v. Wade*[3]), abortion, which involves substantial trade-offs between the welfare of pregnant women and fetuses, would not be allowed. Even if a pregnant woman's life was in danger from the continuation of her pregnancy, this view would support allowing nature to take its course rather than take the life of the fetus to save the life of the mother.

Extreme pro-choice proponents also believe that abortion is a question of equal protection of the laws. They see abortion as involving two classifications of individuals by government: men, who have total control of over their bodies, and women, who should have equal control over their bodies.[4] Pro-choice proponents argue that these two classifications of individuals by government are similar and are entitled to fundamental rights, so that women are entitled to the same control over their bodies as that enjoyed by men. This view supports abortion on demand at any time during pregnancy.

If the abortion question is viewed as an equal protection question, equal protection logic trumps interest balancing through marginal trade-offs. However, equal protection arguments are all-or-nothing arguments—different groups are similarly situated or not, they are entitled to equal rights or not. With respect to abortion rights, equal protection arguments seem consistent with extreme or absolutist preferences; whereas marginal trade-offs as in the context of economic rationality as in *Roe v. Wade* seem consistent with well-behaved preferences that allow a nuanced and particularized response to this complex and most divisive question.

LEXICAL ORDERING

Serial or "lexicographical" ordering of preferences or interests is an ordering or ranking that requires us to satisfy the first principle in the ordering before we can move on to the second, the second before we consider the third, and so on. The term derives from the most familiar example of such ordering: words in a dictionary. As an example of such ordering, substitute numerals for letters, putting "1" for "a," "2" for "b," and so on. Due to the cumbersome nature of the term lexicographical ordering, it is called *lexical* ordering for short. A principle or interest does not come into play until those previous to it are either fully met or do not apply. A serial ordering avoids having to balance interests at all; those earlier in the ordering have an absolute preference with respect to later ones. "We can regard such a ranking as analogous to a sequence of constrained maximum principles. For we can suppose that any principle in the order is to be maximized subject to the condition that the preceding principles are fully satisfied."[5] Recall that the concept of constrained maximization was explained in chapter 2.

FOURTH AMENDMENT RIGHTS

The Exclusionary Rule for Illegally Seized Evidence as a Lexical Constraint

As an example of a lexical constraint in law, consider the exclusionary rule for illegally seized evidence. The question of whether evidence seized in violation of the

Fourth Amendment's prohibition against unreasonable search and seizure must be excluded at trial often involves a controversy over whether the courts should or should not (do or do not) use economic rationality to engage in marginal trade-offs between the costs and benefits of such exclusion. Some judges are and some are *not* willing to employ economic rationality and balance interests via marginal trade-offs.

In *Weeks v. United States*,[6] the Supreme Court held that illegally seized evidence in violation of the Fourth Amendment (that is, evidence gathered by police intrusion in a person's private affairs without a search warrant issued by a magistrate based upon probable cause) must be excluded in a federal trial; it said that to admit such evidence would put a judicial stamp of approval on unconstitutional conduct.[7] The exclusionary rule was extended to state trials in *Mapp v. Ohio*.[8]

However, in *United States v. Leon*,[9] the Supreme Court, 6–3, held that the exclusionary rule should be modified so as not to bar the use of evidence obtained by officers acting in reasonable reliance on a search warrant issued by a magistrate that was ultimately found to be unsupported by probable cause. The majority reasoned that the exclusionary rule is "a judicially created remedy designed to safeguard Fourth Amendment rights generally through its deterrent effect," the applicability of which "must be resolved by weighing the costs and benefits of preventing the use" of illegally seized evidence.[10] The three dissenters questioned whether the exclusionary rule is merely a "judicially created remedy" for Fourth Amendment violations, subject to being narrowed "through guesswork about deterrence" rather than, as indicated in *Weeks*, "a right grounded in that Amendment to prevent government from subsequently making use of any evidence so obtained."[11]

The disagreement between the majority and dissenters can be characterized as a conflict between those judges who are willing to balance interests through marginal trade-offs in Fourth Amendment cases and those who view the Fourth Amendment as a lexical constraint on police behavior.

In Figure 4.1, the vertical axis is labeled Protection Against Search and Seizure; on this axis, 0 represents no right against search and seizure (police can enter a person's home for any reason at any time) and 1 represents absolute prohibition against search and seizure (home is a sanctuary even if the police are in hot pursuit). The horizontal axis is labeled, Probability of Convicting a Guilty Defendant; on this axis, 0 and 1 represent zero probability of conviction and certain conviction, respectively.

In general, many abstract or qualitative variables can be represented by a Euclidean norm or category on an arbitrary scale, usually from 0 to 1; the variable is then said to be normalized. The Gini coefficient, discussed in chapter 1, is an example of a normalized variable. Comparative states of even abstract concepts, such as freedom, can be rank ordered on normalized scales.

Moreover, qualitative variables are sometimes treated as dichotomous variables, which take on only two values, either 0 or 1. For example, the qualitative variable of sex can be described as a dichotomous variable: 0 = male, 1 = female or vice versa. As is well known, super computers, computer generated images (CGI), and streaming video is generated by billions of gigabytes, every one of which is either a 0 or 1.

The absolute placement of the *Weeks-Mapp* and *Leon* cases in Figure 4.1 is unimportant; it is their placement relative to each other that is significant. Assuming that the exclusionary rule against illegally seized evidence, which guarantees

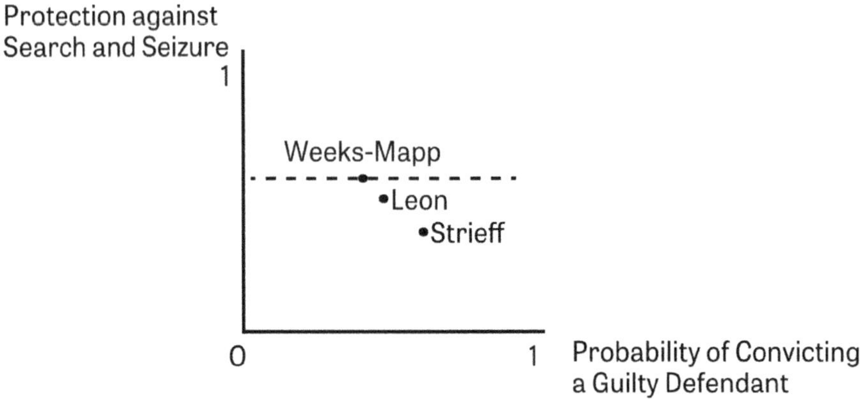

Figure 4.1 Trade-Offs versus Lexical Constraint. *Source*: Created by the author.

the minimum level of protection against search and seizure mandated by the Fourth Amendment, is a significant restriction on police behavior, *Weeks-Mapp* is depicted by a point in Figure 4.1 which has a coordinate on the Protection Against Search and Seizure axis that is closer to 1 than to 0; assuming that the exclusionary rule has at least some negative impact on the probability of convicting a guilty defendant, Weeks-Mapp has a coordinate on Probability of Convicting a Defendant axis of substantially less than 1.

Since the majority in *Leon* are willing to trade-off civil liberties against the cost of letting the guilty go free, economic rationality and interest balancing via marginal trade-offs is relevant for them. The modification of the exclusionary rule announced in *Weeks-Mapp* by the rule in *Leon* means that the majority are willing to reduce Protection Against Search and Seizure somewhat in order to increase Probability of Convicting a Guilty Defendant somewhat. It is clear from the *relative* position of the cases that *Leon* gives a greater relative weight to the Probability of Convicting a Guilty Defendant and a lower relative weight to Protection Against Search and Seizure than does *Weeks-Mapp*. Moreover, for the majority, these two interests are not only commensurate, the marginal trade-off in moving from *Weeks-Mapp* to *Leon* is such that the majority must believe that what is gained by society in terms of the probability of convicting a guilty defendant is as least as great as what is lost in terms of protection against search and seizure; that is, the *sum* of these two social interests in *Leon* is as least as great as that in *Weeks-Mapp*, otherwise the majority would not have modified the rule.

However, for the dissenters, it can be said that the level of protection against search and seizure required by the Fourth Amendment cannot be reduced below the level required by the Fourth Amendment; for them, interest balancing is not relevant. The dissenters' view can be characterized by the belief that the level of protection against search and seizure at the level mandated by the Fourth Amendment is entitled to a lexical priority (represented by the dashed line at the level of *Weeks-Map*). The dissenters focus solely on Fourth Amendment rights and argue that the principle of precedent requires that *Weeks-Mapp* should be followed and that under no circumstances should the level of protection against search and seizure be reduced below that of *Weeks-Mapp*. Moreover, to the extent that the dissenters are willing to consider interest balancing through trade-offs, they appear to believe that the majority's decision in *Leon* could be used by pro-police magistrates to undermine the exclusionary rule substantially, so that this seeming minor exception may substantially diminish rights under the Fourth Amendment.[12]

In *Utah v. Strieff*,[13] a policeman stopped, detained, and asked a person for identification solely on the ground that he was leaving a residence where drug activity was suspected. After determining that the person had an outstanding arrest warrant for an unpaid parking ticket, the policeman searched the person and found illegal drugs. In weighing the benefits of the exclusionary rule in deterring Fourth Amendment violations against the substantial social costs of allowing the guilty to go free, the US Supreme Court (5–3) refused to exclude the evidence on the ground that a valid warrant predated the investigation and was unconnected with the illegal stop. Justice Sotomayor dissented on the ground that "the Fourth Amendment should prohibit, not permit, such misconduct." She wrote that the states and federal government maintain databases with over 7.8 million outstanding warrants, the vast majority for minor offenses, which can be used by police to stop people without cause. In Figure 4.1, *Strieff* is placed lower and to the right of *Leon* because it represents a greater reduction in Fourth Amendment rights and an increase in the probability of convicting a guilty defendant.

The measures for protection of national security that the US government took in response to the terrorist attacks of September 11, 2001, make the constitutional changes in *Leon* seem insignificant. In a recent book, Judge Posner advocates "coercive interrogation," . . . "discretion for public officers to disregard in extreme cases the prohibition against torture," and "defining torture extremely narrowly."[14] Posner justifies these changes by cost-benefit "weighing," which is "usually metaphorical."[15] However, if there are no objective criteria for weighing costs and benefits of torture and no constitutional constraints, then there are potentially there no limits on police behavior. Is it not possible that the logic of legal rationality, which relies on the recognition of similarities and differences between past and present cases, may be more accurate than the logic of economic rationality based on a metaphorical weighing that is totally lacking in objective criteria? Moreover, as is well known, pragmatism as a general philosophy is open to the charge that if the ends always justifies the means, what is forbidden?

Economic rationality and legal rationality have different goals and come from different logical systems. In cases involving fundamental rights, legal and economic rationality are usually untranslatable rather than consistent and isomorphic (chapter 6). When law and economics conflict over fundamental rights, there is no a priori reason to think that economics should trump the law. Such conflicts have to be resolved by asking which abstract logical model, economics or law, more accurately explains experience in the real world. This is a matter of judgment over which reasonable persons can and do differ.

RAWLS' TWO LEXICAL PRINCIPLES

A very well-known use of lexical ordering occurs in John Rawls' *A Theory of Justice*. Rawls' lexical ordering was motivated in part in order to provide an alternative to the utilitarian concepts—Pareto efficiency, Kaldor-Hicks, and cost-benefit analysis—discussed in chapter 1, which concepts sanction a market system that has substantial inequality of income and wealth as one of its byproducts. He set out to develop a set of principles or social contract that individuals would choose as the basic structure of a just society, which he believed would have much less income inequality than currently exists in the United States.

His principles are of constitutional generality. A good analogy is with a group of individuals who are deciding on the set of rules for a game of chance, for example, a card game, in which they will subsequently participate. Another analogy is the familiar cake-cutting problem in which one person is allowed to divide the cake and another person gets to choose the first piece. The cake-cutter's best solution is to cut the cake into equal sized pieces; that is, cut the cake fairly because the cake-cutter does not know in advance which piece he or she will receive. By analogy, one should agree to fair rules for society because one often does not know in advance what one's status in life will be or how it will change over time.

Rawls makes three key assumptions: One, when making basic social rules, individuals must divorce themselves from knowledge of their own personal attributes and original social position by stepping through a *veil of ignorance*. This is the cake-cutting assumption, that is, one assumes that one does not know one's position in society. Two, society is in a state of *moderate scarcity* where it is relatively easy to satisfy the primary wants of all citizens.

Three, individuals care much more about protecting their ability to satisfy primary wants and liberty than they do about receiving gains above primary wants; that is, individuals have a rapidly diminishing marginal utility of additional income once primary wants have been satisfied. This assumption means that individuals are not focused on status and income, do not envy others who have higher incomes, and are extremely risk adverse, which concept is defined in chapter 15. Critics argue that they greatly aid Rawls in ruling out competing principles or conceptions of justice,[16] such as utilitarianism; it is said that the elimination of competing principles is supposed to be a consequence of the analysis rather than a presupposition of the analysis.[17]

Lexical Priority of Liberty

Rawls argues that given the original position behind the veil of ignorance, two lexical principles will be chosen unanimously by individuals when forming a basic social contract or constitution and that these principles will result in a just society. The first lexical principle is the priority of liberty: each person is to have an equal right to the most extensive basic liberty compatible with a similar liberty for all. Rawls justifies giving liberty a lexical priority over economic advantages because of the assumptions concerning moderate scarcity and rapidly diminishing marginal utility of additional income relative to liberty once primary wants have been satisfied.

> Now the basis for the priority of liberty is roughly as follows: as the conditions of civilization improve, the marginal significance for our good of further economic and social advantages diminishes relative to the interests of liberty, which becomes stronger as the

> conditions for the exercise of the equal freedoms are more fully realized. Beyond some point it becomes and then remains irrational from the standpoint of the original position to acknowledge a *lesser* [emphasis added] liberty for the sake of greater material means and amenities of office.[18]

In other words, Rawls does not deny that people who are poverty stricken and starving would trade liberty for food. But he believes that with increasing levels of income, the priority of liberty over other psychological and material needs rises until, at some level of income, liberty takes complete precedence over all other needs. The lexical priority of liberty over income and wealth after primary needs are met is depicted in Figure 4.2. Once primary needs are met, no amount of increase in income and wealth will cause any one to accept a lesser amount of liberty than level L. In other words, after primary needs are met, individuals are assumed to have indifference curves that are horizontal at level L.

If individuals were willing to trade-off liberty for an increase in income and wealth after primary needs are met, such preferences would be expressed by indifference curves that, by allowing marginal trade-offs, would have the usual convex shape; in Figure 4.2, one such indifference curve is depicted by the curve labeled F, where F stands for Faustian preferences; a person with such preferences would accept less liberty for a higher income. However, in Rawls' view, once primary wants are satisfied, no individual would accept a lesser level of liberty than L even if offered a million or a billion dollars! The liberty constraint is an example of a lexical principle not amenable to marginal trade-offs. However, after the terrorist attack of 9/11, it is beyond dispute that citizens of the United States have been willing to trade-off some liberty for some increased security against terrorist attacks.

Rawls' liberty principle, under which each person is to have an equal right to the most extensive basic liberty compatible with a similar liberty for all, is equivalent to the classic definition of liberty championed by John Stuart Mill;[19] as such, it does not have an absolute value because the amount of liberty each person is entitled to is limited by the requirement that it must be compatible with similar liberties for all. This means that the amount of liberty to which each person is entitled may change as social conditions change over time. As the quotation from Rawls makes clear, liberty is not a dimensionless category because he recognizes the possibility of "lesser" states of liberty and by implication greater states of liberty.

Rawls' liberty principle is not operational in the sense that the abstract concept of liberty is itself made up of lesser particular freedoms, such as the right to own land, which freedom, if defined so as to include the right to light, would conflict with the right of adjoining property owners to construct buildings that block access to light. The priority of the liberty principle will do nothing to resolve this particular issue. In order to resolve this conflict between these two freedoms, it is necessary to specify the lexical ordering further or to engage in trade-offs of marginal benefits against marginal costs of constraining each activity. In the United States, the right to construct buildings on one's land is usually given priority over the right to have access to light (see chapter 12). Since courts are concerned with these and similar issues in myriad cases, the level of generality at which Rawls' liberty principle is couched makes it a

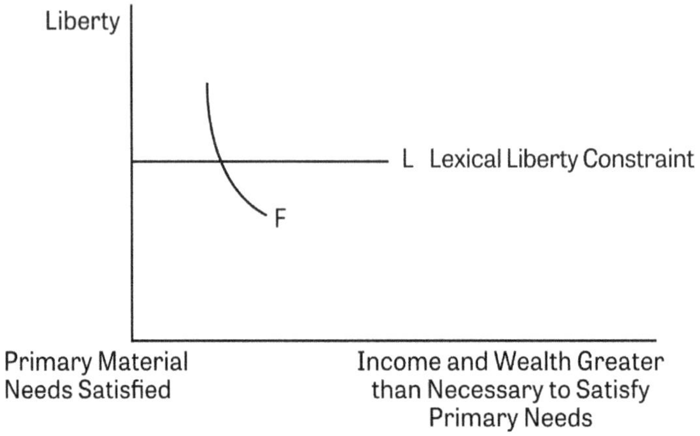

Figure 4.2 Lexical Liberty Constraint. *Source*: Created by the author.

nonoperational guide in many specific cases. However, as discussed in the previous section, the exclusionary rule for evidence seized illegally in violation of the Fourth Amendment is an example of a lexical rule involving liberty.

The Difference or Maximin Principle

Rawls' second lexical principle requires that social and economic inequalities must satisfy two conditions. 1) Social and economic inequalities that come from holding high offices and positions must be open to all under conditions of fair equality of opportunity. Equality of opportunity means that individuals with the same ability and who exert the same effort are entitled to the same educational and economic opportunities without regard to whether they were born rich or poor. 2) Social and economic inequalities are to be of greatest benefit to the worst-off or least-advantaged group; this is Rawls' difference or maximin principle. (The concept of maximin is taken from game theory; see chapter 7.) According to the difference or maximin principle, social and economic inequalities can be justified only if such an inequality improves the welfare of the worst-off individual or group.

As an example of income inequality that can be justified as being of benefit to everyone including the worst-off group, consider the income of a physician. According to Rawls, a departure from income equality is justified because a) few individuals would undertake the many years of arduous medical training without such incentive, and b) such income inequality is necessary to produce medical services that benefit the worst-off group. However, the income of physicians should be raised only the minimum amount above equality necessary to attract qualified individuals into the profession. All other income derived from physicians' services over the minimum needed to attract qualified individuals is to be used to benefit the worst-off group.

The difference principle is based upon the idea that natural endowments, say high IQ or talents that are in great demand by markets (athletic talent), are undeserved.

Rawls says, "In justice as fairness, men agree to share one another's fate."[20] However, this does not mean that everyone should have the equal income. Those who have superior natural endowments are entitled to higher income so long as their being better off also contributes to the wellbeing of the least-advantaged group.

According to utilitarian criteria, either Pareto efficiency or Kaldor-Hicks, an economy that has the greatest total income is the most efficient; however, according to the difference principle, an economy that has the highest income for the least advantage group has a lexical priority, even if that economy has a lower total income.

Critics have pointed out that the ambiguity as to the size and composition of the worst-off category leaves room for significantly different outcomes. Is the worst-off group, say, those with lowest 90% of income and wealth, lowest 10%, lowest 1%, or those in a vegetative state? Would Rawls' difference principle mean that most or all of society's surplus income and wealth would have to be spent on measures of heroic intervention for the elderly at the end of their lives, infants who are born very prematurely and have multiple disabilities, and the like? If the worst-off group is defined broadly, say the bottom 90%, then only the top 10% of incomes would be taxed.

Even assuming that when voting whether to accept the difference principle as the economic foundation of society that one's status in society is unknown due to the veil of ignorance, it has been questioned whether individuals would unanimously agree to the difference principle once they are informed of its implications. Assume for the moment that the worst-off are defined initially to be the lowest 1% of the population. The maximin principle is lexical in the sense that once the income or wealth of the lowest 1% has been raised to the level of the lowest 99% of the population, the maximin principle then applies to the lowest 1% of the resulting income distribution, and so on. However, if the initial worst-off group is defined narrowly, say lowest 1%, then the probability that a person of average income and wealth would be included in either the initial or subsequent worst-off groups in the foreseeable future is extremely small.

Not only is the average person unlikely to receive worst-off group benefits, the average person is not likely to value such benefits highly given that Rawls assumes that society's initial state is one of moderate scarcity such that all the primary wants of individuals are met. An analogy is with being over insured against small risks. Thus, it has been argued that only extremely risk adverse individuals would accept the difference principle. Moreover, contemporary American values do not seem to support the difference principle: even middle income and poor people appear to be against "soak the rich" taxation on the grounds that they or their children may hit the lottery or become wealthy tomorrow even though all rational calculation would heavily discount such possibilities. In other words, US citizens appear to be risk preferrers. (The concept of risk preference is defined in chapter 13.) However, given the fact since 1980, 80% all income economic gains in the United States have gone to the top 1%, American values may be changing. In addition, Rawls' difference principle may be consistent with the values and preferences of, say, Sweden with its extremely high income leveling tax rates.

In sum, Rawls' first lexical principle is political, that of equal citizenship. His second lexical principle concerns economic relations; it answers the question of when deviations from an equal income distribution are justifiable. The evidence of history is that, absent government intervention, Pareto efficiency, although theoretically neutral, tends to favor the rich; Rawls' system favors the poor. What system favors the middle class?

Rawls's Maximin Principle Appears in Two Other Contexts in this Book

Rawls' maximin principle appears two other times in this book. In chapter 8, it is argued that the reason education is financed with compulsory taxation (not private financing) is because the worst-off (poorest) parents would be unlikely to educate their children in private markets. If the poorest are a substantial part of the population, the resulting large number of uneducated persons would threaten democracy and the market system. So, if the reason for public financing is to educate the children of the worst-off group, public funds should not be spent in ways that increase the disadvantage of this group with respect to other groups.

Second, in chapter 10, Rawls's maximin principle is used in the discussion of intergenerational equity and a long-run sustainable economy. Applied intergenerationally, Rawls's principle suggests that the least-advantaged generation should be no worse off than any other generation.

NOTES

1. John Rawls, *A Theory of Justice* (Cambridge, MA: Belknap Press, 1971), 42–43.
2. "No State shall . . . nor deny to any person within its jurisdiction the equal protection of the laws." US Constitution, Amendment XIV, § 1.
3. 410 U.S. 113, 162 (1973).
4. Ruth Bader Ginsburg, "Some Thoughts on Autonomy and Equality in Relation to *Roe v. Wade*," *North Carolina Law Review* 63 (1985): 375, 382–383; Ginsburg, "Speaking in a Judicial Voice," *New York University Law Review* 67 (1992): 1185, 1199–1200.
5. Rawls, note 1, 43.
6. 232 U.S. 383 (1914).
7. 232 U.S. 394.
8. 367 U.S. 643 (1961).
9. 468 U.S. 897 (1984).
10. 468 U.S. 906–907.
11. 468 U.S. 943 (Justice Brennan); Wayne LaFave and Jerold Israel, *Criminal Procedure*, 2d ed. (St. Paul, MN: Thompson West, 1992), 105–111.
12. 468 U.S. 955.
13. 579 U. S. ___, 136 S.Ct. 2056 (2016).
14. Richard Posner, *Not a Suicide Pact: The Constitution in a Time of National Emergency* (New York: Oxford, 2006), 80, 86, 87.
15. Posner, note 14, 24.

16. Dennis Mueller, *Public Choice II*, revised ed. (Cambridge: Cambridge University Press, 1989), 414–423.
17. *Ibid.*, 418.
18. Rawls, note 1, 542.
19. "The liberty of the individual must be thus limited; he must not make himself a nuisance to other people." John S. Mill, *On Liberty* (1859) (Harmondsworth, UK: Penguin, 1974), 119.
20. Rawls, note 1, 102.

Chapter 5

Legal Rationality and Logic Defined

Law professors tell their students they are teaching them to *think like a lawyer* without ever defining what that means. Refusal to define legal rationality is consistent with the Socratic method of teaching law, which is to *show* not *tell*. Law students are expected to and usually do develop a working knowledge of what it means to think like a lawyer through the study of appellate court opinions and replication of their reasoning. However, whether many lawyers can explain precisely what thinking like a lawyer means is doubtful. The purpose of this chapter is to define legal rationality and logic in a way that is parallel to and comparable with economic rationality and logic.

In chapter 2, economic rationality was defined by two behavioral rules, E1(more of a good is preferred to less), E2 (for any two bundles of goods, a rational self-interested individual can determine whether one bundle is preferred to the other or the person is indifferent between them), and constraints, such as budget or resource constraints. In this chapter, legal rationality and logic is also defined by two behavioral rules, which are stated in reverse order, L2 (for any two cases, a judge can determine whether they are like in all material respects, that is, are precedents, or are distinguishable) and L1 (broad legal categories, rules of law, or precedents are preferred to narrow ones); legal rationality is constrained by the principle of precedent or stare decisis, which requires like cases to be decided alike. The next chapter will discuss the conditions under which economic logic and the legal logic are or are not consistent (isomorphic).

Judging is purposeful activity. Judges determine which facts are significant (material or probative) in a case with reference to the purpose or goal they wish to achieve (issue to be decided). They determine whether two cases are like or distinguishable by constructing abstract fact categories, legal concepts, and rules of law which cover the facts of two or more cases. Judges prefer broader legal categories and precedents because they give more guidance to society. The principle of precedent, stare decisis, requires that judges decide like cases alike, that is, consistently; it is the principal constraint that prevents judges from deciding cases broadly, arbitrarily, or inequitably.

The level of generality of rules of law as expressed by judges in cases can be under-inclusive (too narrow), over-inclusive (too broad), or optimum size. This chapter discusses the process by which under-inclusive rules expand and over-inclusive rules shrink, so that rules of law tend toward optimum level of generality. This process by which rules of law expand or shrink incrementally through inclusion or exclusion is law's marginal analysis. I suggest that although economists are trained in the theory of marginal economic analysis, lawyers and judges are actually better at using marginal legal analysis to solve real-world problems; that is why lawyers, as a result of their training in legal logic, are often good arbitrators and negotiators.

The process of marginal broadening or narrowing of rules of law, which occurs by inclusion or exclusion of cases that are within or without a rule, eventually results in optimal size legal categories; this process is illustrated by cases taken from nineteenth- and twentieth-century products liability law, which concern products that cause injury to users or bystanders or cause damage to their property.

Two properties of legal rationality, transitivity and incompleteness, are also explained. The transitivity property in legal rationality is similar to the transitivity property of economic rationality. (Recall the transitivity property of economic rationality in chapter 2: if bundle x is preferred to bundle y, and bundle y is preferred to bundle z, then bundle x must be preferred to bundle z.)

A major difference between economic rationality and legal rationality is that economics is totally ordered, which means that every bundle of goods is strictly ordered in relation to every other bundle (chapter 2); whereas law is incompletely ordered, which means that some, but not all cases are related to each other by precedents.[1] Due to the incompleteness property of law, judges sometimes illegitimately (incorrect reasoning) isolate or confine a specific case to its facts in order to deprive it of precedential value, so that they can make decisions that are inconsistent with other cases (e.g. *Bush v. Gore*, discussed below); these arbitrary decisions are a naked use of judicial power. However, as shown below, if all cases are considered to be nested within broad general legal concepts, such as due process, equal protection of the laws, and horizontal equity, law's incompleteness property is ameliorated and a more consistent, less arbitrary jurisprudence will result.

JUDGING IS PURPOSEFUL ACTIVITY

Judging is purposeful activity because it is not possible to know which facts in a case are material (important) unless one knows the issue for which they are probative. For example, in *O'Brien v. Muskin*,[2] a twenty-three-year-old man jumped from an eight-foot-high garage into a 3 ½ foot deep backyard, above-ground swimming pool and was seriously injured. The manufacturer's warning, "DO NOT DIVE," on the outer wall of the pool, was in letters roughly 1/2 inch high. The young man sued the manufacturer of the swimming pool to recover damages for his injuries.

Under the goal or purpose of deterring careless behavior, a court would be concerned with whether the injured diver, manufacturer, or both were careless (negligent) or reckless. With respect to the manufacturer, was the DO NOT DIVE warning decal, which was only one-half inch high, prominent enough so that a potential diver would expect to see it. With respect to the diver's behavior, the fact that he dove from the roof of an 8-foot-high garage into a pool of water that was only 3 ½ foot deep suggests at least careless if not reckless behavior, with or without an explicit warning.

Under several other related goals or purposes, risk distribution or spreading, or enterprise liability, which emphasize compensation for injured persons, the facts emphasized in the previous paragraph are probably not material. Under these goals, it is significant that the plaintiff is a young person, twenty-three years old, who may not have personal or family health or disability insurance, and that the defendant is a corporation, which may be able to spread the cost of the plaintiff's injury by charging a few dollars more for each pool it sells.

Judges, aided by litigants' lawyers, continually refine and clarify the issue, question, or purpose in a case so as to cause the finder of facts, whether jury or judge, to focus on those facts that are material to the resolution of the dispute. Although, as the discussion below of two centuries of product liability cases makes apparent, there is often a great deal of randomness in individual court decisions, a single judge's freedom of choice among goals, issues, or purposes tends to be circumscribed by legal institutions; a determination by a trial judge that he or she has chosen the proper purpose or issue, selected the material facts, and decided a particular case as required by the law of relevant precedents and legislative statutes, is subject to reversal by appellate judges or new legislation and is also open to possible criticism from the myriad lawyers, legal commentators, and textbook writers who make up the legal community. In sum, case adjudication involves purposeful behavior by all the professional participants in the legal community; lawyers, judges, and legal commentators are expected to engage in a dialogue through which material facts in cases are evaluated in the light of the relevant social goals and law.

HOW CAN CASES THAT HAVE UNIQUE FACTS BE "LIKE"?

A case is a short story of a particular dispute in which the decision of a judge or judges, backed by the coercive power of the state, is authoritative.[3] Treating a case as a short story of a real-world dispute tends to emphasize that the particular facts of each case are unique. Thus, in order to determine whether two cases are like cases or distinguishable, judges must be able to determine the circumstances under which the facts of two cases are like in all material respects; that is, judges must resolve a paradox: are cases that are factually unique nevertheless legally alike?

Consider two hypothetical cases: in one case, *LaFong v. Walpole*, a man named Harry Walpole, a red head, was driving blue Toyota, serial number 348973, on the road between Plainville and Birlington on October 22, 2012, at 11:15 AM, a clear and sunny day, when he turned his head to look at his wife, Sandra, a passenger in the back seat, to whom he was speaking. In the process, the car swerved off the road and hit Jay LaFong, who was walking on the sidewalk, breaking his hip.[4]

In a second hypothetical case, *Chadra v. Locaste*, a woman named Doris Locaste, a blond, was driving a black Chevrolet pick-up truck with white upholstery at 75 mph, 20 mph in excess of the 55 mph speed limit, in the rain on Route 80 near Gloversville,

November 4, 2012, at 3:30 p.m. When the car in front of her slowed down due to road construction, Ms. Locaste applied the brakes, but due to the unreasonable speed at which she was traveling, she was not able to prevent a collision between her automobile and the rear end of a red Saturn automobile driven by Sam Chadra. Although Mr. Chadra was uninjured, his car's rear bumper and trunk were severely damaged.

Although the facts in each case are unique, if facts are considered not as concrete facts but in terms of general categories, cases that have different concrete facts can have like abstract fact categories. Llewellyn said each concrete fact is "the *representative* of a much wider abstract *category* of facts, and it is not in itself but as a member of the category that you attribute significance to it."[5]

If *LaFong v. Walpole* is compared to *Chadra v. Locaste* on a factual level, a blue Toyota is not like a black Chevrolet pick-up truck; on a slightly more general level, an automobile or motor car is still not the same as a truck; however, on a still more general level, both are automotive vehicles. What is the appropriate level of generality? Before answering that question, consider the similarities and differences among some other concrete facts and fact categories in the two cases: a pedestrian with a broken hip is not like damage to an automobile's rear bumper and trunk; on a slightly more general level, personal injury is still not like damage to one's property;[6] on a still more general level, both plaintiffs are victims who have suffered tangible economic damage. What is the appropriate level of generality? A rainy day is not like a sunny day; however, in both cases, the collisions took place during the daytime when there was good visibility. Judges often find the appropriate level of generality by comparing similar facts in several cases in order to determine if there is a common denominator; that is, the appropriate level of generality at which to consider facts is often arrived at inductively.

The question of which facts are material is a function of both the purpose and appropriate level of abstraction at which such facts should be considered in all relevant cases. Consider the two hypothetical collision cases discussed above. In the first case, Harry Walpole's automobile went off the road and onto the sidewalk due to his momentary inattention, causing it to strike a pedestrian, Jay LaFong. In the second case, Doris Locaste's truck, which was exceeding the speed limit, collided with the rear end of Lois Chadra's automobile. On the concrete factual level, these cases are different (distinguishable).

Although each of the hypothetical cases, *LaFong v. Walpole* and *Chadra v. Locaste,* has unique facts, if the level of generality of the facts in both cases is selected for the purpose of achieving the social goal of deterring careless or inefficient behavior and in reference to the facts selected in the other case, the cases can be seen as similar (like). Llewellyn said: "The court can decide the particular dispute only according to a general rule which covers a whole class of like disputes."[7] Both *LaFong* and *Chadra* can be covered by a general rule of law: a *driver* of an *automotive vehicle*, which was being *driven carelessly (negligently)* on a *public road* and *causes* a *collision* that results in *injury* to a *victim,* is legally liable for the *economic damage*. The level of generality of facts and the rule of law in each of these two cases should be selected by making similar comparisons with all possible cases that involve negligent driving.

THE TWO BEHAVIORAL RULES OF LEGAL RATIONALITY

As discussed in chapter 2, economic rationality assumes that self-interested individuals will maximize utility or satisfaction. Individual utility or satisfaction, which is a subjective concept, was given concrete meaning by revealed preferences or choices that individuals make among bundles of competing goods or ends. Similarly, in legal rationality, judges, who are impartial or disinterested rather than self-interested (this assumption, which is justified by game theory, is discussed in chapter 7), are assumed to maximize social utility or social welfare as they see it. Social utility or social welfare, which is also a subjective concept, can be given concrete meaning by judicial decisions that reveal judicial preferences among the sometimes competing goals of efficiency, equity, liberty, order, majority rule, and so on.

The model of legal rationality will be stated in a way that is parallel to and comparable with the model of economic rationality. Legal rationality has two principles or behavioral rules, L1 and L2, which are similar to the two rules, E1 and E2, of economic rationality, which were restated at the beginning of this chapter. Rule L1 states that broader legal categories and rules of law are preferred to narrower ones. Rule L2 states that for any two cases, judges can determine whether they are like or distinguishable. Legal rationality is constrained by the principle of precedent also known as stare decisis.

For ease of exposition, Rule L2, which states the relationship between two cases, that is, defines what a precedent is, will be presented before Rule L1, which states that judges prefer broader precedents to narrower ones. The reason for stating L2 before L1 is that the definition of what a precedent is should be stated before statements about what type of precedents are preferred.

> Rule L2: For any two cases, a judge can determine whether they are like in all material respects (precedents) or unlike (distinguishable).

Rule L2 is supported by a) conclusions from the discussion of the hypothetical *LaFong* and *Chadra* cases, b) Llewellyn's statement quoted above that "[t]he court can decide the particular dispute only according to a general rule which covers a whole class of like disputes," and c) Rule E2 of economic rationality to which Rule L2 is parallel.

As depicted in Figure 5.1 below, there are only five ways two cases can be related to each other in terms of binding precedents: 1) cases x and y are mutual precedents; 2) case x is a precedent that includes case y; 3) case y is a precedent that includes case x; 4) case x and case y are related but distinguishable; and 5) case x and case y are unrelated.[8] Rule L2 says that for any two cases, judges are able to determine which of these five figures describes their relationship.

Although the level of generality at which legal categories and rules of law are stated is capable of great variation, Rule L1 states that, other things equal, judges prefer broader rules of law or precedents to narrower ones.[9] The rationale for Rule L1 is that broader legal rules of law and precedents provide greater social guidance to society than do narrower rules of law. If a decision in a specific case is applied to that case only, rules of law in cases would provide no social guidance in similar

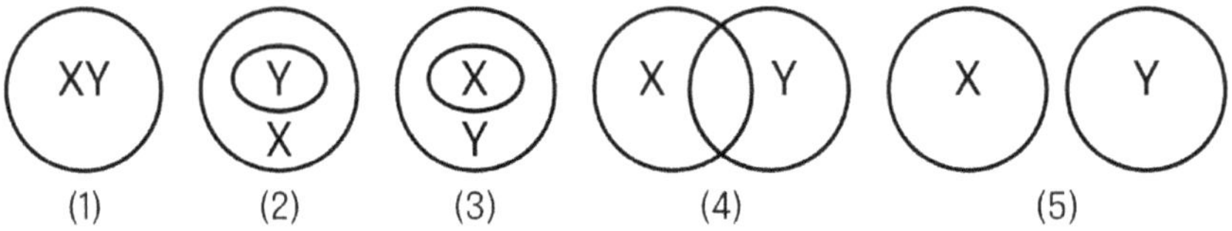

Figure 5.1 Two Cases Related (or not) by Precedent. *Source*: Created by the author.

circumstances or cases, and judges would be free to decide cases inconsistently and inequitably.

Rule L1: Broad rules of law or precedents are preferred to narrow ones.

Over time, Rule L1 tends to eliminate under-inclusive rules of law. What keeps rules of law from becoming over-inclusive, that is, too broad? Just as economically rational individuals are assumed to prefer more of a good to less but cannot have an unlimited amount of goods because their choices are constrained by budget and resource restrictions, so too, judges, who are assumed to prefer more general legal categories or rules of law to narrow ones, confront constraints on over-inclusive rules, the principal one being the principle of precedent or stare decisis, as described below.

HARD CASE SELECTION HYPOTHESIS

The case selection hypothesis is that hard cases are more likely to be selected for litigation than easy cases. Hard cases involve conflicting principles, are over- or under-inclusive rules of law, or are "interstitial"; that is, they fall in the "cracks" between rules of law. Hard cases usually reach the highest appellate courts for decision.[10] Easy cases fall clearly within an uncontroversial legal category or rule of law, are usually not litigated, or are disposed of at trial and usually do not reach the highest appellate courts.

How do the two rules of legal rationality, the case selection hypothesis, and the principle of precedent or stare decisis tend to expand under-inclusive and constrain over-inclusive legal categories and rules of law, so that rules of law tend toward optimum level of generality? Discussion of the evolution of products liability law over the last two centuries, which concerns personal injuries and property damage arising from the use of consumer products, will illustrate this process.

In general, during the last two centuries, products liability law has changed from near caveat emptor (buyer beware) to near venditor emptor[11] (seller beware). Products liability cases from 1830 to 1960 are used to illustrate the tendency of under-inclusive rules of law to expand; three of these cases are also used to explain the transitivity property of legal rationality. Other cases from 1960 to 1980 illustrate the tendency of over-inclusive laws to be constrained by the principle of precedent that like cases must be decided alike.

RULE L1 CAUSES UNDER-INCLUSIVE RULES OF LAW TO EXPAND

Judges often realize that a rule of law is *under-inclusive* when they must decide a new case, say case y, which does not fall within the rule of law as currently formulated in case x, but which they believe should be decided consistently with case x. In order to provide a rationale for the decision based upon a general rule of law, judges must either reformulate the rule of law in case x more broadly or interpret its language more broadly to include both case x and case y. Rule L1, which states that broad rules of law are preferred to narrow ones, implies that under-inclusive rules will expand over time through the process of marginal inclusion.

EXPANSION OF IMMINENTLY DANGEROUS PRODUCTS CATEGORY

During the nineteenth century, courts were concerned with the question of whether sellers should incur legal liability for injuries caused by negligently made products to persons other than the buyer; that is, did sellers incur legal liability where there was no *privity of contract* (contractual relationship) between the seller and the injured person? Nineteenth-century products liability cases chronicle the gradual erosion of the legal rule that privity of contract between the seller and the injured party was a legal prerequisite for suit. The privity rule was arbitrary in the sense that the right (standing) to sue depended upon the entirely fortuitous circumstance whether the injured party was the buyer or merely the ultimate user of the product that caused injury. However, in the opinion of the courts, the privity rule at least had the virtue of limiting the number of lawsuits; without privity, courts thought there would be an "infinity of actions."[12] An alternative rule would also have to be able to confine the scope of liability within manageable bounds. The courts chose to expand liability for negligently made products by allowing suits for injuries by victims without privity for products that are within the *imminently dangerous* products category, which category expanded during the second half of the nineteenth century.

In *Thomas v. Winchester* (1852),[13] defendant's employee negligently labeled and sold a jar of extract of belladonna, a poison, as extract of dandelion, an innocuous medicine. Plaintiff's physician prescribed a dose of dandelion and her husband purchased what was believed to be medicine from a druggist. After taking a small quantity of medicine, his wife experienced "coldness of the surface and extremities, feebleness of circulation, spasms of the muscles, giddiness of the head, dilation of the pupils of the eyes, and derangement of mind." Although her life was thought to be in danger for a short time, she subsequently recovered. The court said that the question was whether the suit could be maintained despite lack of privity between plaintiff and defendant, who was "a remote vendor." The suit was permitted notwithstanding lack of privity because "[t]he defendant's negligence had put human life in imminent danger."

In *Devlin v. Smith* (1882),[14] the deceased, an employee painter, was killed when a ninety-foot scaffold used in painting the rotunda of a court house collapsed due to negligent construction. The court held that the defendant who had contracted to build the scaffold for the painting contractor could be sued by the deceased's legal representatives. Relying on *Thomas,* the poison case, the court, which ruled for plaintiff, supported the distinction between two classes of cases, not imminently dangerous class, where a seller is only liable for negligence to the person with whom he contracted, and imminently dangerous class, where lack of privity does not bar suits against negligent sellers.

In *Statler v. Ray* (1909),[15] three large, connected coffee urns in a restaurant, which exploded and killed a man, were held to be in the imminently dangerous category. The court explained that when an article that is *inherently dangerous* is manufactured negligently, it becomes *imminently dangerous.* Suit and recovery by plaintiff was allowed.

In the nineteenth-century products liability cases discussed above, *Thomas, Devlin* and *Statler*, the category that courts used to expand liability—imminently dangerous products—remained unchanged, but they continually broadened the class of products to which it applied. This literal view of precedents "results in thinking and arguing exclusively from language found in past opinions and in citing and working with that language wholly without reference to the facts of the case that called the language forth."[16] This way is often used as a "springboard" to expand precedents that seem helpful.

TRANSITIVITY PROPERTY OF LEGAL RATIONALITY

In the context the model of legal rationality, the expansion of the imminently dangerous category for negligently made products is an example of applying Rule L1 to an under-inclusive products liability category. Although the nominal legal category remained the same, the implicit rule of law underlying the legal category expanded over time like those Russian nesting dolls (matryoska), where each smaller doll is contained in a larger doll. The nesting Russian doll image also captures the consistency criterion or *transitivity* property of legal rationality,[17] which will be explained both verbally and with Venn diagram in Figure 5.2, below.

In order to prevent against an infinity of actions, the court in *Thomas v. Winchester,* the poison case, could have limited the legal category of the imminently dangerous articles by a rule of law that defined the term to include *articles whose function is to destroy*; that is, the court could have defined the concept very narrowly. If imminently dangerous is defined in this way, consistency requires that poison be included within the imminently dangerous category but that scaffolds and coffee urns be excluded. In order to broaden the definition of imminently dangerous articles to include the scaffold case, *Devlin v. Smith*, the court could have stated a rule of law that defined the term to include *articles dangerous in ordinary use*. If imminently dangerous is defined in this way, consistency requires that poison and scaffold be included within the imminently dangerous category but that coffee urns be excluded. In order to broaden the definition of imminently dangerous articles still further to include the coffee urns case,

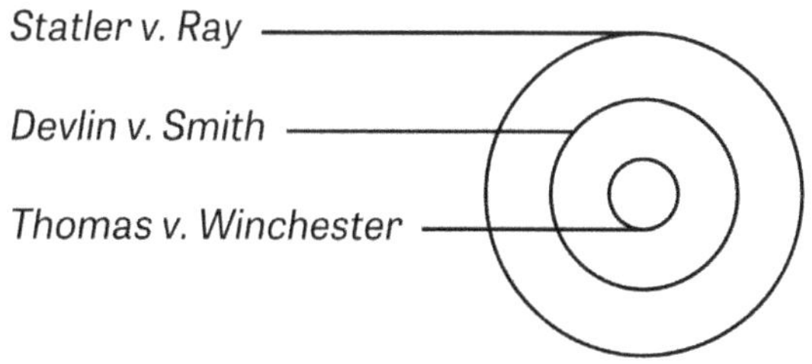

Figure 5.2 Transitivity in Law. *Source*: Created by the author.

Statler v. Ray, the court could have stated a rule of law that defined the term to include *articles capable of being dangerous*. If imminently dangerous is defined in this way, consistency requires that poison, scaffold, and coffee urn all must be included within the imminently dangerous category.

The consistency or transitivity property of legal rationality is illustrated by a general or inclusive rule of law, which contains a narrower, less inclusive rule, which in turn contains a still less inclusive rule. By transitivity, *if Statler*, which is a broadly defined legal category, is a precedent that includes *Devlin*, which is a medium-sized legal category; and, *Devlin* is a precedent that includes *Thomas*, which is a narrowly defined legal category; *then*, *Statler* is a precedent that must include or cover *Thomas*. Notice that transitivity runs in one direction only: *Statler* includes *Devlin* includes *Thomas*; the converse is not true; it would be logically inconsistent.

TWENTIETH CENTURY EXPANSION OF PRODUCTS LIABILITY LAW

In *MacPherson v. Buick* (1916),[18] the plaintiff was driving a car manufactured by Buick when one of its wheels, made of defective wooden spokes collapsed, causing him to be thrown from the car and to be injured. The plaintiff had bought the car from an automobile dealer and so did not have privity with the manufacturer-defendant. Judge Cardozo eliminated the more than half-century old distinction between the products which are or are not imminently dangerous: he held that if a negligently made product is reasonably certain to cause injury or death, it is a "thing of danger"; manufacturers of such products are liable to injured parties "irrespective of contract."

In *MacPherson,* Cardozo expanded the imminently dangerous so that the exception swallowed up the whole; privity of contract was eliminated as a prerequisite for suit by users who were injured by a defective product.

Writing in 1948, Levi said that the imminently dangerous products category was an example of three stages of a legal concept: creation, development, and breakdown.[19] However, from the vantage point of the twenty-first century, although the concept of imminently dangerous products was no longer legally significant, the expansion of legal liability for defective products that caused injury was far from over; after *MacPherson,* products liability law continued to expand for another fifty years via other legal concepts because it was still an under-inclusive category. Why were other concepts necessary for the continued expansion of products liability?

After *MacPherson*, a buyer or ultimate user of a defective product that caused injury could sue based upon the tort of negligence. However, in order to win the suit, the injured party had to prove that the product 1) was dangerously defective, and 2) was negligently (carelessly) made. Assuming that a defective product is not destroyed when the injury occurs (e.g., in an automobile collision, the part that made automobile defective could be destroyed along with the automobile), proving that a product is dangerously defective is usually not too difficult. However, proving that the defect was due to negligence or carelessness in the manufacturing process often necessitates knowledge of conditions in the factory at the time of the product's manufacture, information usually wholly controlled by the manufacturer and difficult or impossible for the plaintiff to find out, especially if the product was manufactured in some distant country and went through a long chain of production and distribution.

During the first half of the twentieth century, courts used concepts other than negligence, such as implied and express contract warranties and strict liability in tort, to expand manufacturers' liability for defective products that cause injury to persons and property, a still under-inclusive category, and searched for ways to ground and confine such expanded liability within limits.

In *Mazetti v. Armour Co.* (1913),[20] which predated *MacPherson*, a restaurateur was permitted to recover for damage to his business that resulted when a customer was served defective canned food manufactured by defendant. The court stated that the nineteenth-century rule of law requiring privity was based on the assumption that a retailer of food had opportunity to inspect food bought from others before selling it to the public and was accordingly charged with knowledge of its defects. However, under modern conditions, it "recognize[d] that the contents of a can are sealed up, not open to the inspection or test, either of the retailer, or of the customer, until they are opened for use; and then not susceptible to practical test, except the test of eating." When a manufacturer markets goods in this form, it represents to purchasers that the contents of the can are suitable for eating, the same as if an express representation was printed upon a label. The court grounded its theory of "implied warranty" upon "the demands of social justice."

In *Baxter v. Ford Motor Co.* (1932),[21] the plaintiff was driving an automobile manufactured by defendant, when a pebble, which was thrown up by the wheels of a passing car, struck the windshield of his car, which shattered, causing small pieces of glass to fly into his left eye, resulting in the loss thereof. The issue in the case was whether the trial court erred in refusing to admit into evidence defendant's brochure which represented that the glass was shatterproof. The court held that the brochure was improperly excluded from evidence by the trial court. *Baxter* grounded liability on a contract theory of "express warranty." The court said that manufacturers use advertising on radio and printed media to create demand for products; having created demand, manufacturers should not be allowed to evade liability for injury to consumers from products that do not have the advertised qualities.

In *Henningsen v. Bloomfield Motors, Inc.* (1960),[22] buyer bought an automobile as a Mother's Day present for his wife. While driving the car, she was badly injured when "something went wrong" with the steering gear, and the car turned sharply to the right and collided with a wall. The issue in the case concerned the effect that should be given to the contract, which contained an express warranty that sought to limit the manufacturer's liability to replacement of defective parts and which disclaimed all other warranties, express or implied.

The court stressed the "inequality of bargaining position between the parties" and noted that this warranty was not only standard with manufacturer but was the uniform warranty of the Automobile Manufacturers Association. The court held that manufacturer's "attempted disclaimer of an implied warranty of merchantability [average quality] and of the obligations arising therefrom is so inimical to the public good as to compel an adjudication of its invalidity."

By the 1970s, products liability law was reasonably clear: manufacturers were strictly liable for injuries and property damage caused by dangerously defective consumer products to buyers, users, or bystanders, and such liability could not be disclaimed by contract. Proof of negligence or carelessness was not required.

PRINCIPLE OF PRECEDENT (STARE DECISIS) CONSTRAINS OVER-INCLUSIVE RULES

Just as economically rational individuals are assumed to prefer more of a good to less but cannot have an unlimited amount of goods because their choices are constrained by budget and resource restrictions, so too, judges, who are assumed to prefer more general legal categories or rules of law to narrow ones, confront constraints, the principal one being stare decisis.

Judges often realize that a rule of law is over-inclusive when they must decide a new case, say case y, which falls within the rule of law as currently formulated in case x, but which they believe should be decided differently from case x. An over-inclusive rule of law includes cases that a court would prefer to distinguish but cannot due to the principle of precedent or stare decisis. In order to provide a rationale for the decision to distinguish cases x and y based upon a general rule of law, a judge must reformulate the rule of law in case x more narrowly to exclude case y from the rule of case x by the process of marginal exclusion. Thus, due to the principle of precedent, which requires courts to decide like cases alike, courts are constrained from adopting over-inclusive rules of law.

PRODUCTS LIABILITY LAW SPLIT INTO MANUFACTURING DEFECTS AND DESIGN DEFECTS

As shown above, during the twentieth century courts expanded products liability, which was still an under-inclusive legal concept, by contract warranties, either express or implied, and by strict liability in tort. However, beginning in the 1970s, cases show a gradual awareness by some courts that the category of strict liability in tort was becoming over-inclusive, so that the concept had to be confined (constrained) to a product with a manufacturing defect, which is a product that differs from all other products that a manufacturer produces by the same process (as, for example, an automobile with a defective steering mechanism). For products with design defects, which are all products manufactured by the same process that subject consumers to unreasonable risks (as, for example, all automobiles with unpadded steering wheels, chapter 3), liability is not strict but should be based on fault or negligence as determined by the Judge Hand test or a risk-utility analysis, as discussed in chapter 3. Other courts disagreed and tended to hold sellers absolutely liable for all injuries in any way related to a product's use, even if the cost of product safety was prohibitive or there was product misuse by the injured party. This latter tendency seemed to gather momentum; that is, some courts seem to adopt the doctrine of absolute liability or venditor emptor. The following cases illustrate these developments.

In *Greenman v. Yuba Power Products* (1963),[23] plaintiff brought an action for damages against manufacturer and retailer of a combination saw, drill, and wood lathe. While the plaintiff was using this lathe to turn a large piece of wood into a chalice, it suddenly flew out of the machine and struck him on the forehead, inflicting serious injuries. The California Supreme Court held: "A manufacturer is strictly liable in tort when an article he places on the market, knowing that it is to be used without inspection for defects, proves to have a defect that causes injury to a human being."

The Court appeared not to notice that previous product liability precedents involved manufacturing defects, as for example, *MacPherson*, in which the defective automobile wheel that caused the injury differed from all other automobile wheels manufactured by the same process. However, in *Greenman*, the plaintiff's expert witnesses testified that injury was caused by a "defective design" because inadequate set screws were used to hold material being worked upon in the machine. In a design defect case, the issue is whether all products manufactured by the same process, the entire product line, subjects consumers to unreasonable risks. In other words, in *Greenman,* the court unknowingly conflated manufacturing defects with design defects and created an over-inclusive rule of law. This distinction between manufacturing and design defects, which was overlooked in *Greenman*, would be seen as crucial beginning in the 1970s.

In *Phillips v. Kimwood Machine Co.* (1974),[24] the plaintiff was injured while operating a sanding machine that did not hold its grip on a piece of fiberboard, but suddenly expelled it, striking plaintiff in the abdomen. There was evidence that for a relatively small expense there could be installed on the machine a line of metal teeth, which pointed in the direction the fiberboard progresses through the machine and which would press lightly against the sheet, but in case of attempted regurgitation, would be jammed into it, thus stopping its backward motion. The court recognized that "[t]he problem with strict liability of products has been one of limitation. No one wants absolute liability where all the article has to do is to cause injury." It said: "A test for unreasonable danger is vital." It proposed a distinction between defective products and defective designs and suggested that there should be strict liability for defective products whereas liability for defective designs should be grounded in negligence. In other words, the court realized that applying strict liability in tort to both manufacturing defects and design defects was over-inclusive.

In *Prentis v. Yale Mfg. Co.* (1984),[25] the court essentially agreed with the *Phillips* court; it said that although many courts have insisted that in defective design cases "they are not applying negligence tests because their focus is on the *product* rather than the manufacturer's *conduct* . . . the distinction on closer examination is semantic."[26] In other words, defective design cases are de facto negligence cases and should be decided according to the Judge Hand test. Surprise! In the 1970s, negligence-like liability, which was thought to have been eliminated in products liability cases, came back again in design defect cases.

However, other courts continued to expand product liability toward caveat venditor either by blurring the standard between negligence and strict liability or by openly holding that manufacturers should bear the cost of any injury related to product use, regardless of user fault or whether safer alternatives exist. In *Beshada v. Johns-Manville Products Corp.* (1982),[27] workers whose health suffered from use and handling of asbestos products sued asbestos manufacturers. Since the court accepted as a fact that the medical profession did not recognize the health hazards from using insulation materials made from asbestos until the 1960s, the defendants maintained that they could not have warned of such danger before then because they did not know of its existence. The unanimous court disagreed, stating that to allow a state-of-the-art defense would be to reintroduce a negligence test. Relying on the semantic product-conduct distinction, it said: "Strict liability focuses on the product, not the fault of the manufacturer." (Other courts have allowed a state-of-the-art defense.[28]) In *O'Brien v. Muskin Corp.*, discussed above, the case in which plaintiff was severely injured when he dove from the roof of an 8-foot-high garage into a 3 ½-foot deep above-ground swimming pool, the court held that the plaintiff need not prove the existence of alternative or safer designs; it said that even if there are no alternative methods of the making bottoms of above-ground pools, the jury might have found that the risk posed by the pool outweighed its utility.

By the middle of the 1980s, legal commentators began to argue that the caveat venditor line of cases was leading to an insurance crisis because manufacturers were finding it increasingly difficult to pass on to consumers the cost of ever increasing insurance premiums and ever expensive safety measures.[29] As a solution to the insurance crisis, they suggested legislative tort reforms. Many states responded by passing laws that placed caps on noneconomic damages, so-called pain and suffering awards. In cases in which plaintiffs seek punitive damages in addition to compensatory damages, some states placed caps on punitive damages, raised the plaintiff's burden of proof from the preponderance of evidence standard to a clear and convincing evidence standard for punitive damages, or required that defendants be guilty of gross negligence or recklessness before punitive damages could be awarded. Other state tort reforms allowed the defense of contributory negligence or applied comparative negligence in product liability actions, regulated attorney's contingent fees, and changed the law in product liability cases from joint to apportioned liability in order to protect defendants who were not substantially at fault but were sued merely because they had deep pockets.

With respect to the common (judge made) law of products liability, *Restatement (Third) of Torts: Products Liability* (1998) has resolved the disagreement among courts in favor of those courts that recognize two categories of products liability: strict liability for manufacturing defects and a negligence-like standard for design defects.[30] (The theoretical case for maintaining this distinction is given in chapter 18.)

Manufacturing defects are essentially a question of efficient quality control, for which a manufacturer should be strictly liable. However, since it is almost always possible to make any product safer, if a plaintiff in a design defect case only had the burden of proving that a better design is possible, regardless of cost, the burden on manufacturers would be crushing. In a design defect case, courts now generally require plaintiff not only to prove that a safer design is feasible, but also that it is cost efficient; in other words, whether stated or not, in design defect cases, courts now generally require the use of the Judge Hand test, risk-utility test, or the similar negligence-like test.

NESTED RULES OF LAW PREVENT ISOLATION OF INCONSISTENT CASES

Unlike economic rationality, which has the property of completeness, legal rationality has the property of incompleteness; that is, unlike economics, in which every bundle is related to every other bundle, in law not every case is related to every other case by binding precedent. The incompleteness property of law raises the following question: Are the many rules of law in cases disparate and isolated due to legal rationality's incompleteness property, or are they bound together by fundamental principles such as due process, equal protection, and horizontal equity into a coherent whole? The answer to this question determines whether the principle of precedent is a weak or a strong constraint on decisions by judges.

The incompleteness property of law allows judges to "distinguish a case to death" or "confine a case to its particular facts,"[31] so as to decide cases inconsistently by disingenuously isolating such cases under rules of law which purport to be separate and discrete or sui generis. For example, a latter court could decide that the hypothetical decision in *LaFong v. Walpole,* discussed above, holds only for red-headed individuals named Walpole who drive blue Toyotas. Llewellyn said that if this is said of a case, it is overruled because it has no precedential value.[32] In practice, this technique is a way for either diminishing unwelcome precedents or making arbitrary, inconsistent decisions based on the naked power of courts, as in *Bush v. Gore*, discussed below, the legitimacy of which decision has been seriously questioned. This sleight of hand is possible because in legal rationality not all cases are related to each other by precedent due to its incompleteness property.

However, if legal categories and rules of law are nested (like the example of Russian dolls discussed above) within broader fundamental principles, such as due process, equal protection, and horizontal equity, law's incompleteness property will be less able to mask inconsistent decisions.

An example discussed by Dworkin is a good place to begin discussion. He defines a legal system bound together by fundamental principles as having the virtue of integrity. As an example that violates legal integrity, he discussed the rule that although other professions were liable for damages caused by their negligence, British barristers had legal immunity. This rule violates legal integrity because special treatment for barristers cannot be justified in principle. Dworkin said that although the exemption for barristers has been curtailed, integrity will not be satisfied until the exemption is entirely eliminated.[33]

Having one rule for negligent barristers and a different rule for other professionals who are negligent exemplifies disparate, isolated, and local legal rules that take advantage of law's incompleteness property. However, if narrow, local rules are treated as being nested within broader, inclusive legal rules, the broadest of which have the status of fundamental principles and standards, such as due process, equal protection, and horizontal equity, legal rationality begins to look more like a coherent whole. The process Dworkin describes may reflect what Lord Mansfield had in mind when he spoke of "the law working itself pure."[34] Dworkin conceives of the two rules of law, one for barristers and another for other professions, as being nested[35] in a broader legal category, which he calls integrity that requires the same standard of care for all professionals, but which could be called horizontal equity. Horizontal equity is the judicial analogue to equal protection of the laws that applies to legislative enactments.

Taking advantage of law's incompleteness property, that is, isolating a case within a narrow discrete rule of law to avoid explaining why the case is not like other cases, all of which are nested within the same fundamental principles, weakens the principle of precedent, which is the law's principal constraint on arbitrary exercises of judicial power.

Consider the following two examples of attempts by the United States Supreme Court to isolate a rule of law in a disparate category, as opposed to interpreting rules as being nested within fundamental principles that make law a coherent whole.

In *Griswold v. Connecticut*,[36] the Supreme Court struck down a law prohibiting the use of contraceptives by married persons on the ground that it infringed a fundamental right to privacy. However, in *Bowers v. Hardwick*,[37] the Court upheld a Georgia state sodomy statute that made consensual homosexual activity between adults in private a crime on the ground that sexual activity is not a fundamental right. This case conflicted with *Griswold,* which was grounded on the fundamental right of privacy. A rule of law that can distinguish the right to privacy of married couples from the right to privacy of homosexual couples without being based upon suspect, unreasonable classifications is hard to imagine. *Griswold* and *Bowers* are contained or nested within the fundamental right to privacy. *Bowers* was subsequently overruled by the US Supreme Court in *Lawrence v. Texas*.[38]

In the controversial case *Bush v. Gore*[39] that decided the 2000 presidential election, the 5–4 conservative majority overtly rejected the principle of precedent. The legitimacy of the decision has been seriously questioned.[40] The Supreme Court's conservative justices intervened in Florida's electoral process twice and held that a planned manual recount violated the right to equal protection even though in many past cases involving manual recounts they had not raised the equal protection issue. The majority said that the equal protection right was unique to this case and that it would not be bound in future cases by any principles stated in *Bush*.[41] The five conservative justices used law's incompleteness property to isolate *Bush* and make it sui generis. In sum, *Bush* is an example of naked, unprincipled use of judicial power.

The principle of precedent in strong form mandates that courts should not isolate cases within a disparate rule of law. Before promulgating a rule of law in a case, an appellate court ought to look for conflicts between the proposed rule of law and every fundamental legal principle within which the case is nested. If the court finds inconsistencies, then one or more cases has been decided incorrectly.

NOTES

1. The next chapter shows that when the rules of economics, E1 and E2, and law, L1 and L2, are stated as simple binary relations, transitivity, completeness, and incompleteness are mathematical properties of these rules. The conditions under which economic and legal logics are consistent (isomorphic) are also stated.
2. 94 N.J. 169 (1983).
3. Steven Burton, *An Introduction to Law and Legal Reasoning* (New York: Aspen, 1985), 11–13.
4. This hypothetical case, slightly altered, is taken from K. N. Llewellyn, *The Bramble Bush* (New York: Oceana Publications, 1960), 47.
5. *Ibid.*, 48.
6. In criminal law, an intentional injury to person is much more harshly dealt with than theft of property.
7. Llewellyn, note 4, 41.
8. Figure 5.1 is adapted from Joseph Tussman and Jacobus tenBroek, "The Equal Protection of the Laws," *California Law Review* 37 (1949): 341, 347, which is discussed in the next chapter.
9. For advocacy of a minimalist approach to rules of law, see, Cass Sunstein, *One Case at a Time: Judicial Minimalism in the Supreme Court* (Cambridge, MA: Harvard University Press, 2001).
10. Ronald Dworkin, *Law's Empire* (Cambridge, MA: Belknap Press, 1986), 106.
11. H. Craig Petersen, *Business and Government,* 4th ed. (New York: HarperCollins, 1993). 450.
12. *Winterbottom v. Wright* (1842) 152 ER 402 (Ex Ct) (Lord Abinger's opinion).
13. 6 N.Y. 397.
14. 89 N.Y. 470.
15. 195 N.Y. 478.
16. Llewellyn, note 4, 74.
17. In chapter 6, the transitivity property of legal rationality will be derived from a mathematical property of Rule L2.
18. 217 N.Y. 382.
19. Edward Levi, *An Introduction to Legal Reasoning* (Chicago: The University of Chicago Press, 1948), 8–9, 14.
20. 75 Wash. 622.
21. 168 Wash. 456.
22. 32 N.J. 358.
23. 59 Cal.2d 57.
24. 269 Or. 485.
25. 421 Mich. 670.
26. 421 Mich. 687–88.
27. 90 N.J. 191.
28. *Schneider v. Chrysler Motor Corp.*, 401 F.2d 549 (8th Cir. 1968).
29. George Priest, "The Current Insurance Crisis and Modern Tort Law," *Yale Law Journal* 96 (1987): 1521, 1521–22; Robert Rabin, "Indeterminate Risk and Tort Reform," *Journal of Legal Studies* 14 (1985): 633–43.
30. "A product . . . is defective in design when the foreseeable risk of harm posed by the product could have been reduced or avoided by the adoption of a reasonable alternative design by the seller or other distributor, or a predecessor in the commercial chain of distribution, and

the omission of the reasonable alternative design renders the product not reasonably safe." *Restatement (Third) of Torts: Products Liability* (St. Paul, MN: American Law Institute, 1998), Section 2(b).

31. Llewellyn, note 4, 72–73.

32. *Ibid.*

33. Dworkin, note 10, 219–20.

34. Lon Fuller, "Forms and Limits of Adjudication," *Harvard Law Review* 92 (1978): 353, 377.

35. Dworkin, note 10, 412.

36. 381 U.S. 479 (1965).

37. 478 U.S. 186 (1986), overruled by *Lawrence v. Texas*, 539 U.S. 558 (2003). [*Bowers v. Hardwick*].

38. 539 U.S. 558 (2003). After *Bowers* but before *Lawrence,* in *Powell v. State*, 510 S.E.2d 18 (Ga. 1998), Supreme Court of Georgia reversed its *Bowers* decision and held that the Georgia statute violated the fundamental right of privacy under the state constitution.

39. 531 U.S. 98 (2000) (per curiam).

40. See, for example, David Cole, "Symposium: Assessing the Rehnquist Court's Parting Words on Criminal Justice: The Liberal Legacy of *Bush v. Gore*," *Georgetown Law Journal* 94 (2006): 1427–31.

41. 531 U.S. 109 ("Our consideration is limited to the present circumstances, for the problem of equal protection generally presents many complexities").

Chapter 6

When Are Economics and Law Consistent (Isomorphic)?

> A foolish consistency is the hobgoblin of little minds, adored by little statesmen and philosophers and divines. With consistency a great soul has simply nothing to do.
>
> —Ralph Waldo Emerson, "Self-Reliance" (1841)

Chapter 2 defined economic rationality and logic by two behavioral rules, E1, more of a good is strictly preferred to less, and E2, for any two bundles, an individual can determine whether he or she prefers one bundle to the other or is indifferent between them, plus constraints, for example, budget constraints. Chapter 5 defined legal rationality and logic by two parallel behavioral rules, L1, broad legal categories, rules of law, or case precedents are preferred to narrow ones, and L2, for any two cases, a judge can determine whether they are like in all material respects or unlike (distinguishable), as constrained by the principle or precedent or stare decisis. These two chapters answered the question, how does thinking like an economist differ from thinking like a lawyer?[1]

How do these two decision processes relate to each other? Under what conditions are these two logics consistent? In order to determine when law and economics are consistent (isomorphic), the first step is to translate economic and legal rules, which thus far have been stated verbally, into simple logical relations; the next step is to determine whether the logical relations of economics can be "mapped onto" the logical relations of law so as to preserve the mathematical properties and results of both disciplines. When such mapping is possible, economics and law are consistent or isomorphic; when such mapping is not possible, economics and law are inconsistent; and, if inconsistent, the question, which logic should take precedence or trump the other, must be answered.

An additional benefit from translating the verbal rules of economics and law into logical relations is that two significant characteristics of economic rationality, transitivity and completeness, and two significant characteristics of legal rationality, transitivity and incompleteness, which were discussed in chapters 2 and 5, are shown to be mathematical properties of the rules stated as binary relations.

In this chapter, the term binary relation is defined; it simply refers to two entities that are related to each other by an operation; for example, 2 + 3 is the binary relation of addition.

The two verbal rules of economics, Rules E1 and E2, are translated into binary relations; when so stated, Rules E1 and E2 have the mathematical property of transitivity and Rule E2 has the mathematical property of completeness.

Tussman & tenBroek's (T&t) logical model of equal protection of the laws is presented because it is a precursor to this book's general model of legal rationality. In T&t's model, equal-protection categories of constitutional law are described with Venn diagrams (in two-dimensional space). The T&t equal protection model makes the general model of legal rationality easier to understand because their Venn diagrams can be translated into binary relations and generalized easily into a theory of law.

The two verbal rules of law, Rules L1 and L2, are translated into binary relations; when so stated, they have the mathematical properties of transitivity and incompleteness. Three internal processes in economics and law that cause rules of law to tend toward some optimum are explained. Three case selection hypotheses, which conclude that some cases are more likely to be selected for litigation than others, are discussed. Two of these hypotheses are from economics. The third case selection hypothesis from law, which was discussed in the previous chapter, is briefly reprised; it states that both under- and over-inclusive rules of law are likely to be litigated, so that rules of law tend toward optimal level of generality.

Two examples of the relationship between economic and legal rationality in case law are discussed: one example in which economics and law are consistent (isomorphic) and one in which they are not. One tentative conclusion is that in judge-made law or in judicial interpretation of statutory ambiguities, economic rationality and legal rationality have two usual relationships: when judges use economic rationality as the dominant rationality, they will adopt the goal of efficiency (or efficiency as constrained by equity) and balance competing interests through marginal trade-offs. Under these circumstances, economic rationality and legal rationality will be isomorphic or close to it because judges will tend to interpret legal concepts to preserve the properties and results in both disciplines.

However, if judges employ legal rationality as the dominant rationality, legal rationality and economic rationality will rarely be isomorphic or consistent, because legal concepts, such as fundamental rights, are based upon goals that are often inconsistent with efficiency. The last section contains conclusions.

BINARY RELATIONS EXPLAINED

Math phobic persons are often fearful of the term binary relation, even though, like Molière's bourgeois gentlemen,[2] who was surprised to find out that he had been speaking prose all his life, they have been using binary relations most of their lives. For example, 3 + 2, which in prose reads, "three plus two," is the binary relation of addition. A binary relation is defined as two elements that are related by an operation, in this case addition. The result of performing the operation of addition on the binary relation 3 + 2 equals (=) 5. The binary relation 4 × 6 reads "four times six"; the result of performing the operation of multiplication equals 24. Binary relations can be stated more abstractly in symbols: addition is $x + y$ and multiplication is xy, where x and y stand for any two numbers. $x = y$ is also a binary relation, which expresses equality between two entities or elements.

In the text below, we will be discussing binary relations or economics and law and their mathematical properties. Since grade school, you are familiar with three mathematical properties of addition and multiplication even if you cannot name them; most people cannot. First, $3 + 2 = 2 + 3$, and $4 \times 6 = 6 \times 4$: the commutative property means that numbers can be added or multiplied in any order. Second, $2 + (3 + 5) = (2 + 3) + 5 = 10$, and $2(3 \times 5) = (2 \times 3)5 = 30$ (binary operations within parentheses are performed first): the associative property means that if the order of numbers does not change, grouping of numbers related by addition or by multiplication can change without changing the result. Third, $2(3 + 5) = 2(3) + 2(5) = 16$ (multiplication distributes over addition): the distributive property means a number can be taken through the parentheses or can be factored out of the parentheses.

The rules of economics, E1 and E2, and law, L1 and L2, discussed in chapters 2 and 5, will be formally stated in terms of binary relations and their mathematical properties.

BEHAVIORAL RULES OF ECONOMIC RATIONALITY AS BINARY RELATIONS

Stated verbally, Rule E1 of economics states that more of a good is strictly preferred to less, and Rule E2 states that for any two bundles of goods, an individual can determine whether he or she prefers one bundle to the other or is indifferent between them. Rule E1, stated as a binary relation between any two bundles x and y, is:

> Rule E1: if bundle y is a bundle of goods, and bundle x is a bundle of goods with at least as much of all goods as bundle y and more of one good, then bundle x $\succ$ bundle y (where $\succ$ is read "is strictly preferred to").

For example, bundle x (40 hours of work/week, \$80,000) is strictly preferred to bundle y (40 hours of work/week, \$60,000) because bundle x has the same amount of work but more income than bundle y; that is, bundle x $\succ$ bundle y.

Recall from chapter 2 that Rule E1 cannot be used to decide an economic preference between two bundles when one bundle has more of one good and another bundle has more of another good. For example, Rule E1 cannot be used to determine preference between bundle y (40 hours of work/week, \$60,000) and bundle z (60 hours of work/week, \$90,000) because the former bundle has more leisure and the latter bundle has more income.

Rule E2 eliminates indeterminacy between any two bundles. Rule E2 states that for any two bundles x and y, an individual must be able to determine whether he or she prefers bundle x to bundle y, prefers bundle y to bundle x, or is indifferent between them. Rule E2, stated as a binary relation between any two bundles x and y, is:

Rule E2: For any two bundles x and y, an individual can determine whether x ≽ y, y ≽ x, or x = y (where ≽ is read "is weakly preferred to" or "at least as good as").

For any two bundles x and y, one of three binary relations must be true: bundle x ≽ bundle y, bundle y ≽ bundle x, or bundle x = bundle y. Not everyone will have the same preferences. Those who prefer leisure to income may prefer bundle y (40 hours of work/week, $60,000) to bundle z (60 hours of work/week, $90,000); those who prefer income to leisure may have the opposite preferences; or, a person may be indifferent between the bundles.

Transitivity (Consistency) Property of x ≻ y, x ≽ y, and x = y

The binary relations of Rules E1 and E2, x ≻ y, x ≽ y, and x = y (strict preference, weak preference, or indifference) have the mathematical property of transitivity, which is defined as a relationship between any three bundles, such that if bundle x is preferred or equal to bundle y, and bundle y is preferred or equal to bundle z, then bundle x must be preferred or equal to bundle z. Transitivity holds for the binary relations of strict preference, weak preference, and indifference. Stated symbolically:

Strict preference transitivity property: If x ≻ y, and y ≻ z, then x ≻ z.
Weak preference transitivity property: If x ≽ y, and y ≽ z, then x ≽ z.
Indifference transitivity property: If x = y, and y = z, then x = z.

Transitivity is the consistency criterion of economic rationality. For example, with respect to the strict preference transitivity property, if a person prefers bundle x (50 hours of work/week, $80,000) to bundle y (60 hours of work/week, $70,000) because bundle x has both more leisure and more income, and prefers bundle y (60 hours of work/week, $70,000) to bundle z (70 hours of work/week, $60,000) for the same reasons, then he or she must prefer bundle x to bundle z. Although a real-world person may have intransitive preferences, such a person is inconsistent in terms of economic logic. Transitivity gives economic rationality its rigid logical structure; it is the backbone of economic reasoning.

Completeness Property of x ≽ y

Although the binary relation of Rule E1, x ≻ y, x is strictly preferred to y, is transitive, it is incomplete because it cannot be used to determine preference between two bundles where one bundle has more of one good and the other bundle has more of another good; that is, not all bundles of competing goods are related to each other by strict preference relations. However, the binary relation of Rule E2, x ≽ y, weak preference, is both transitive and complete: for any two bundles whatsoever, one of the three binary relations that express an individual's weak preferences, x ≽ y, y ≽ x, or x = y, must be true. The completeness property is important because it means that in economic rationality every bundle is related to every other bundle by Rule E2. The completeness property also justifies one of economic rationality's fundamental principles, opportunity cost. Since the cost of any bundle depends upon the cost of the next best alternative bundle, and the completeness property relates every bundle to every other bundle, opportunity cost of any bundle can always be determined.

Chapter 3 contains examples in which due to economic rationality's completeness property, all cases that balanced the same competing interests were shown to be completely ordered in relation to each other.

Rule E2, $x \succcurlyeq y$, Is a Total or Linear Order Relation

The binary relation of Rule E2, $x \succcurlyeq y$, is defined for all bundles of competing goods; these bundles can be represented by real numbers in two or three spatial dimensions or by mathematical vectors for four or more goods (recall the discussion in chapter 2). The binary relation, $x \succcurlyeq y$, has four mathematical properties (only the last two of which are significant in economics): it is reflexive, antisymmetric, transitive, and complete; a binary relationship having these four properties is called a total order relation.[3] A total order relation is also suggestively called a linear order relation because it guarantees that all bundles can be *rank ordered* in relation to each other in spatial diagrams or on a line (examples were given in chapters 2, 3).

The transitivity property of the binary relations in Rule E1, $x \succ y$, and Rule E2, $x \succcurlyeq y$, and completeness property of the binary relation in Rule E2, $x \succcurlyeq y$, are essential to understanding what thinking like an economist means because they provide the logical foundation for balancing competing interests via marginal trade-offs, which was used in the theory of efficient (optimal) choice in chapter 2 (recall that Rule E2 was used to construct indifference curves and marginal trade-offs along the indifference curves).

Maximizing Satisfaction Subject to a Budget Constraint

In chapter 2, maximizing satisfaction or utility subject to a budget constraint, which is called the theory of optimal choice, was explained. Rules E1 and E2 make possible a complete rank ordering of an individual's choices over all possible bundles of competing goods. In that chapter, an individual's efficient or optimal choice among bundles of work-leisure and income depended upon the individual's preferences between leisure and income in relation to the market for her or his skills as expressed by the job possibilities curve. Each individual trades-off leisure and income according to her or his preferences as constrained by availability of jobs in the market.

Thinking like an economist means employing the two economic rules and the theory of optimal (constrained) choice rigorously (respecting transitivity) to any and every possible pair of the infinitely variable choices (completeness).

TUSSMAN & TENBROEK MODEL OF EQUAL PROTECTION

Tussman and tenBroek's (T&t) article, "The Equal Protection of the Laws," describes constitutional equal protection categories with Venn diagrams.[4] The T&t model provides a convenient bridge to the model of legal rationality because their Venn diagrams can be translated into binary relations and generalized easily into a theory of law.

When legislation concerns general matters of economic or social welfare, the law need only relate rationally to a legitimate governmental purpose. However, when legislation classifies persons for differing advantages or obligations, it must conform to the Fourteenth Amendment, which requires that no person shall be denied "equal protection of the laws."

In reviewing any classification, a court must determine whether persons classified by law for similar treatment are similarly situated or dissimilar.[5] In order to determine whether a law that classifies different groups for different benefits or burdens is legitimate, that is, is based upon constitutionally permissible criteria, a court must first look to the purpose of the legislation. The purpose of a law is legitimate if it either eliminates a public mischief or achieves some positive public good. This focus on the purpose of legislation is consistent with the notion that a rational decision-making process must be purposeful (chapter 2).

Once a court has determined that the governmental purpose is legitimate, the relation between legitimate purpose and actual classification must be considered. For example, in time of war, Congress can legitimately pass a law to meet the dangers of sabotage. But when the actual legislation identifies saboteurs by ethnic origin, as in *Hirabayashi v. United States*,[6] a court must analyze whether the classification violates equal protection.

T&t use Venn diagrams to illustrate the five possible relations between a classification based upon a legitimate governmental purpose and the law's actual classification of individuals. These five relations are reproduced as Figure 6.1. Let L stand for a classification based on a legitimate governmental purpose, and A stand for the actual government classification defined in law.

The reasonableness of any actual government classification, A, defined in law depends upon its relation to the classification based on a legitimate governmental purpose, L. 1) The law is perfectly reasonable: the general classification based on a legitimate governmental purpose includes every member of the actual class defined in the law; the converse is also true. 2) The law is perfectly unreasonable: the general classification based on a legitimate governmental purpose does not include any member of the actual class defined in the law.

3) Represents under-inclusiveness: the classification based on a legitimate governmental purpose includes everyone in the actual class defined in law, but the actual class excludes some people who are included in the legitimate governmental purpose. Although this relationship is not perfectly reasonable, the Supreme Court has upheld under-inclusive legislation to allow the legislature to deal with a general problem in a piecemeal manner.

4) Represents over-inclusiveness: the actual class defined in law includes individuals who are outside the classification based on a legitimate governmental purpose. Because "over-inclusive classifications reach the innocent bystander, the hapless victim of circumstance or association," the case against over-inclusive classifications is stronger than the case against under-inclusiveness. Nonetheless, over-inclusive classifications have been sustained in emergency situations, such as wartime.[7]

5) Represents a law that is both under-inclusive and over-inclusive. For example, *Hirabayashi*, the legitimate purpose was to deter saboteurs, but the actual class was

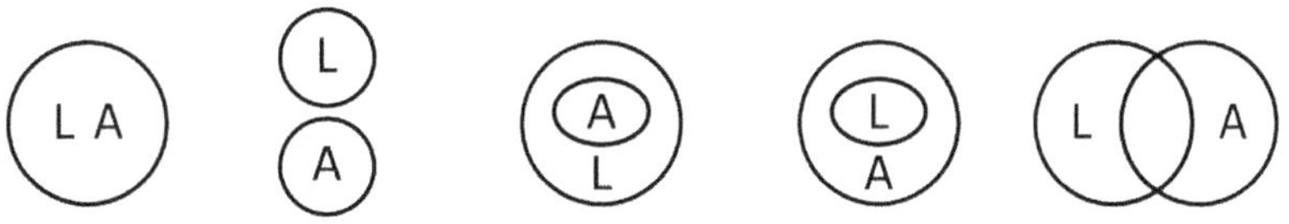

Figure 6.1 Equal Protection of the Laws. *Source*: Created by the author.

defined as Japanese-Americans, which was over-inclusive since not all Japanese-Americans were traitors and is under-inclusive because German-Americans and Italian-Americans were excluded.

The Venn diagrams used by T&t can be represented symbolically in terms of binary relations. Translating the five Venn diagrams into binary relations demonstrates the essential bridge between their model and the general model of legal rationality, which uses these binary relations and is presented below. The translation shows that the T&t model is merely a specific example of the more general model of legal rationality. In T&t's Venn diagram (1) both $L \supset A$ and $A \supset L$, where the binary relation $\supset$ is read "contains." In Venn diagram (2) L and A are disjoint, so the relation $\supset$ is not defined. In diagram (3) $L \supset A$; diagram (4) $A \supset L$; and, diagram (5) L and A intersect, so the relation $\supset$ is not defined.

T&t's diagrams can also depict the five possible relations between any two cases through the principle of precedent. Diagrams (1), (3), and (4) depict the three possible ways that two cases can relate to each other as *binding* precedents. In diagram (1), the two cases are like in all material respects; they are mutual precedents. In diagram (3), case L is a binding precedent that includes case A; in diagram (4), case A is a binding precedent that includes case L. In diagrams (2) and (5), the cases L and A are distinguishable; that is, unrelated by binding precedent. Thus, Tussman and tenBroek's specific model of equal protection of the laws is part of the general theory of legal rationality.

BEHAVIORAL RULES OF LEGAL RATIONALITY AS BINARY RELATIONS

The model of legal rationality, which was presented in chapter 5, will be stated as binary relations that are parallel to and comparable with the rules of economic rationality. Legal rationality also has two behavioral rules: Rule L1 states that broad legal categories and rules of law are preferred to narrow ones. Rule L2 states that for any two cases, judges can determine whether they are like (in all material respects) or unlike (distinguishable).

Legal rationality is constrained by the principle of precedent or stare decisis. Whether the principle of precedent is a strong or weak constraint depends upon whether legal rationality's incompleteness property (not all cases are related to each other by precedent), which allows courts to arbitrarily isolate and confine a case to its facts, is ameliorated by the requirement that all cases be nested within and be consistent with fundamental principles, such as due process, equal protection, and horizontal equity.

The two rules define legal rationality or what it means to think like a lawyer. As in chapter 5, for ease of exposition, Rule L2, which states the relationship between two cases, that is, defines what a precedent is, will be presented before Rule L1, which states that judges prefer broader precedents to narrower ones. The reason for stating L2 before L1 is that the definition of what a precedent is should be stated before statements about what type of precedents are preferred. Rule L2 can be stated as a binary relation between any two cases x and y:

> Rule L2: For any two cases, x and y, case x ⊃ case y, case y ⊃ case x, both (case x = case y), or the binary relation ⊃ is not defined (where ⊃ is read "contains" or "is a binding precedent on")

Rule L2 says that the only way that the binary relation ⊃ can be satisfied is if case x is a *binding* precedent on case y, case y is a *binding* precedent on case x, or both (case x = case y). Any two cases for which the binary relation ⊃ is not defined are unlike or distinguishable, that is, unrelated by binding precedent.

Rule L1 can be stated as a binary relation between any two cases x and y:

> Rule L1: If case x ⊃ case y, then case x ≻ case y.

Rule L1 states that if case x is a binding precedent that includes case y, then the rule of law in case x is strictly preferred to the rule in case y.

Just as economic rationality assumes that budget constraints prevent individuals from acquiring an unlimited amount of goods, one important constraint judges confront that limits the scope of legal categories despite their preference for greater inclusiveness is the principle or precedent or stare decisis, which is discussed below.

The Transitivity (Consistency) Property of $x \supset y$

Transitivity property: if case x ⊃ case y, and case y ⊃ case z, then case x ⊃ case z.

An example of legal rationality's transitivity property was given in chapter 5's discussion of three nineteenth-century, "imminently dangerous" products liability cases as depicted by the Venn diagram, Figure 5.2: *Statler v. Ray* (the exploding coffee urn case) ⊃ *Devlin, v. Smith* (the collapsing scaffold case) ⊃ *Thomas v. Winchester* (the poison case).

If *Statler* ⊃ *Devlin*, and *Devlin* ⊃ *Thomas, then Statler* ⊃ *Thomas.*

However, the reverse order is not transitive because the transitivity principle in law works like a set of Russian dolls (*matryoska*) that go in one direction only; larger dolls include smaller dolls, not vice versa.

Incompleteness Property of $x \supset y$

The binary relation x ⊃ y also has the mathematical property of incompleteness, which means that not every two cases are related to each other by precedent. In chapter 5, the incompleteness property figured prominently in the discussion of whether the principle of precedent is a strong or weak constraint on judicial decisions. As discussed there, due to the incompleteness property, if cases are not nested within fundamental legal principles such as due process, horizontal equity, and equal protection of the laws, judges can, by expressly confining a case to its particular facts or distinguishing a case to death, decide cases inconsistently by disingenuously isolating such cases under rules of law that they purport to be separate and discrete or sui generis.

$x \supset y$ Is a Partial (Incomplete) Order Relation

The binary relation $x \supset y$, which is defined on sets which can be represented in two-dimensional space by Venn diagrams, has four mathematical properties: it is reflexive, transitive, antisymmetric, and incomplete; a binary relation that has these four mathematical properties is called a partial (incomplete) order relation.[8]

THREE THEORIES OF OPTIMAL RULES IN LAW

Both economics and law have case selection hypotheses, which posit that some cases are more likely to be selected for litigation and appeal than others. Two case selection hypotheses from economics and a third from law will be discussed: hypothesis one, inefficient rules of law are more likely to be litigated than are efficient rules of law; hypothesis two, the handicapping model in which cases which are litigated and appealed will tend to reflect the known bias of appellate court judges (e.g., US Supreme Court); and hypothesis three, discussed in chapter 5, that under- or over-inclusive rules of law are more likely to be litigated than are optimal size rules of law.

Inefficient Rules of Law Are Selected for Litigation

Law and economics scholars have constructed a hypothesis that judge-made law tends toward economic efficiency over time even if judges do not consciously strive for this goal.[9] The theory concludes that judge-made law becomes more efficient through the process of selective litigation. This theory has two key assumptions: first, inefficient rules of case law are more likely to be selected for challenge than efficient laws. The theory assumes that efficient laws generate more economic benefits (e.g, greater gross domestic product, GDP) than inefficient laws. If the allocation of a legal entitlement is inefficient, the gain to those who would benefit from a change to an efficient law will exceed the loss to those who are favored by the currently inefficient law. Thus, the former will file more lawsuits to change inefficient laws and spend more on such suits than those who are favored by the inefficient status quo. Conversely, if the allocation of a legal entitlement is efficient, those who would benefit from a change to an inefficient law will have less to gain than the cost to those who are favored by an efficient law. Thus, the former will file fewer lawsuits to change an efficient law and spend less on such suits than those who are defending the efficient status quo. In sum, litigants will spend more on litigation to overturn inefficient laws than to overturn efficient laws.

The second assumption of the hypothesis is that judges are not hostile to efficiency. As long as judges are at worst indifferent but not hostile to efficiency, the theory will be accurate. It follows as a conclusion from the two assumptions that even if judges decide cases randomly, say by flipping a coin, judge-made law will become more efficient as long as inefficient laws are litigated more than efficient laws. In other words, the worst case version of the efficient case selection hypothesis is based upon probability.

Exceptions to the theory have been noted: repeat players, that is, litigants, who expect to be in court multiple times (e.g., cigarette companies), will often spend more on a particular lawsuit than it is worth due to the strategic significance of one adverse decision. However, if the amount in controversy or stakes is high for both parties, the percentage of cases won by appellants will be closer to the distribution of suits as a whole. Second, those who litigate on principle often have noneconomic motives. Third, the more lawyers err in their assessments of the probable outcome of appeals and the more irrational the process, the less likely is the efficient result. Fourth, if courts have known political or economic biases, the efficient outcome might not be achieved. For example, economic decisions of the Supreme Court under Chief Justice Warren were substantially different than those of the Court under Chief Justice Roberts. This observation has spawned the handicapping model of selective litigation, which is discussed next.

Handicapping Hypothesis of Selective Litigation

Given the known bias of a court, cases litigated and chosen for appeal will tend to reflect that bias.[10] The process can be analogized to gambling on sporting events where there is a point spread. For example, if a football game is to be played between two teams, the Zebras and the Giraffes, where the Zebras are heavily favored to win, few people will be willing to bet that the Giraffes will win the game; however, many people may be willing to bet that the Zebras will not win by more than a stipulated number of points. For example, someone who wants to bet on the Giraffes will commonly say, "I'll take the Giraffes and thirteen points." Thirteen points is the point spread. If the Zebras win by less than fourteen points, the person who bet on the Giraffes wins. Although the probability that the Zebras will win the game is considerably greater than 50 percent, if the point spread is set so that the bet is a fair gamble, the probability that the Zebras' score will exceed the point spread will be 50 percent.

Consider the point spread analogy in the context of the probable decisions in cases litigated and selected by the United States Supreme Court under its certiorari jurisdiction. In terms of the linear description of the political spectrum, where left represents liberalism and right represents conservatism, the Warren Court of 1960s was to the left of center and the current Roberts Court is to the right. In the context of handicapping case selection hypothesis, cases litigated and selected for hearing by the Warren Court should be skewed to the left of the spectrum, and cases litigated and selected for hearing by the Roberts Court should be skewed to the right. Moreover, since lawyers will be aware of a court's biases, if their goal is assumed to be solely that of winning cases, the cases they choose (select) to appeal will be skewed to reflect the court's bias; only cases in which the court's decision is difficult to predict will be appealed. Other things being equal, if lawyers make accurate decisions concerning which cases to appeal, the case selection hypothesis predicts that appellants should win 50 percent of the cases. For example, litigation of the question whether the Affordable Health Care Act was constitutional, which would have been unlikely under a liberal court, was nearly held to be unconstitutional under the Roberts Court. Of course, other things are not always equal; the several caveats or complicating factors that affect the efficient case selection hypothesis, discussed above, repeat players, suits based upon principle regardless of the probability of winning, and attorney error, suggest that observed winning percentage of appellants will often differ from 50 percent.

Over- and Under-Inclusive Legal Rules Are Selected for Litigation

The legal theory, which concludes that rules of law progress toward rules of optimum level of generality, neither under- nor over-inclusive, was presented in chapter 5. It has two key assumptions: First, through the process of selective litigation, hard cases are more likely to be litigated than easy cases. Hard cases are interstitial, that is, they fall in the cracks between rules of law or involve arguably over-inclusive, under-inclusive, or inconsistently isolated categories or rules. Easy cases, which fall clearly within an uncontroversial legal category or rule of law, usually are not litigated or if litigated, are disposed of at trial and usually do not reach the highest appellate courts.

Second, in the model of legal rationality, Rule L1, broad legal categories and rules are preferred to narrow ones, implies that under-inclusive law rules will tend to be eliminated over time by selective litigation. The principle of precedent, stare decisis, like cases ought to be decided alike, implies that over-inclusive law rules will be constrained or eliminated in order for judges to retain freedom to distinguish future cases. The process by which rules of law expand or shrink incrementally through inclusion or exclusion is law's marginal analysis.

However, as the discussion of two centuries of products liability law in chapter 5 shows, it may take considerable time and myriad cases for the optimal level of generality to be achieved; moreover, the optimal level of generality may itself change over time. In addition, there are some legal fields, such as business law, which require consistency and uniformity, where the process of reaching an ideal rule of law through selective litigation in multiple jurisdictions has been thought to be too slow or to produce inconsistent results. These legal fields are ripe for legal codification by legislative enactment, such as the Uniform Commercial Code. In other fields of law, such as torts, the fact that it has not been codified means either that the common law process of selective case adjudication produces acceptable results, or there is a legislative stalemate among contending interests.

The most problematic assumption of the legal model of selective litigation is that all cases are nested within fundamental principles and must be decided consistently with these principles. Experience show that some judges will take advantage of the incompleteness property of legal rationality so as to give themselves freedom to distinguish and isolate cases inconsistently or illegitimately; for example, *Bush v. Gore*, which was discussed in chapter 5.

TWO EXAMPLES ON WHETHER ECONOMIC AND LEGAL RATIONALITY ARE CONSISTENT

This section contains two examples that illustrate the relation between economic rationality and legal rationality in judicial decisions: in one example economics and law are consistent or isomorphic; in the other example they are inconsistent. A tentative conclusion as to the relationship between economic rationality and legal rationality in judicial decisions is offered. The question of which logic should prevail when economics and law conflict is also discussed.

In order for law and economics to be consistent or isomorphic, it is necessary to show that the binary relations of economics and law can be translated or mapped onto each other so as to preserve the mathematical properties and results in both disciplines. Recall that the two economic rules, E1 and E2, whether stated verbally as in chapter 2 or as binary relations as in this chapter, are defined for bundles of competing goods, which can be depicted by real numbers in two or three spatial dimensions or as mathematical vectors for four or more goods; and, that the two legal rules, L1 and L2, whether stated verbally as in chapter 5 or as binary relations in this chapter are defined on sets, which are collections of things that can be depicted spatially in two dimensions by Venn diagrams. In order for economics and law to be isomorphic, it must be possible to map the dimensions of economics space onto legal space so as to preserve mathematical properties and results; and conversely.

First, the crashworthy cases discussed in chapter 3 are restated in terms of binary relations as an example in which economic rationality and legal rationality are consistent or isomorphic.[11] The second example, in which economics and law are not consistent, concerns the exclusionary rule for evidence seized in violation of the Fourth Amendment, which was discussed in chapter 4 in the context of the conflict between cost-benefit analysis and lexical constraints. A third interpretation of this example, which involves a conflict between economic and legal rationality, is also given.

Crashworthy Automobile Cases and Judge Hand Test

Recall or review chapter 3's discussion of crashworthy automobile cases, which were depicted and arranged in Figure 3.1 according to economic rationality's Judge Hand test. In that figure, cases in which automobiles were negligently designed, that is, where $B < PL$, were depicted above the 45° line, and cases in which designs met the due care standard, that is, where $B \geq PL$, were depicted on or below the line. The set of all cases, in which, $B < PL$, contains the set of all cases that belong to legal category of design defect; the converse is also true.

$(B < PL) \supset$ negligent design; the converse, negligent design $\supset (B < PL)$, is also true.

Similarly, the set of all crashworthy cases that met the due care design standard according to the Judge Hand test, $B \geq PL$, contains the set of all cases that belong in the legal category of due care design.

$(B \geq PL) \supset$ due care design; the converse, due care design $\supset (B \geq PL)$, is also true.

Thus, the Judge Hand test from economic rationality and the categories of negligent design and due care design from legal rationality are isomorphic (consistent).

Note that the Judge Hand test, which is a marginal trade-off test, belongs to economic rationality. Legal rationality is concerned exclusively with determining similarity and difference, like from unlike (distinguishable) cases; trade-offs have no meaning in economic rationality because they are not defined. However, notwithstanding their different binary relations, an isomorphic relationship exists between law and economics if binary relations from one discipline can be mapped onto the other in a way that preserves the mathematical properties and results of operations in both disciplines.

Three Interpretations of Unreasonable Search and Seizure Cases

In chapter 4, cases involving the interpretation of the exclusionary rule for evidence seized in violation of the Fourth Amendment were discussed; these cases involved the question whether or not to make trade-offs between personal liberty and public safety. *United States v. Leon* (6–3),[12] involved a conflict between the majority, who weighed the costs and benefits of preventing the use of illegally seized evidence at trial, and dissenters who believed that *Weeks v. United States*[13] and *Mapp v. Ohio*[14] established a minimum (lexical) level of Fourth Amendment protection against unreasonable search and seizure, which cannot be reduced or traded-off without violating the US Constitution. In *Utah v. Strieff* (5–3),[15] where a policeman stopped and detained a person for identification without probable cause, the evidence so obtained was allowed at trial on the ground that defendant was subsequently found to have an outstanding arrest warrant. The balance in *Strieff* between Fourth Amendment rights and convicting the guilty was tilted even further toward the latter goal than in *Leon*. Justice Sotomayor dissented on the ground that "the Fourth Amendment should prohibit, not permit, such misconduct."

Another interpretation of the conflict between the majority and the dissenters in these cases is one where the majority employed economic rationality and the dissenters employed legal rationality, in which trade-offs are not defined. For the dissenters, *Weeks-Mapp* stands for a rule of law: evidence seized in violation of the Fourth Amendment prohibition against unreasonable searches and seizures must be excluded from trial. In *Leon* and *Strieff*, the warrants allowing the seizure were without probable cause; therefore, the seizure violated the Fourth Amendment and must be excluded. In other words, for the dissenters in these cases, *Weeks-Mapp* can be interpreted as a binding precedent that includes the facts of the other two cases, that is, *Weeks-Mapp* $\supset$ *Leon; Weeks-Mapp* $\supset$ *Strieff* and mandates that the tainted evidence be excluded.

In cases involving fundamental rights, legal rationality and economic rationality are usually untranslatable rather than consistent and isomorphic. When law and economics conflict over fundamental rights, there is no a priori reason to think that economics should trump the law or vice versa.

CONCLUSION

Once one acknowledges that economic rationality and legal rationality are separate and distinct rational decision processes and that the two models of rationality have different values embedded in their logics, the question that naturally arises is whether these two logics are consistent. One conclusion of this chapter is that in judge-made law or judicial interpretation of statutory ambiguities, economic rationality and legal rationality have two usual relationships: First, when judges use economic rationality as the dominant rationality, they will adopt the goal of efficiency (or efficiency as constrained by equity) and balance competing interests through marginal trade-offs; economic rationality and legal rationality will usually be consistent or isomorphic, or close to it, because economic concepts can be mapped onto relevant legal concepts; judges will tend to interpret legal concepts so as to preserve the properties and results in both disciplines.

Second, if judges employ legal rationality as the dominant rationality, legal rationality and economic rationality will rarely be isomorphic or consistent, because legal concepts, such as fundamental rights, are based upon goals that are often inconsistent with efficiency. If economic rationality and legal rationality are inconsistent and imply conflicting policy recommendations, a third question must be answered: which logic takes precedence or trumps the other? Such conflicts have to be resolved by asking which abstract logical model, economics or law, more accurately explains experience in the real world and is more consistent with our values. This is a matter of judgment over which reasonable persons can and do differ.

NOTES

1. A version of the material in chapters 5 and 6 first appeared in, John Cirace, "When are Law and Economics Isomorphic," *Golden Gate University Law Review* 39 (2009): 183–220.
2. *Le Bourgeois Gentleman* (1670).
3. Rosemary Hirschfelder and John Hirschfelder, *Introduction to Discrete Mathematics* (Pacific Grove, CA: Brooks/Cole, 1991), 77, 78, 89, 90.
4. Joseph Tussman and Jacobus tenBroek, "The Equal Protection of the Laws," *California Law Review* 37 (1949): 341; Exposition of the T&t model follows John Nowak and Ronald Rotunda, *Constitutional Law*, 4th ed. (St. Paul, MN: Thomson West, 1991), 569–73.
5. Whether an individual properly falls within a specific classification concerns procedural due process.
6. 320 U.S. 81 (1943).
7. *Korematsu v. United States*, 323 U.S. 214 (1944).
8. Hirschfelder and Hirschfelder, note 3, 89.
9. Paul Rubin, "Why Is the Common Law Efficient?" *Journal of Legal Studies* 6 (1977): 51; George Priest, "The Common Law Process and the Selection of Efficient Rules," *Journal of Legal Studies* 6 (1977): 65; John Goodman, "An Economic Theory of the Evolution of the Common Law," *Journal of Legal Studies* 7 (1978): 393.
10. George L. Priest, "Selective Characteristics of Litigation," *Journal of Legal Studies* 9 (1980): 399; George Priest and Benjamin Klein, "The Selection of Disputes for Litigation," *Journal of Legal Studies* 13 (1984): 1.
11. Chapter 3's discussion of Lon Fuller's hypothetical contract cases could also be used as an example of when law and economics are consistent (isomorphic).
12. 468 U.S. 897 (1984).
13. 232 U.S. 383, 394 (1914).
14. 367 U.S. 643 (1961).
15. 579 U. S. ___, 136 S.Ct. 2056 (2016).

Part II

PROVIDING INFRASTRUCTURE AND CONTROLLING EXTERNALITIES REQUIRES RESOLVING THE CONFLICT BETWEEN INDIVIDUAL SELF-INTEREST AND COOPERATION

Part II consists of four chapters. Chapter 7 contains an introduction to game theory. The Prisoners' Dilemma game shows that unrestricted self-interested behavior usually produces antisocial results. Competitive markets, which are good for buyers but bad for sellers, are an example of the Prisoners' Dilemma. The Chicken Game, Coordination Game, zero-sum games, and dominant strategy and Nash equilibrium concepts are also discussed.

Chapter 8 is concerned with legal institutions (infrastructure) that are preconditions for markets; these institutions must solve or ameliorate the three market failures: commons problems, externalities, and public goods, which are Prisoners' Dilemmas. Government failure is concerned with reasons why government attempts to ameliorate market failures may go awry; some government failure problems, such as the voting paradox and special interest groups, are also Prisoners' Dilemmas.

Chapter 9 explains the theory of supply and demand, the necessary requirements for competitive markets, and the inefficiencies of monopoly in order to determine the circumstances in which constraints on markets are necessary in order for individual self-interest to be consistent with social welfare.

Chapter 10 is concerned with whether decisions made by short-lived, self-interested individuals are consistent with long-run (intergenerational) social welfare. The concept of present discounted value and how it lays waste to the future is explained. When current decisions have intergenerational ramifications, for example, those involving wildlife, virgin forests, global warming, and climate change, the present discounted value approach is contrasted with a sustainability approach.

Chapter 7

Prisoners' Dilemma and Introduction to Game Theory

Economists have shown that within the context of competitive markets, Adam Smith's "invisible hand," which states that if all individuals are motivated solely by self-interest society as a whole will benefit, is true. However, the validity of the invisible hand is extremely sensitive to legal and institutional context. The Prisoners' Dilemma, game theory's most famous game demonstrates that outside of the context of competitive markets and similar institutions in which self-interested behavior is narrowly channeled and severely constrained (as in the law's adversarial system where judges set limits on the behavior of self-interested lawyers), unrestricted self-interest is more likely to lead to social disasters typical of American urban ghettos, third world countries, and the former Soviet Union. "The sad reality is that Adam Smith's invisible hand has a relatively small span."[1] Thus, unless appropriately constrained by social institutions, self-interested behavior is more likely to result in a Hobbesian state, "a war . . . of every man against every man" in which "the life of man is solitary, poor, nasty, brutish and short."[2]

Surprisingly, invisible hand of competition and Prisoners' Dilemma are related because the former is a subset of the latter; that is, the invisible hand of competition is just one of many Prisoners' Dilemmas. As shown in this chapter, in most Prisoners' Dilemmas, individual self-interested behavior is usually bad for all concerned. A competitive market is one of the few Prisoners' Dilemmas where positive social consequences outweigh negative consequences: competition is bad for sellers, who would all be better off if they could stop competing against each other and make agreements to charge high prices, that is, form cartels, but is good for consumers, who benefit from sellers' inability to cooperate in the form of low prices.

The fact that the invisible hand of competition is a subset of the Prisoners' Dilemma also exposes the "fallacy of composition[3] behind the belief of those who argue that if self-interest is socially beneficial in competitive markets, self-interest must be beneficial for society in all contexts, many of which are Prisoners' Dilemmas.

If laws and institutions are perceived as variables, the invisible hand theorem is clearly a special case. Analysis of the common law and government regulations for the purpose of determining which laws and institutions are necessary in order to constrain self-interested behavior so as to promote social welfare is of prime importance in the interdisciplinary field of law, economics, and game theory.

This chapter contains an introduction to game theory. There are three main reasons for studying game theory. One reason is so that the reader will know and understand the many and complex relationships between self-interest and cooperation that are necessary for a modern society to function and which provide a justification for infrastructure and government regulation of externalities discussed in subsequent chapters. The second reason is that much law involves one-on-one adversarial relationships that are often more accurately modeled and understood by game theory, which is largely concerned with one-on-one conflict relationships, than by competitive market theory, which requires many interdependent participants. Third, the concepts developed in this chapter will be employed throughout the book.

Other games and concepts presented in this chapter are: the Repeated Prisoners' Dilemma, which shows that individuals can learn from experience that cooperation often produces greater social welfare than narrow self-interested behavior; the Chicken Game, in which, like it or not, we all engage in from time to time, and which figures prominently in both bargaining theory and torts involving negligent driving; zero-sum games and mixed or randomized strategy games are also discussed. The solution concepts of dominant strategy and Nash equilibrium are explained and emphasized.

ADAM SMITH'S INVISIBLE HAND

In Adam Smith's famous book, *An Inquiry into the Nature and Causes of the Wealth of Nations* (1776), he clearly stated his belief that individuals are motivated by self-interest: "It is not from the benevolence of the butcher, the brewer, or the baker that we expect our dinner, but from their regard to their own self-interest."[4] The presumed positive relationship between individual self-interest and competitive markets was also clearly stated by Smith: Every individual employs his resources so as to produce greatest profit without intending to "promote the public interest. . . . [H]e intends only his own gain, and he is in this, as in many other cases, led by an *invisible hand* to promote an end which was no part of his intention."[5]

Many of Smith's followers have taken his argument out of the context of competitive markets where individual self-interested behavior is severely constrained and have argued that individual self-interested behavior will serve the public interest in all contexts and in all conditions. A moment's reflection shows that in many real-world contexts, unregulated individual self-interested behavior, which, in addition to hard work, thrift, foresight and industry, also includes murder, theft, duress, fraud, opportunism, polluting, and free riding, will not maximize social welfare. If one takes the self-interest assumption seriously, one must ask not, why is there so much theft, duress, fraud, etc., but why isn't there much more of this behavior? If all individuals are self-interested, why doesn't everyone cheat, litter, pollute, and steal?

PRISONERS' DILEMMA

The Prisoners' Dilemma, game theory's most famous game, demonstrates that self-interested behavior outside of competitive markets usually makes everyone worse off. The verbal paradigm of the Prisoners' Dilemma is based on an age-old interrogation technique. Two suspects, who are carrying concealed weapons, are arrested at the entrance to a bank before they draw their guns and are charged with two crimes: conspiring to rob the bank and carrying concealed weapons. The prisoners are interrogated by police in different rooms so that they cannot communicate with each other. If one confesses and agrees to testify against the other at trial and the other does not confess, the one who confesses and testifies for the prosecution goes free, and the other, "the sucker," is sentenced to ten years; if both confess and agree to testify against each other, both get reduced sentences of five years; if neither confesses, each is sentenced to one year for illegal possession of a weapon. Both prisoners are told that this is a one-time offer to confess and that the same offer with the same consequences is being made to the other prisoner.

The Prisoners' Dilemma is presented formally in the game matrix of Figure 7.1. Each prisoner must choose between two strategies: Confess or Do Not Confess, where a *strategy* specifies an action a player can take in a game. A player's *strategy set* consists of all possible actions a player can take in a game. The prisoners must choose their strategies simultaneously and only get to play the game one time. The prisoners are called Row and Column because Row's strategy set is either a top row, Do Not Confess strategy, or a bottom row, Confess strategy; and, Column's strategy set is either a left column, Do Not Confess strategy, or a right column, Confess strategy. Since there are two prisoners and each prisoner has two strategies, there are four possible strategy combinations, the outcomes of which are represented in the four cells in the game matrix. Note that the prisoners cannot choose a cell or outcome; each must choose from row strategies or column strategies. The cell or outcome that results is due to the interaction of their strategies. The first number in any matrix cell is the sentence of Row, the second number is the sentence of Column. For example, if Row adopts a Do Not Confess strategy, and Column adopts a Confess strategy, the outcome is given in the cell in the upper-right corner of the game matrix: Row is sentenced to ten years in jail and Column goes free (−10, 0).

No matter which strategy the other prisoner adopts, a *self-interested* prisoner is always better off choosing a Confess strategy. Consider both choices in the context of the *best response to the other's strategy method* for solving a game. First, analyze the game from Row's point of view: if Column were to choose a Confess strategy, Row's best response is a Confess strategy, so as to get a sentence reduction from ten to five years (circle Row's best response, the left most 5 in the lower-right cell); if Column were to choose a Do Not Confess strategy, Row's best response is a Confess strategy so as get a reduction in sentence from 1 year to no sentence at all, 0 (circle Row's best response, 0, in the lower-left cell).

Since this game is symmetric (not all games are), the same reasoning holds for Column. If Row were to choose a Confess strategy, Column's best response is a Confess strategy, so as to get a sentence reduction from −10 to −5 years (circle Column's best response, the right most −5 in the lower-right cell); if Row were to choose a Do Not

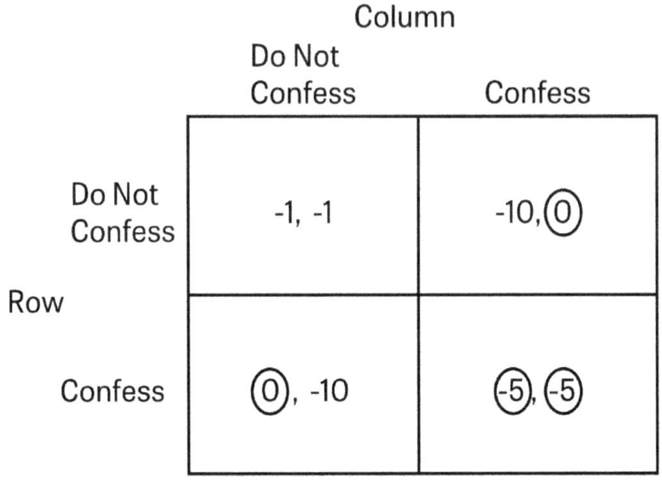

Figure 7.1 Prisoners' Dilemma. *Source*: Created by the author.

Confess strategy, Column's best response is a Confess strategy, so as to get a reduction in sentence from −1 year to no sentence at all, 0 (circle Column's best response, 0, in the upper-right cell).

Thus, each prisoner has a *dominant strategy* which is defined as a strategy that does better against every other strategy of the other player.[6] Each player's dominant strategy is to Confess and inform on the other. Row's two circles show that Row always plays the bottom row, Confess strategy; Column's two circles show that Column always plays the right column, Confess strategy.

In a Prisoners' Dilemma, no matter which strategy the other prisoner adopts, a self-interested prisoner is always better off *not* choosing a Do Not Confess strategy: If Row chooses a Confess strategy, Column should not choose a Do Not Confess strategy; if Row chooses a Do Not Confess strategy, Column should not choose a Do Not Confess strategy. The same reasoning holds for Row. Thus, each player has a *dominated strategy*, which is one that a player will never choose because some other strategy is always better choice.[7] A player will never adopt a dominated strategy regardless of the other player's strategy. Do Not Confess is a dominated strategy. If a player has a dominant strategy, then all other strategies are dominated. However, the converse is not true; having a dominated strategy does not imply that there must be a dominant strategy.

In game theory, the absolute (cardinal) value of numbers is usually of little or no importance; games are defined by the relative relationship among numbers in the matrix.[8] (Similarly, microeconomics or price theory is mostly concerned with relative, not absolute, price.)

The intuitive reason why both self-interested prisoners will chose a Confess strategy is that although the outcome in the cell where both prisoners adopt Do Not Confess strategies (−1, −1) is in their *joint* self-interest because their combined or collective sentence is lower than when they both adopt Confess strategies (−5, −5), the Do Not Confess strategies are unstable because there is always a temptation for one prisoner to double cross the other, Confess, get immunity, and go free [(0, −10) or (−10, 0)], coupled with the fear that the prisoner who does not adopt a Confess strategy will be "left holding the bag." This is the Prisoners' Dilemma: a choice of strategy based on individual self-interest results in an outcome in which both are worse off than if they had chosen cooperative strategies that are in their joint or collective interest; but, adopting the cooperative strategy exposes a prisoner to a potential double cross by the other prisoner.

It might be argued that the dilemma would disappear if the prisoners made a prior agreement not to confess in the event of capture so that they might mutually benefit from cooperation. However, if the pact is not enforceable, a new dilemma arises: each prisoner must decide whether to keep or break a pact. In other words, after a pact, a metagame, which is another Prisoners' Dilemma, results; as before, each Prisoners' self-interest is to break a pact regardless of whether the other adheres to it. Therefore, if cooperative agreements are to make a difference in the choice of strategies in a Prisoners' Dilemma, there must be enforceable sanctions for breaking the agreement. It is well known that among criminals, the sanctions used by organized crime against its members who are informers are very harsh so as to overcome the Prisoners' Dilemma.

Self-interested rational players may have different attitudes toward risk (chapter 15). Consider a negative attitude toward risk: assume an individual is *risk adverse*, that is, the individual dislikes choices where there is a significant possibility that a large loss will result. This type of person often buys a lot of insurance against negative outcomes, like substantial home and property damage insurance. Risk adverse individuals may choose a maximin strategy.

A *maximin* strategy is one where the lowest possible payoff is at least as high as the lowest payoff of any other strategy. In the Prisoners' Dilemma, a Confess strategy is the maximin strategy for both players. In Figure 7.1, if Row adopts a bottom row, Confess strategy, −5 is Row's lowest possible payoff, whereas if Row adopts a top row, Do Not Confess strategy, −10 is the lowest possible payoff; therefore, the maximin strategy for Row is Confess, which allows Row to avoid the worst possible outcome, ten years in jail, −10. The same reasoning holds for Column.

In sum, in the Prisoners' Dilemma, the Confess strategy is both the dominant strategy and the maximin strategy; Do Not Confess strategy is a dominated strategy.

Recall that in chapter 4's discussion of Rawls' *A Theory of Justice*, his difference or maximin principle, which stated that income inequality should only be allowed if it benefited the worst-off individuals in society, was described as a principle that would be accepted unanimously by all individuals only if they were very risk adverse; that is, individuals who accept Rawls' difference principle want to guard against the risk that they would be in the worst-off group. Rawls said that game theory's maximin strategy, which is designed to protect a player against the worst that can happen, was part of the inspiration that led him to advocate the difference principle.

TWO-PERSON, REPEATED PRISONERS' DILEMMA

In an iterated or Repeated Prisoners' Dilemma, where two players play the game against each other a fixed number or infinite number of times, for smaller payoffs rather than one long-term jail sentence, the cooperative (joint self-interest) solution can dominate the uncooperative (individual self-interest) solution if payoffs in future iterations are not small relative to the payoff in a current game (that is, the future is not heavily discounted) and both players learn from experience that they will be better off if they cooperate.

Backward Induction Solution

Using the argument of *backward induction*, some game theorists argue that if the two-person Prisoners' Dilemma is repeated a limited number of times as opposed to an infinite number of times, the uncooperative solution is *the* rational solution. The argument is as follows: suppose the Prisoners' Dilemma is played a limited number of times in succession, say 100. Whatever strategy players played on the first 99 plays, on the 100th and last play, the players are back in a single-game Prisoners' Dilemma. Since it is clear that a Confess strategy is the dominant strategy for a Prisoners' Dilemma that is played a single time, both players will adopt a Confess strategy on the 100th and last play. But if the 100th outcome is known and if nothing can be done about it, then the 99th outcome becomes effectively the last game, and the same reasoning applies to it, and so on back to the first game. In the context of this hyper-rationalism, that is, thinking 100 games in advance of the first game, the cooperative behavior which appears to make strategic sense in a Repeated Prisoners' Dilemma is said to collapse like a row of dominoes. The conclusion is that two players who are pursuing their individual self-interest ought to adopt a Confess strategy all 100 times in succession.

Even theorists who accept this reasoning admit that it does not apply to a Prisoners' Dilemma that is repeated an infinite number of times, because there is no last game, nor where the total number of games is unknown. Experiments, discussed below, have not verified the backward induction conclusion of total lack of cooperation.[9] Thinking strategically, both players will realize that it is in their self-interest not to adopt a Confess strategy until somewhere near the last game. At what point or game does one stop cooperating? That is the question?

Two-Person, Repeated Prisoners' Dilemma Tournaments and Group Selection

In two tournaments for game theorists and social scientists, in which each player was paired with every player for 200 plays of the Prisoners' Dilemma (as well as playing against its own strategy and a program that cooperated and defects with equal probability), the question was whether individual self-interested (uncooperative) behavior or joint self-interested (cooperative) behavior would be more successful.[10] Both tournaments were won by the same player who used a simple strategy that was initially cooperative and thereafter practiced reciprocity, called tit-for-tat: that is, start off with a cooperative strategy in the first round; in every subsequent round, adopt your opponent's strategy in the previous round. Tit-for-tat differs from the golden rule, "do unto others as you would have them do unto you," which advises unconditional cooperation; that is always play Do Not Confess.

A strategy of reciprocity makes exploitative strategies less successful because a player, who is initially cooperative but who will retaliate swiftly in response to uncooperative behavior, sends a clear signal that such behavior will not be tolerated. Cooperation can be promoted by making interactions more frequent. "The foundation of cooperation is not really trust, but the durability of the relationship."[11]

Whether the results of a two-person, Repeated Prisoners' Dilemma will be cooperative or uncooperative depends upon two factors. First, are the payoffs in future iterations small or large relative to the payoff in a current game. If the future is discounted at a sufficiently high rate so that uncooperative behavior in a current game has a low future cost, uncooperative behavior will be profitable; on the other hand, if there is significant future cost to current defecting, the cooperative strategy will be profitable. Either outcome is possible.

Second, will the two paired players meet enough in the future so that a cooperative player would be able to retaliate in the future for current treachery? If a large number (pool) of players are paired together randomly, the uncooperative strategy will tend to dominate because the likelihood of future retaliation for current treachery is small. For example, consider the difference in automobile driving behavior by persons in a small town in which all persons know each other versus driving behavior in New York City where the likelihood that any driver will encounter the same driver at another intersection or entrance ramp is almost zero. Thus, in two-person, Repeated Prisoners' Dilemmas where the opposite player is chosen randomly, social sanctions or enforcement are almost always necessary in order for the cooperative solution to result.

When a large pool of persons play Repeated Prisoners' Dilemmas in pairs, simulations of evolution to determine whether cooperative or uncooperative behavior dominates over time tend to produce cyclical results: if most people cooperate, uncooperative players can exploit them; however, if uncooperative players become known or "marked men," then cooperation among the known cooperators may be the superior strategy (because of mutual score raising); as the number of cooperative players increases, they become ripe for exploitation anew by uncooperative players and another cycle begins. However, if cooperative players are clustered, they can protect themselves from exploitation by adopting a territorially based social structure.

In the past, evolutionary theory concentrated on individual and kin selection (the selfish gene). However, recent research suggests that natural selection works on multiple levels: individual, group, culture, and complex interactions among the three levels. In group selection simulations based upon two-person Prisoners' Dilemmas, a self-interested individual within a group can take advantage of and prosper at the expense of cooperative individuals; however, in competition between groups, the more cooperative and cohesive group will defeat the less cooperative group. In the most extreme example of group competition, war between groups, self-interest and even the instinct for self-preservation must be suppressed in order for one group to prevail over another group (what other rationale is there for hazing in military "boot camp"?). Group selection works best if there are many small groups and less well if there are a few large groups.[12]

N-PERSON PRISONERS' DILEMMA

An N-Person Prisoners' Dilemma refers to a Prisoners' Dilemma that is played *simultaneously* by an indefinitely large number of players, where "N" stands for any number. In terms of the verbal paradigm of the Prisoners' Dilemma, one can think of the technique of individual crime interrogation as being applied to an entire criminal organization. The basic conclusion is both simple and clear: If an N-Person Prisoners' Dilemma is played only once, as the number of players increases, the likelihood that at least one player will adopt a Confess strategy, sending all others to jail, increases and approaches certainty.

With respect to group interaction by large numbers of individuals, the invisible hand of competition is less widely applicable than the N-Person Prisoners' Dilemma because the invisible hand is an example (subset) of the Prisoners' Dilemma. Many forms of antisocial behavior where individual self-interest leads to a decrease in social welfare, such as theft, pollution, or litter can be modeled as N-Person Prisoners' Dilemmas. Consider three examples: First, consider littering: if no one else litters, a self-interested person will litter in order not to incur the slight cost of walking to a waste basket to dispose of waste; if everyone else litters, a self-interested person will still litter not to incur cost of waste disposal; thus, all self-interested persons will litter to the detriment of all. Second, if toxic waste from factories is substituted for litter, an N-Person Prisoners' Dilemma with much more serious consequences results.

Third, whether to vote in an election can also be modeled as an N-Person Prisoners' Dilemma. In any election where there are at least two candidates or issues, for example, candidate x and candidate y, if all other persons who favor candidate x vote, a self-interested person who favors candidate x will not vote because that person's vote is almost certain not to be decisive in the election; if no other person who favors candidate x votes, a self-interested person will not vote because that person's vote certainly will not be decisive in the election; thus, a self-interested person derives no benefit from voting; but, if no one votes, all are worse off. From the view point of N-Person Prisoners' Dilemmas, individuals should not vote! However, as shown in the next section, if voting is thought of as an N-Person, Repeated Prisoners' Dilemma, individuals have an incentive to vote.

Voting as N-Person, Repeated Prisoners' Dilemma

If voting is analyzed as an N-Person, Repeated Prisoners' Dilemma, a persuasive argument can be made that rational individuals ought to vote, which reverses the conclusion concerning voting by modeling voting as a one-time N-Person, Prisoners' Dilemma. The reason is that if an identifiable age, ethnic, gender, racial, or interest group repeatedly votes in lower percentages than other groups, and they get a reputation for low voter turnout, politicians will tend to discount or ignore their wishes; they will be "off the radar screen." Thus, if members of a socially recognizable group to which you belong do not vote, they not only harm themselves, they harm you as well.

HOW MANY FUTURE MOVES DOES A RATIONAL PERSON ANTICIPATE?

Whether individuals in N-Person, Repeated Prisoners' Dilemmas will act in a hyper-rational manner using backward induction or act strategically is an example of a crucial unresolved problem in economics: how many moves or steps into the future should a rational person look?

One extreme view as to how individuals react in the present with respect to the future is known as the classical, neoclassical, or *rational expectations* view. This view postulates that individuals look many steps into the future and react immediately to any new information. The hyper-rational backward induction solution to the N-Person, Repeated Prisoners' Dilemma is an example of this view. Moreover, this view assumes that all individuals not only react to new information but also react to anticipated information. For example, if a rise in the interest rate by the Federal Reserve is expected in the near future, the stock market traders will interpret such action as a move to slow the economy in the future, and stock market values will decline in anticipation of the Federal Reserve's action. Another example of this view is known as the *efficient market hypothesis*, it postulates that no individual can beat the stock market consistently because as soon as information becomes known or is anticipated, myriad individuals act on it at the same time, so no one can get a consistent advantage. In extreme form, even insider trading on corporate information will be of little avail to the insiders because market participants will know the financial state of all corporations and will anticipate self-dealing by insiders.

The other extreme view is identified with John Maynard Keynes, the great twentieth-century economist, who argued that individuals either react slowly or belated to change (prices and wages are "sticky") and that markets often exhibit herd-like behavior as a result of attempts by rational individuals to out-anticipate each other. An example of such behavior is his famous analogy of the stock market to a beauty contest.

> It is not a case of choosing those which, to the best of one's judgment, are really the prettiest, nor even those which average opinion genuinely thinks the prettiest. We have reached the third degree where we devote our intelligences to anticipating what average opinion expects the average opinion to be. And there are some, I believe, who practice the fourth, fifth and higher degrees.[13]

Although thinking one or two moves ahead of the average person often gives one an advantage, research in behavioral economics shows that an individual will often lose if she thinks either too few or too many moves in the future with respect to the average opinion.[14] A person can be either too far behind or ahead of her time.

The general question of how far it is rational to look into the future is of paramount importance in economic theory. Even if two theorists agree on the relevance of specific variables and even agree upon a particular model, if they postulate different reactions or rates of change in variables dealing with information about the future, they can reach diametrically opposite conclusions.

CHICKEN GAME

In this section, the Chicken Game is explained; it is used later in this chapter to explain one-on-one bargaining; and, its sequential version is used in chapter 18's *last clear chance* model. The verbal paradigm of the Chicken Game assumes that two automobiles start at opposite ends on a straight track. The goal for each driver in this game is to reach the other's point of origin first. If both drivers insist on staying in the center of the track, a crash results and both suffer severe damage. If each moves to the right as they approach each other, that is, they cooperate to avoid a crash, a draw results. But, if one intimidates the other by threatening to crash rather than cede the center of the track and the other yields, the intimidator wins and the other, the loser, is a "chicken." The Chicken game explains why not only aggressive business persons and lawyers but tough guys and gals of all types affect a "don't mess with me" attitude, so that their opponents will turn chicken and back down in one-on-one conflict situations like contract bargaining or disagreements that have the potential to get physical. These individuals try to create the impression that nobody is crazier or "badder than me."

In daily life, people engage in chicken games quite often, sometimes involuntarily. Any time a person is walking on a sidewalk or a path and becomes aware that he is on a potential collision course with someone coming from the opposite direction, a potential chicken game ensues. By hostile facial expression, truculent body language, or by attire, for example, a Hell's Angels outfit, one or both can signal that he or she is an intimidator. In this latter situation, a chicken game ensues. If neither player turns chicken and yields the center of the path to the other, a collision and possible fight will ensue.

If two pedestrians who are coming toward each other on the same path are polite, by *convention*, each will move slightly to her right as they approach each other in order to avert a collision. Although it is not important whether the convention is that everyone moves right or left, common or public knowledge of the convention is crucial: whether one drives on the right in the United States or on the left in Great Britain is not important; what is important is that everyone drives on the same side.

Pedestrians and drivers also engage in chicken games with each other: when crossing a street, pedestrians sometimes either refuse to look or at least pretend not to look at an oncoming vehicle, so as to put the driver in the position of either having to yield the right of way or run them down; for their part, drivers can attempt to drive fast enough so that they cannot stop in time to avoid a pedestrian who is trying to cross the street in order to put the pedestrian in the position of choosing between not crossing or being run over. Other examples of vehicular chicken games involve situations in which two vehicles approach on opposite sides of a one-lane bridge, or where a vehicle is attempting to make a left hand turn into oncoming traffic at an intersection, or where a vehicle on an entrance ramp of a highway is trying to break into a stream of traffic.

A formal statement of the Chicken game is presented in Figure 7.2. First, analyze the game from Row's point of view. If Column chooses a Concede strategy, Row's best response is to choose an Intimidate strategy, so as to increase its payoff from 1 to 2 (circle Row's best response, 2 in the lower-left cell); if Column chooses an Intimidate strategy, Row's best response is to choose a Concede strategy, so as to decrease

		Column	
		Concede	Intimidate
Row	Concede	1, 1	-2, 2
	Intimidate	2, -2	-5, -5

Figure 7.2 Chicken Game. *Source*: Created by the author.

it loss from −5 to −2 (circle Row's best response, −2 in the upper-right cell). Analyzing the game from Column's point of view is symmetric. If Row chooses Concede, Column's best response is Intimidate (2 is more than 1); if Row chooses Intimidate, Column's best response is Concede (−2 is less bad than −5).

There are similarities and differences between the Prisoners' Dilemma and the Chicken game. The Chicken game is like the Prisoners' Dilemma in that the players will be jointly better off if they adopt mutually cooperative strategies rather than uncooperative strategies. In the Prisoners' Dilemma, both are jointly best off if both adopt Do Not Confess strategies so that the result is (−1, −1); in the Chicken game, both are jointly best off if both adopt Concede strategies so that the result is (1, 1). A second similarity between the two games is that self-interested players are tempted to defect from the cooperative outcome: in the Prisoners' Dilemma, a self-interested player would adopt a Confess strategy if he or she knew the other player will adopt a Do Not Confess strategy; in the Chicken game, a self-interested player would chose an Intimidate strategy if he or she knew the other player was going to adopt a Concede strategy.

There are, however, three distinct differences between the Prisoners' Dilemma and Chicken game. First, the worst result a player can experience in the Prisoners' Dilemma occurs when both players adopt different strategies: if one adopts a Do Not Confess strategy while the other player adopts a Confess strategy, the former will get a ten-year sentence while the latter will go free (−10, 0). In the Chicken game, the worst result occurs when both players adopt the same strategy, an Intimidate strategy, so that both lose five (−5, −5).

Second, in the Prisoners' Dilemma, there is both a dominant strategy, always Confess, and a dominated strategy, never Do Not Confess. Recall that a dominant strategy exists if there is a strategy that is the best strategy which a player can adopt regardless of the strategy the other player adopts; a dominated strategy exists if there is a strategy a player would never play. In the Chicken game, there are neither dominant nor dominated strategies because there is neither a strategy that a player should always play nor a strategy that a player should never play.

There is a third significant difference between the Chicken Game and the Prisoners' Dilemma: In the Prisoners' Dilemma, a maximin strategy by both players produces a stable outcome; however, in the Chicken Game, the adoption of a maximin strategy by both players is unstable. In Figure 7.2, if Row adopts a top row, Concede strategy, its lowest possible payoff is –2, whereas if it adopts a bottom row, Intimidate strategy, its lowest possible payoff is –5; therefore, a Concede strategy is Row's maximin strategy. However, if both Row and Column were both to adopt a maximin strategy in the Chicken Game, the outcome would be unstable because the best response by a player who knows that the other player is going to adopt a maximin or Concede strategy is to adopt an Intimidate strategy, not a Concede strategy. (Maximin strategies in the Chicken Game do not result in a Nash equilibrium, which concept is discussed below.)

Due to extrinsic circumstances (not part of the game), it may be obvious which player will choose the Intimidate strategy and which will choose the Concede strategy. For example, if two male pedestrians, who are walking toward each other on a side walk, become involved in a Chicken Game for the right of way, where one is 6'5", 270 pounds and dressed in scruffy attire, while the other is 5'7", 140 pounds, and dressed in a business suit, it is likely that the big man will adopt an Intimidate strategy and the small man will adopt a Concede strategy. The obvious reason is that if both adopt an Intimidate strategy so that a collision occurs, possibly leading to a fight, it is likely that the small man will get the worst of it. In other words, payoffs in Figure 7.2 may not be the same for both players; in the asymmetrical case, if both men adopt Intimidate strategies, the payoffs in the lower-right box would not be (–5, –5) but may be (–1, –8) instead, where –1 represents the payoff to the big man and –8 represents the payoffs to the small man, which would cause the small man to adopt a Concede strategy.

NASH EQUILIBRIUM

This section is concerned with the concept of a Nash equilibrium, named for the mathematician, John Nash. It is arguably the most important concept in game theory. *A Nash equilibrium is strategy combination or cell in the game matrix such that given the strategy choice of the other player, neither player has an incentive to switch strategies.* This concept is an important concept in game theory for three reasons: 1) Nash equilibrium formalizes the process of reasoning inherent in games, where the strategy one adopts is greatly influenced by the strategy one's opponent is expected to adopt. 2) All finite games in which the players have a finite number of strategies with finite outcomes must have at least one Nash equilibrium, which equilibrium may involve mixed or randomized strategies (discussed below). 3) A Nash equilibrium is *self-enforcing* in the sense that neither party has an incentive to change its strategy.[15]

An easy method for finding a Nash equilibrium is to use the best response method discussed above. Any cell in a game in which payoffs of both players are circled is a Nash equilibrium. For example, in Figure 7.1's Prisoners' Dilemma, the strategy pair, Confess-Confess (–5, –5), each of which is circled, is the Nash equilibrium. None of the other three cells is a Nash equilibrium. For example, the lower-left cell, Confess-Do Not Confess (0, –10), is not a Nash equilibrium: although Confess is Row's best response given Column's Do Not Confess strategy, Do Not Confess is not Column's best response given Row's Confess strategy (–10 is worse than –5). Given that Column chooses a Confess strategy, Row's best response is to change to a Confess strategy. Thus, the only Nash equilibrium strategies in a Prisoners' Dilemma are the strategy combination Confess-Confess because neither player can benefit by a change of strategy. As discussed above, the best response method can also be used to show whether either or both players have a dominant strategy.

In Figure 7.2's Chicken game, there are two Nash equilibria, the strategy pairs Concede-Intimidate and Intimidate-Concede. The cell (1, 1) which results when both players choose Concede strategies is not a Nash equilibrium: given that one player chooses a Concede strategy, the other player's best response is an Intimidate strategy, not a Concede strategy. Because there is more than one Nash equilibrium in the Chicken game, game theory cannot determine which of the two Nash equilibria will result. Extrinsic circumstances and institutional context, like who the parties are and what they know about each other, may help but are not definitive. One shortcoming of game theory is that although it can analyze the structure conflict situations, it often cannot determine which of several Nash equilibria strategies a player should adopt.

COOPERATIVE OUTCOMES IN NON-COOPERATIVE GAMES

Prisoners' Dilemma and Chicken games are non-cooperative games. However, even a non-cooperative game can have a *cooperative* outcome, which occurs if there is at least one cell or outcome in which a) the interests of the players are not strictly opposed, and b) the sum of their payoffs in this cell is at least as large as payoffs in cells in which their interests are opposed. (Any game in which there is at least one cooperative outcome is a nonzero-sum game; zero-sum games, those in which there are no cooperative outcomes, are discussed below.)

In the Prisoners' Dilemma, the strategy combination (Do Not Confess-Do Not Confess) is a cooperative outcome because interests of the players are not strictly opposed and the outcome (−1, −1) is greater in the sense of being less negative than the other three non-cooperative results. In the Chicken game, the strategy combination (Concede-Concede) is a cooperative outcome because interests of the players are not strictly opposed, and the sum of the outcomes (1, 1) is greater than the sum of other three non-cooperative results. Although Prisoners' Dilemma and Chicken game have cooperative outcomes, neither of these cooperative outcomes is a Nash equilibrium, which means that they are possible but, unstable outcomes.

COORDINATION GAME AKA BATTLE OF THE SEXES

Two or more individuals may not only be better off if they cooperate, they may want to *coordinate* their activity. For example, if two individuals want to lift an object that is too heavy for either to lift alone, cooperation is not enough; they must know how to coordinate their individual efforts.

Consider the third canonical game, the Coordination Game, also known as Battle of the Sexes. Assume a couple is trying to determine which of two entertainments, opera or football, to attend. Row prefers to go to the opera; Column prefers to go to a football game; and both prefer going to an entertainment together rather than alone.

		Column	
		Opera	Football
Row	Opera	3, 2	1, 1
	Football	0, 0	2, 3

Figure 7.3 Cooperation Game. *Source*: Created by the author.

The player who goes to her or his favored entertainment together with the other gets the highest psychic gain; second best outcome is to go to a player's less favored entertainment together with the other; third best outcome is to go to the player's favored entertainment alone; and, worst outcome is going the player's less favored entertainment alone. These relationships are depicted in Figure 7.3.

In the Battle of the Sexes game there is no dominant strategy and two Nash equilibria (Opera, Opera and Football, Football). How do players choose which of these two Nash equilibria to adopt? Before answering this question, consider the conversations that are likely to occur prior to playing Prisoner's Dilemma, Chicken Game, and Battle of the Sexes. In the Prisoners' Dilemma, before being arrested both players pledge loyalty and swear they will not confess, only to double cross each other when the game is played. In the Chicken Game, before getting in their autos, one of both players profess that he or she will never back down, preferring a fiery death to defeat and humiliation. Such bravado may or may not be a bluff; only playing the game will reveal the truth. However, in the Coordination Game, since both players want to attend an entertainment together, they can voluntarily agree and commit to a cooperative-coordinated strategy. To which Nash equilibrium will they agree? They could decide which entertainment to attend based on a coin flip, with or without the stipulation that on the next occasion they will go to the other entertainment. Possibly one is in the habit of deferring to the other. The point is that since they want to coordinate their choices, honoring a commitment is in the self-interest of both; prior agreement will be respected because neither wishes to sabotage either the agreement or their relationship.

BARGAINING GAMES AND COMPETITIVE MARKETS

Two-Person Bargaining Games

A two-person bargaining game, as for example when a buyer and a seller attempt to negotiate the sale of a used automobile, is a Chicken Game with a significant difference; prior to playing the game, it is possible for both players to make a legally binding commitment or *contract* as to which strategy will be played because such contracts can be enforced in a court of law. Consider a simple bargaining game in which two people are negotiating over the sale of a used automobile. Assume that minimum price seller will accept for the automobile is $4,000; this is seller's *threat value*. Below this amount, seller would keep the automobile or will search for another buyer rather than take less. Maximum price buyer will pay for the automobile is $6,000; this is buyer's threat value. Above that amount buyer will not buy this automobile and will search for another seller. A player's threat value is greatly affected by whether a party has alternative third parties with whom he or she can negotiate. For example, in a competitive market, where there are many alternative buyers and alternative sellers, so that a buyer can say to a seller, "You're not the only seller of widgets," seller will have a lower threat value (minimum sale price) than will seller who is a monopolist, that is, one who has no competitors.

If buyer and seller can agree on a price between $4,000 and $6,000 so that the automobile is transferred from seller to buyer who values it more, $2,000 in value will be created. The value created as a result of transferring resources from where they are less valued to where they are more valued produces a *cooperative surplus*. If buyer and seller cannot come to an agreement, the non-cooperative outcome results, no bargain, and there is no cooperative surplus. Bargaining involves negotiation over the distribution of the cooperative surplus. The strategic bargaining game involving the automobile is depicted in game matrix in Figure 7.4.

Assume buyer and seller each have two strategies: Easy bargaining or Hard bargaining. If both buyer and seller adopt Hard bargaining strategies, that is, seller demands at least $6,000, and buyer insists on paying at most $4,000, bargaining will be unsuccessful, so there is no cooperative surplus. The payoffs for non-cooperative result, ($0, $0), are depicted in the lower-right cell, which is bordered in bold. This cell, which represents no sale and no cooperative surplus, is equivalent to a crash in the Chicken Game in which both players adopt Intimidate strategies. If seller adopts a Hard bargaining strategy and buyer adopts an Easy bargaining strategy, there will be an agreement to sell the automobile for $6,000; as depicted the lower-left cell; seller gets the entire cooperative surplus ($2,000, $0). This is equivalent to a Chicken Game in which Seller adopts an Intimidate strategy and Buyer adopts a Concede strategy. If their selling strategies are reversed, so that Seller adopts an Easy bargaining strategy and Buyer adopts a Hard bargaining strategy, there will be an agreement to sell the automobile for $4,000; as depicted in the upper-right cell, buyer will get the entire cooperative surplus of ($0, $2,000). This is equivalent to a Chicken Game in which Seller adopts a Concede strategy and Buyer adopts an Intimidate strategy.

If both players adopt Easy bargaining strategies, there will be an agreement to sell the automobile for $5,000; buyer and seller will split the cooperative surplus each getting $1,000 as depicted in the upper-left cell ($1,000, $1,000). This is equivalent to a Chicken Game in which both Seller and Buyer are cooperative and adopt Concede strategies. Although, like the Chicken game, the cooperative solution is not a Nash equilibrium in this bargaining game, the players can make a binding commitment or contract enforceable at law to play the cooperative strategy before the game is played.

Which of the four outcomes in the game matrix is most likely, the cooperative result or the uncooperative results? If both parties have *complete information*, that is, each party can see Figure 7.4 and knows each other's threat value, the cooperative solution

		Buyer	
		Easy Bargaining	Hard Bargaining
Seller	Easy Bargaining	$1,000, $1,000	$0, $2,000
	Hard Bargaining	$2,000, $0	No Sale, No Coop. Surplus $0, $0

Figure 7.4 Bargaining as Chicken Game. *Source*: Created by the author.

where the parties share the cooperative surplus equally (the top-left cell) is the most likely outcome because strategic behavior like bluffing, which requires uncertainty, is ineffective in a complete information setting. However, in the real world, there is usually incomplete information in bargaining situations because buyers and sellers know their own threat values, but they usually do not know each other's threat value, so that bluffing occurs when either seller tries to convince buyer that her threat value is higher than it actually is or buyer tries to convince seller that her threat value is lower than it actually is. With incomplete information (discussed in chapter 16), say where each party knows its own threat value but is uncertain as to its opponent's threat value, predicting which outcome will occur is not possible. If either buyer or seller engages in bluffing or demands more than half of the cooperative surplus from agreement, failure to come to an agreement, the non-cooperative result, may occur.

A two-person bargaining game is like a Chicken Game with two significant differences: one, the players can make a binding commitment or contract to play a cooperative strategy before the game is played; and two, there is usually incomplete information as to each player's threat value; that is, seller usually does not know maximum price buyer will pay and buyer usually does not know minimum price seller will take.

In sum, in two-person bargaining there is range of possibilities. With respect to the negotiations concerning the automobile, a sale may or may not be made, and, if made, price will be in a range between $4,000 and $6,000. The price in two-person bargaining is said to be *indeterminate*; that is, there is a range of possible prices.

Competitive Markets Are a Subset of the Prisoners' Dilemma

This section explains how the range of prices in two-person bargaining, that is, an indeterminate price, becomes the *law of one price* in a competitive market where there are many buyers and sellers. As the number of buyers and sellers increases, the range of indeterminacy, that is, the spread between the highest and lowest prices diminishes, as shown by extending the previous example.

In two-person bargaining described above, seller has a minimum price of $4,000 and buyer has a maximum price of $6,000. Now assume the market consists of two sellers and two buyers, the original seller and buyer and a second seller whose minimum sale price is $5,000 and a second buyer whose maximum price is $5,500. Consider bargaining in this used automobile market that consists of two sellers and two buyers with the aid of Table 7.1.

In Table 7.1, if two buyers offer $4,000 for used automobiles, only one seller wants to sell at that price. So, when the one seller realizes that two buyers want the one automobile he or she has to sell, seller can raise price until it is $5,000. At $5,000, there are now two willing sellers and two willing buyers, so automobiles could be sold at that price because supply and demand are in equilibrium. What if because of the hard bargaining of sellers, price rises to $5,500? At this price, there are still two willing sellers and two willing buyers, so supply and demand are still in equilibrium. What if sellers bargain still harder and demand $5,600? One of the buyers drops out of the market because price has now exceeded her or his maximum price. In Table 7.1, if two sellers demand $5,600, there is only one buyer remaining, so supply exceeds demand. The remaining buyer realizes that he or she can force the sellers to compete against each other and reduce price.

Compare the results of two-person bargaining with four-person bargaining: In Figure 7.4, where there was one seller and one buyer, the range of indeterminacy was between $4,000 and $6,000. With the addition of another seller who has a higher minimum sale price, $5,000, and another buyer who has a lower maximum buying price, $5,500, the range of indeterminacy decreases to between $5,000 and $5,500. As more sellers enter a market with different minimum prices (some firms are more efficient and can for less, some less efficient and must charge higher prices) and different buyers enter the market with different maximum prices, the range of indeterminacy diminishes until when there are many buyers and many sellers, a competitive market results, where the law of one price governs and price is *determinate* rather than indeterminate.

In sum, in markets with one buyer and one seller, there is a range of prices. As the number of sellers and buyers increase, the range of price indeterminacy diminishes until, when there are many buyers and sellers, the market is competitive and the law of one price governs (see chapter 9).

In the context of game theory, the reason that there is one price in a competitive market is that sellers are in N-Person Prisoners' Dilemma. Although sellers would be better off if they all agreed to restrict output so as to be able to charge a higher price, each will fear that other sellers will not honor the restrictive output agreement, which is both unenforceable and illegal in the United States, so that sellers who adhere to the agreement and restrict output will lose sales and income to those who do not honor the agreement and sell additional output. The result of the N-Person Prisoners'

Table 7.1 Bargaining among Two Buyers and Two Sellers

Price	*Used Automobile Supply*	*Used Automobile Demand*
$4,000	1	2
$5,000	2	2
$5,500	2	2
$5,600	2	1

Source: Created by the author.

Dilemma is that no seller will be willing to restrict output for fear that it will lose sales and profit to other sellers, so the price of products will be driven down to the cost of production.

The invisible hand of competition is one of many Prisoners' Dilemmas; that is, competition is a subset of the more general concept of Prisoners' Dilemma. Moreover, a competitive market is one of the few Prisoners' Dilemmas that has positive as well as negative social consequences. In most Prisoners' Dilemmas, the inability of self-interested individuals to make binding commitments to engage in cooperative behavior leads to antisocial consequences for everyone in society; in competitive markets, the inability of sellers to make a binding commitment to each other is bad for the competitors but is socially beneficial for consumers, who can buy products and services at prices driven down to the cost of production by competition.

The relation between competitive markets and Prisoners' Dilemma, the former is a subset of the latter, exposes the fallacy of composition behind the belief of those who argue that if self-interest is socially beneficial in competition, self-interest must be beneficial for society in all contexts (Prisoners' Dilemma itself is an example of the fallacy of composition). Self-interested behavior in most Prisoners' Dilemmas greatly reduce social welfare.

SIMULTANEOUS VS. SEQUENTIAL GAMES

Prisoners' Dilemma, Chicken Game, and Coordination Game aka Battle of the Sexes described above are games in which both players 1) know who the other player is, 2) know all possible outcomes in the game matrix, that is, they have complete information concerning the game matrix, and 3) must choose their strategies *simultaneously*, that is, both have no advance knowledge what strategy their opponent will play. Sequential games are games in which one player moves first: examples are chess, tennis, and card games like bridge and gin rummy. In Chapter 18 on negligence law (torts), a sequential version of the Chicken Game known as *last clear chance* will be discussed; this sequential game is a Chicken Game that has a single Nash equilibrium instead of two as in the Chicken Game played simultaneously.

Ultimatum Game

The Ultimatum Game is a simple two-person, sequential game that explores whether individuals are motivated strictly by self-interest as is usually assumed in economics or whether they are also motivated by distributional equity and fairness. In this game, there is a stipulated sum of money which is to be divided between two players. The 1st mover, called the Proposer, is allowed to make one and only one proposal as to how the sum of money is to be split between the two players; the 2nd mover, the Responder, can either accept the offer, in which case the Proposer's distribution is accepted, or reject the offer, in which case both players get nothing. This is a one-shot game in which there is a cooperative surplus to divide if the players can agree and no cooperative surplus if they do not agree.

The game has attracted attention because experimental results differ so dramatically from the results that are expected if the players are motivated solely by individual self-interest. Assume that the amount for which the game is played is $10. If both players are totally *self-interested* income maximizers, and both players are mutually aware of this assumption, Proposer should offer a penny or the smallest unit of currency available, and the Responder should accept. Instead, offers typically average about 30–40 percent of the total, with a 50–50 split often the mode. Offers of less than 20 percent are frequently rejected.[16]

In experiments with children for a dollar or a pile of M&M candies, kindergartners accepted minimal offers (one penny or one M&M) about 70% of the time, compared to about 40% for third- and sixth-graders, and probably much less for typical college student subjects. The study concluded that the tendency to reject low offers is probably learned over time as one's social sense of fairness or distributional equity develops.

The highest value for which ultimatum games have been played and reported is $100 per game. The proportions offered by Proposers in both $10 games and $100 games were not significantly different. Query: What if the Ultimatum Game is played for $10 million or $10 billion? If you are the Responder in a $10 billion game, would you reject the offer of a measly $1 billion?

ZERO-SUM GAMES

A game in which one player's gain is always matched by another's loss is called a *zero-sum* game. For example, if we flip a coin so that if heads comes up, you win $1 from me, and if tails comes up, I win $1 from you, this is a zero-sum game. Chess, card games like poker, and tennis are zero-sum games. In two-person, zero-sum games, the interests of the players are strictly opposed so cooperation is ruled out; one player is a winner, the other is a loser.[17] Zero-sum games are unlike the Prisoners' Dilemma and Chicken game in which there are both cooperative and non-cooperative outcomes. The Prisoners' Dilemma and Chicken game each has one cooperative outcome (in the Prisoners' Dilemma if both players choose Do Not Confess strategies and in the Chicken game if both choose Concede strategies) and three uncooperative outcomes (the three other cells in these games).

Hotelling Game

Consider a zero-sum game, called the Hotelling game, after its author, in which two sellers of ice cream must decide where to locate on a beach on a summer's day.[18] Buyers, each of whom will buy one ice cream cone, are arrayed at uniform intervals along a beach, which can be visualized as a straight line. Assume that sellers must choose a stationary location and that buyers will travel to and buy ice cream from nearest seller and that buyers will buy ice cream no matter how far they must walk. Since either seller can supply enough cones for all buyers in this market, this is zero-sum game, one seller's gain is the other's loss.

Where should the two sellers locate? Both sellers should locate next to each other at the center of the beach because the center solution is a Nash equilibrium. If one seller were to locate anywhere else, say one seller located ¼ of the distance from the left end of the beach, the other seller would locate just to the right of the first seller and get ¾ of the business. Competition would force the sellers to locate next to each other at the center of the beach, even though the locations that would minimize the distance buyers have to travel are where one seller located ¼ of the distance from the left end of the beach and the other located ¼ of the distance from the right end of the beach. In this market, competition does not result in the socially optimal location.

This simple zero-sum game has also been used to explain the tendency in winner-take-all elections for two political parties to move to the center of the political spectrum, on which left is liberal and right is conservative, even though people's political preferences would be more accurately reflected if the two parties' were to move from the center to the first and third quartile locations. In political science this is known as the *median voter theorem.*

There are several problems with the political predictions of this model: First, since political extremists often feel disenfranchised by the prospect of having to choose between two centrist parties, they sometimes sit out elections (low voter turnout) or support splinter or third parties in order to force the two parties to move from the center. In terms of economics of the model, this qualification may be expressed by assuming that buyers limit the distance they are willing to travel to a seller. Second, buyers or voters may not be uniformly distributed on the beach or over the political spectrum; they may be clustered near the extreme ends of the spectrum. Third, there may be more than two candidates; with three candidates the model is unstable because the candidate in the middle may attempt to leap frog one of the others so as to put that candidate in the middle. (There were three candidates in the presidential election of 1992: President George H. W. Bush, Governor Bill Clinton, and Ross Perot.) Fourth, there may be more than one political issue or dimension to an election. Fifth, there may be primaries in which moderate or centrist candidates are eliminated. Sixth, since candidates cannot commit themselves to positions, candidates may not be truthful as to their location on the political spectrum.

Randomized (Mixed) Strategies in Zero-Sum Games

In two person, zero-sum games, *randomized* or *mixed strategies* are often an integral part of the game because an unpredictable opponent is usually harder to beat in these games than is a predictable opponent. A randomized or mixed strategy is one where a player chooses a strategy based on some random method. For example, in the game of matching fingers, two players simultaneously thrust out either one or two fingers. If both players show the same number of fingers, the matching player wins, say $1, and if players show a different number of fingers, the nonmatching player wins $1. If both players play one finger 50% of the time and two fingers 50% of the time, they should break even over many games, that is, both players will on average neither win nor lose money; this is the optimal random or mixed strategy equilibrium for this game. A player might choose between a one or two finger strategy by flipping a coin, so that heads means choose a one finger strategy, tails means choose a two finger strategy.

However, if one player knows that the other player chooses either one finger or two fingers more than 50% of the time, it can exploit that knowledge to its advantage. For example, if the matching player knows the other player chooses one finger 60% of the time, the matching player should choose one finger all the time, so that the sum will be even 60% of the time and it will win 60% of the time. Of course, this is not a Nash equilibrium because the other player will likely notice that the other player is playing one finger all the time. But, if the matching player notices that the other player is playing one finger 60% of the time, he or she should take advantage of that bias by playing one finger enough to get a winning edge, say 60 or 70% but not enough to make the other player aware of her or his bias in response to the other player's bias.

If a game has three possible strategies and a player could choose among them by rolling a six-sided die such that if 1 is on top, strategy A is followed, if 2, 3, or 4 are on top, strategy B is followed, and if 5 or 6 are on top, strategy C is followed: strategy A has a 1/6 probability of being chosen, strategy B has a 1/2 probability of being chosen, and strategy C has a 1/3 probability of being chosen.

In many games in which it is optimal to follow a mixed strategy, the best proportion is often not 50/50. Consider the game of tennis. The server must choose whether to serve to the receiver's forehand or backhand. Typically both the server and the receiver know that the receiver's forehand returns are more accurate than backhand returns. However, if the server always serves to the receiver's backhand, the receiver will prepare in advance by shifting her or his stance toward that side of the court and by adopting a backhand grip, with the result that the receiver will be more successful in returning serve than if the server served to the backhand most of the time but not all the time. The optimal proportion of serves backhand/forehand depends upon the relative strength of the receiver's backhand and forehand returns and the effectiveness of the serve to the left or right of the service box. "The basic idea is that one resorts to chance as the means of keeping the other player from exploiting any systematic behavior on your part."[19]

An easy way to determine whether randomness is efficient in a zero-sum game is to consider whether the other player will benefit if he or she knows what your move will be before you make it. In the card game of poker, if a player never bluffs, her opponents will know that any time she bets a large sum she must have a very good hand, so she would seldom win large pots because no one will ever match her large bets; on the other hand, if she always bluffs, the other players would always call her bluff and she would lose most of the time. The optimal poker strategy requires a mix between bluffing and not bluffing, that is, a randomized or mixed strategy. Maximin strategies also figure prominently in zero-sum games. For example, in the game of poker, a maximin strategy may be to put a limit on the amount one is willing to lose in any game or over the course of play.

AT LEAST ONE NASH EQUILIBRIUM IN ANY FINITE GAME

John Nash's great achievement, for which he won a Nobel prize, was to prove that for any finite game, that is, a game with a finite number of players and finite payoffs, whether or not it is a zero-sum game, there must at least be one Nash equilibrium, which equilibrium may consist of either pure or random (mixed) strategies.

NOTES

1. Avinash Dixit and Barry Nalebuff, *Thinking Strategically: The Competitive Edge in Business, Politics, and Every Day Life* (New York: Norton, 1991), 223.
2. Thomas Hobbes, *Leviathan* (London: Melborne, 1914), pt. I, ch. 13.
3. The error of assuming that what is true of a member of a group is true for the group as a whole. Prisoners' Dilemma itself is an example of the fallacy of composition.
4. *Wealth of Nations*, Modern Library ed. (New York: Random House, 1937), 14.
5. *Ibid.*, 423 (emphasis added).
6. The definition in the text is for a strictly *dominant* strategy. A weakly dominant strategy is one that does as least as well against every strategy of the other player and does better against some.
7. The definition in the text is for a strictly *dominated* strategy. A weakly dominated strategy is a strategy which has a worse outcome than another strategy while for all other strategies the two strategies have the same payoffs.
8. Although the absolute values of the numbers in Figure 7.1 are arbitrary, in order for the game to be a Prisoners' Dilemma, the numbers in the matrix must have a strict relationship to each other: where the lower-right cell has numbers (a, a), lower-left cell (b, c), upper-right cell (c, b) and, upper-left cell (d, d), they must have the following relation to each other: $b > d > a > c$. In Figure 7.1, Prisoners' Dilemma, the numbers are $(b = 0) > (d = -1) > (a = -5) > (c = -10)$.
9. "[B]ackward induction receives little support from evolutionary game theory." George Mailath, "Do People Play Nash Equilibrium? Lessons from Evolutionary Game Theory," *Journal of Economic Literature* 36 (1998): 1347, 1349.
10. Robert Axelrod, *The Evolution of Cooperation* (New York: Basic Books, 1984).
11. *Ibid.*, 182.
12. Martin Nowak with Roger Highfield, *Super Cooperators: Altruism, Evolution, and Why We Need Each Other to Succeed* (New York: Free Press, 2011), 93.
13. John Maynard Keynes, *The General Theory of Employment, Interest, and Money* (New York: Palgrave Macmillan, 2007), ch. 12.
14. Colin Camerer, *Behavioral Game Theory: Experiments in Strategic Interaction* (Princeton, NJ: Princeton University Press, 2011).
15. The definition in the text is for a strict Nash equilibrium. A weak Nash equilibrium is a strategy such that *given* the strategy choice of the other player, the player's response is at least as good as every other response and best for some.
16. Colin Camerer and Richard Thaler, "Anomalies: Ultimatums, Dictators and Manners," *Journal of Economic Perspectives* 9 (1995): 209, 210.
17. Three person, zero-sum games are not necessarily strictly competitive if coalitions (cooperation) of two against one are possible.
18. Harold Hotelling, "Stability in Competition," *The Economic Journal* 39 (1929): 41–57.
19. Dixit and Nalebuff, note 1, 180.

Chapter 8

Market and Government Failures as Prisoners' Dilemmas

The theory of supply and demand and perfect competition, both of which are discussed in the next chapter, are dependent upon the existence of legal institutions that support the market system and make it possible. It is important to know what legal institutions are necessary preconditions for self-interested behavior to be channeled into markets where it will promote social welfare rather than spill over into antisocial consequences.

Economists discuss the preconditions necessary for competitive markets to exist and be efficient under the headings of *market failures* and *market imperfections*. Market imperfections, such as monopoly and incomplete information, will be discussed in the next chapter. This chapter will focus on market failure, that is, reasons why markets may either fail to exist or produce socially unacceptable outcomes, as well as on remedies for market failures, many of which are known as *social infrastructure*. The three market failures, commons problems, externalities, and public goods, which are overlapping concepts, can be modeled as Prisoners' Dilemmas.

Because economists are generally more concerned with analysis of markets and market outcomes rather than with the assumptions underlying markets, discussion of market failures is often relegated to the last chapters in many economics textbooks. However, in the field of law and economics, discussion of the necessary legal and institutional assumptions underlying markets and social infrastructure should be discussed before markets.

This chapter is also concerned with government failure, that is, reasons why government attempts to ameliorate market failures may go awry, some of which involve Prisoners' Dilemmas, such the voting paradox and special interest groups.

COMMONS PROBLEM

The market failure known as the common property resource problem or, in short, *commons problem*, takes its name from grazing land held in common in agricultural villages in England and colonial New England towns. This system worked well for centuries because the number of cattle was small relative to the size of the commons. However, as herds become ever larger, eventually a point is reached where the commons is destroyed because all grass is either eaten or trampled. A similar scenario applies to unrestricted whaling, fishing, and the drilling of multiple oil wells in a single oil field. Air pollution, which can be thought of as waste disposal of greenhouse gases in an air commons, threatens the earth with substantial global warming and climate change. (Air pollution and its control will be discussed under the second market failure, externalities.)

The commons problem is an example of N-Person, Prisoners' Dilemma. Each herdsman realizes that the benefit from adding one more cow to the commons accrues entirely to the cow's owner whereas the cost of the grass eaten by an additional cow is shared by all other herdsmen and is therefore negligible to the cow's owner. If no other herdsman increases his herd, each herdsman's self-interest is to increase his herd; if all other herdsmen increase their herds, each herdsman's self-interest is to increase his herd. Even if every herdsman is aware of the impending disaster, he knows that if he does not add one more cow to the commons, someone else will. If the words whales and whalers are substituted for cows and herdsmen, the analysis is the same.

Commons Solutions: Private Property, Regulation, or Government Ownership

There are three solutions to commons problems: 1) the institution of private property, 2) a system of rules restricting entry and/or regulating use, or 3) government ownership.

Private Property

Private property in land was introduced centuries ago in order to solve the commons problem in farming; no one would plant crops if they could not prevent others from reaping where they did not sow. The eighteenth century enclosure movement in England solved the commons problem inherent in animal grazing by turning most commons over to private owners. The nineteenth-century "range wars" in the American west between the cattlemen and farmers, "sod busters," portrayed in epic motion pictures, concerning whether cattle should roam freely on open ranges or whether land should be fenced, also involve a commons problem dispute.

When someone takes private property from someone else by force or misrepresentation, say steals an automobile, a coat, or money, we call that activity theft or fraud. In market failure terms, thieves attempt to make a commons out of private property. A police force is necessary to prevent private property from becoming a commons and to ensure that transfers of property take place through voluntary exchange in markets as opposed to coercive transfers from victim to wrongdoer. In advanced societies, commons problems are usually solved by the legal institution of private property, a court system which settles conflicting claims to property, and a police force which enforces the legal determination of property rights.

Regulation or Government Ownership

With respect to hunting in forests, fishing in small bodies of water, and the right to divert and appropriate water from rivers, the usual solution to the commons problem in the past has not been private property but to place legal restrictions on the number of people who can exploit the resource, the times it can be exploited, the technology that can be used—limiting the catch, kill, or amount of water appropriated, so that the resource can be sustained through reproduction or additional rainfall. Sometimes the legal regulations become de facto property rights for the hunters, fishers, or water users.

In the eastern states of the United States, where the supply of water is relatively plentiful in relation to demand, water rights are held more or less in common: the basic concept is that of *riparian rights*, where those who own land adjoining a body of water are entitled to make reasonable use of the water, reasonable use being defined as that which does not unduly interfere with downstream uses of others who also have riparian rights. However, in western states where the supply of water is scarce relative to demand, the eastern states' approach would be unsustainable, so a system of property rights in water is based on first use or appropriation. However, due to the increasing scarcity of water in the western United States from droughts and climate change, the appropriation system, which gives priority based upon historical usage regardless of current efficiency considerations, has come under attack and water rationing has been imposed.

With respect to the right to drill wells in a common oil field (a depletable, as opposed to sustainable, resource), some countries have opted for private property, others solve the commons problem with government ownership. However, private ownership of surface rights does not solve the commons problem with respect to an underground pool of oil that is beneath (common to) many individually owned surface plots. Unless the amount of oil that each individual well produces is limited by agreement or government regulation, individual self-interest would cause each well owner to produce as much oil as possible as fast as possible, thereby depleting the oil pool at a rate which is inefficiently fast. The situation is analogous to one in which several people have straws in and are sipping from the same bottle filled with a beverage. If one person is impolite, a race to the bottom is on. The race ends when your hear "Slurp, Slurp" at the bottom of the bottle!

If the number of broadcasters over the electromagnetic spectrum were not limited, the resulting congestion and noise would make listening to radio and watching television nearly impossible since the demand for radio and television broadcast frequencies exceeds the limited supply of this resource. In 1934, Congress created the Federal Communications Commission (FCC) in order to deal with this congestion or commons problem. The FCC was authorized to grant three-year renewable licenses for a nominal charge to applicants who would use the broadcast frequencies in the public interest. Once obtained, broadcast licenses have rarely been revoked, thus becoming de facto property rights; the licensee is usually able to transfer the license along with the physical assets of the station for much more than the physical assets are worth, thus reaping a windfall gain.

Initial applicants for licenses spend a great deal of money on lawyers and lobbyists in an attempt to win FCC approval. The total spent by all applicants on lawyers and lobbyists on a broadcast frequency or other government granted monopoly reduces and may even exceed the value of the monopoly. This inefficiency, called *rent-seeking* behavior, refers to competition among individuals for a monopoly on the basis of spending that is mutually frustrating and may be collectively in excess of the value of the monopoly.[1] For example, the search for a sunken ship, which contains a $1 million treasure, by 10 teams, each of which spends $150,000 to find the treasure, results in a $500,000 net social loss, even if the treasure is found. (A patent race, whereby several inventors duplicate each other's research can also be characterized as rent-seeking behavior.)

For reasons of efficiency, economists generally advocate that government ought to solve the commons problem where possible by auctioning off scarce or monopoly resources, such as the electromagnetic spectrum and off-shore oil drilling rights, to the highest bidder. The auction approach has twin benefits: if the highest bidder is given a property right in scarce resources, which includes the right to transfer that property to others, that is, if a market is permitted, resources tend to gravitate toward their most valuable uses. Second, if scarce resources owned by government are auctioned off to the highest bidder rather than given away to fortuitous individuals or distributed based on political favoritism, society as a whole will get the windfall benefit from scarce resources. Currently, the federal government uses the auction method for allocating licenses for the electromagnetic spectrum devoted to cell phones.

Population Growth and Global Warming as Commons Problems

In a well-known article, Garrett Hardin said that since the earth is limited in size, human freedom to reproduce is the ultimate commons problem.[2] He argued that government controls on the right of individuals to reproduce are necessary in order to prevent social disaster due to over-population of the commons we call the earth. Since he wrote, the government of China instituted such controls but is now relaxing them. On the other hand, in countries, which have educated women and have high income per capita, the rate of population growth has declined dramatically.

Another commons problem related to the earth as a whole is the greenhouse effect, which refers to the role that polluting the unowned atmosphere plays in warming the earth's surface. Much of the incoming solar radiation, which penetrates the atmosphere, is held within the earth's atmosphere by gases such as carbon dioxide, methane, and ozone. It has been estimated that increasing amounts of greenhouse gases produced by burning of fossil fuels could eventually result in a global warming of perhaps 2° C (3.6° F) by the year 2100.

Commons as Prisoners' Dilemmas

Commons problems that are international in scale, like over-fishing in international waters and greenhouse gases that pollute the atmospheric commons or acidify the ocean commons, are not only ownership and regulation of usage problems, they are Prisoners' Dilemmas. For agreements regulating usage of these commons to be successful, not only must all nation states be parties to agreements, there must be enforcement mechanisms so that rogue states and private actors who violate these agreements cannot do so with impunity.

EXTERNALITIES

Externalities are a second market failure. Pollution is a common example of an externality. If a steel manufacturer pollutes air and streams with smoke and slag that are incidental byproducts of the process of producing steel, individuals who live in the vicinity of the factory will suffer damage because of reduced air and water quality. The social cost of such damage, which is not considered in the market for steel by either the manufacturer or its customers, is said to be external to the market or an externality. An externality represents a discrepancy between private and social cost in the sense that products are priced at less than their true cost to society, which social cost should include the lesser of cost of damage or cost of preventing such damage. (Prices of products that cause significant negative externalities are not Pareto efficient by definition; chapter 1.)

Specifically defined, an *externality* occurs when an activity undertaken by one party, or a transaction between two parties, imposes a cost or confers a benefit on another party as an incidental byproduct that is not charged or credited directly to the originating activity through the price system. An externality affects the price system indirectly rather than directly. For example, although air pollution from a steel mill does not affect the price of steel directly, it has indirect effects on the market system, such as an increase in medical costs consequent on higher rates of asthma in the population, more expenditure on paint as a result of the more frequent painting of houses and buildings, and increases in the cost of food due to greenhouse gases.

In a sparsely settled agrarian economy typical of the United States up to the end of the nineteenth century, where many individuals lived on isolated, single-family farms which were largely self-sufficient and where one person's activities had minimal effect on others, externalities were negligible. However, in modern, densely populated metropolises, where individuals and their activities are highly interdependent, externalities are manifold: urban traffic congestion, noise, smog, and automobile accidents, in which a person is injured or property is damaged as a result of another's activity, are examples of negative externalities. If unchecked by laws and institutions, negative externalities resulting from self-interested behavior that interferes with the rights of others would destroy both society and markets. The greater prevalence of externalities, both positive and negative, in urban, as opposed to suburban or rural, areas may have political ramifications in terms of an individual's political party affiliation. Urban dwellers are more likely to be receptive to government or cooperative action than are rural dwellers (those who live in "empty spaces").

Externalities may be positive and confer benefits as well be negative and impose costs. If a person builds a park next to another's property, which raises its value, the park builder confers a positive externality on the neighboring land owner. Communications systems are another example of positive externalities, which are called *network externalities*: if one person is added to an existing communications network (e.g., Facebook), that person receives external benefits that are substantially greater than the cost to her or him of joining the network. Police, courts, education, banking systems, roads, libraries, public health measures, such as immunization and water purification, are other examples of activities that produce positive externalities or eliminate negative externalities, which are collectively called social infrastructure.

Public Financing of Education Is Required by Externalities

Why is everyone compelled (through taxes) to contribute to the education of children? Why don't we rely on individual self-interest of parents to educate children? Public education is an example of both positive and negative externalities. Laws mandating public funding of education and mandatory school attendance by children are justified by positive externalities. Educating one child benefits everyone else; democracy and the market system would collapse if a substantial part of the population were allowed to choose not to educate their children.[3] Mandatory education of children produces large positive externalities. Looked at from the negative side, uneducated persons often become a charge on the state so that neglect of education results in large negative externalities.

If education was not publically financed through taxes but was financed and provided solely by private markets, who would probably not educate their children? Answer: it is likely to be Rawls's worst-off (poorest) group. So, if publically financed education is required in order to eliminate the negative externalities that would occur if worst-off group did not educate their children, the disbursement of compulsory taxes should be consistent with Rawls's maximin principle. In other words, one can argue, as discussed later in this chapter, that when evaluating plans for school vouchers, charter schools, and public schools, such plans should not be to the detriment of the most disadvantaged group.

Externalities as Prisoners' Dilemmas

Like the commons problem, externalities can also be analyzed as N-Person, Prisoners' Dilemmas. Consider the problem of littering (or spitting gum on sidewalks where it turns to black spots), which is a form of pollution. Every purely self-interested individual will litter whether or not anyone else litters: even if no one else litters, a self-interested person will litter because it saves her or him the cost of proper disposal of trash; if everyone litters, a self-interested person will still litter in order to avoid the cost of disposal. If all self-interested individuals litter, society as a whole will be worse off. Consider another example: if a production process produces toxic waste as an unintended byproduct, dumping waste in a public sewer at midnight is in a manufacturers' self-interest due to lower costs and higher profit; paying to dispose of the toxic waste in a manner that does not threaten injury to the public is more expensive. The greenhouse effect, which causes global warming and climate change and was discussed above as a commons or ownership problem, can also be analyzed as an unwanted byproduct of production, that is, as a negative externality or activity problem.

When the cost of a negative externality like pollution is fully *internalized*, that is, included within a product's price, the discrepancy between private and social costs is eliminated, and prices are Pareto efficient. There are three ways of using law to internalize negative pollution externalities so that the price of products includes the lesser of either the cost of eliminating an externality or the social cost of damage it causes: 1) property right approach, 2) performance standards, 3a) carbon taxes, 3b) sale of pollution rights, 3c) cap and trade (tradable emissions permits), which are three different versions of one method.

Property Rights Approach to Externalities

An externality can be dealt with as if it were a commons problem, that is, an ownership problem instead of an activity problem. For example, air pollution may be thought of as occurring because no one owns the air. If individuals had a property right or entitlement to clean air, they would be entitled to a court-mandated injunction, which forbids any individual or business entity from engaging in an activity that pollutes the air, unless the entitlement to clean air is transferred (sold) from its owner to polluter by means of a contract voluntarily entered into by both parties. This approach, known as the Coase Theorem, will be more fully discussed in chapter 12. However, even the staunchest adherents of the property rights method for dealing with externalities admit that it is not a feasible solution for greenhouse gases from carbon dioxide (CO_2) emissions that are a byproduct of burning fossil fuel for four reasons.

1) Transaction costs, which are involved in making contracts among huge groups of persons, most of whom suffer small individual damage, are prohibitive; 2) proving causation, which factory caused damage to which individuals, is not possible; 3) determining the amount and extent of damage to specific individuals and property is speculative; and, 4) when a property right, for example, the right to clean air, is owned by multiple parties and unanimous agreement among owners is required in order to sell the right to a business which wants to buy the right to engage in limited pollution, any one individual, and by extension every individual, can hold-up or prevent sale of the right unless an exorbitant price is paid to that person. Conversely, when a group of individuals wish to buy a property right from business which has a right to pollute, any one person can be a free rider, that is, refuse to contribute her or his fair share; by extension all persons can be free riders. Reason 4) is an example of an N-Person Prisoners' Dilemma. In the context of externalities involving large numbers of persons and covering large geographic regions, the carbon tax, cap and trade, or performance standards approaches discussed below are more feasible alternatives than is the property rights approach.

Carbon Taxes, Sale of Pollution Rights, and Cap and Trade

Government can place a tax on an activity that generates negative externalities. For example, bottle return laws, under which a consumer pays a five- or ten-cent deposit on each bottle, can, or container of beverage bought, internalize the externality caused by litter from beverage containers so that activities of manufacturing and consuming beverages will bear the cost of litter or its removal. Consumers of beverages sold in containers are given a choice between either incurring the opportunity cost in time and effort required to sort, save, and return containers to redemption centers or not incurring such cost and forfeiting bottle deposits, which is equivalent to paying a tax equal to the social cost of having someone else remove the litter. In either case, the price of beverages will include both its private cost (materials and distribution cost) and its social cost (removal cost). Each individual will choose the option that is least costly for her or him: individuals with high opportunity cost of time will choose not to return containers; those with lower opportunity cost of time will return containers; those with still lower opportunity cost, say children and the unemployed, will search for unreturned containers and return them for refunds. If bottle return laws do not prevent litter at the prevailing tax per container, the remedy is to raise the tax per bottle until litter disappears.

Consider another example: a tax on smoke emission by steel manufacturers is an attempt to internalize the smoke externality so that the cost of pollution or its removal will be internalized in the price of steel. Manufacturers of steel are given a choice between either paying a tax or investing in methods to prevent smoke pollution. Each firm will choose its least cost option. In either case, the price of steel will reflect both its private cost (materials) and its social cost (pollution damage cost or cost of abatement). Newer firms, or those currently being planned, which can inexpensively install equipment to prevent smoke pollution, will decide that it is cheaper to prevent pollution than to pay the tax. Old firms with antiquated technology will often find the cost of installing pollution control technology prohibitive in the sense that such expenditure would cause profitable firms to incur losses, driving them out of business; such firms will choose to pay the pollution tax, which is cheaper than the cost of preventing pollution. A higher pollution tax means less pollution, but at a cost of higher prices for products. The level of tax should be determined via marginal cost-benefit analysis so as to balance, that is, trade-off, the marginal benefits from reducing pollution against the marginal cost of higher prices for products.

In order to make the processes described by the bottle tax and carbon taxes clear, a simple model of the tax approach to externalities is presented in Figure 8.1. The annual cost per ton of carbon dioxide (CO_2) reduction is depicted on the vertical axis. The cost of CO_2 reduction is an annual cost because pollution control, such as switching to electric generation fired by gas rather than coal, requires a large capital expenditure that must be amortized over the life of the equipment, just as the cost of a home mortgage must be amortized over a period of time, say thirty years.

The amount of CO_2 per year, which is being produced by factories as an unintended and unwanted byproduct, is depicted on the horizontal axis. Assume the current amount of CO_2 per year is a total of 8 tons emitted by four businesses, A, B, C, and D, which burn fossil fuel. Firms C and D are old and have antiquated technology, so that for them switching to cleaner fuel is much more costly than for newer businesses, A and B, for which switching is less expensive. In Figure 8.1, firms A and B each produce 2 tons annually of CO_2, firm C produces one ton annually of CO_2, and firm D produces 3 tons annually of CO_2 for a total of 8 tons. The height of each line or step represents annual cost *per ton* of pollution reduction by one of the four firms as measured on the vertical axis; the width of each line represents the number of tons of CO_2 that each firm is currently polluting. The annual cost of pollution reduction is the

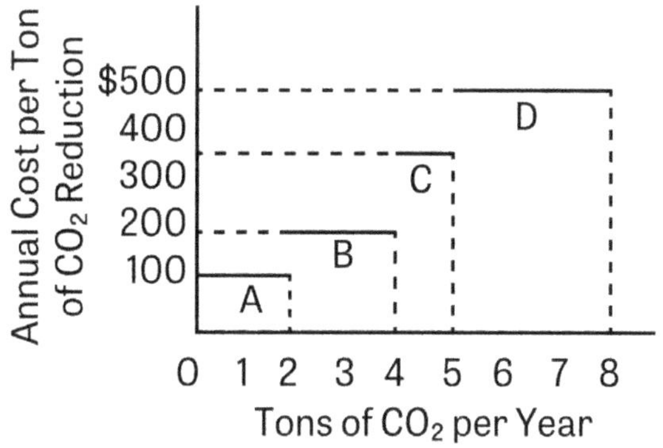

Figure 8.1 Pollution Abatement Costs of Four Firms. *Source*: Created by the author.

cost of clean air per ton. Notice that the annual cost of pollution reduction for firm D is \$500 per ton, which is significantly higher than for firm A, which is \$100 per ton.

Next, compare the cost of pollution reduction by a regulation requiring each of the 4 firms to reduce pollution by the same amount, say 1 ton each, to the cost of same total amount of pollution reduction, 4 tons, via a carbon tax, where each firms decides whether to clean up or pay the carbon tax.

Equal Pollution Reduction by All Firms

Assume a pollution control law is enacted such that each of the 4 firms is required to reduce CO_2 emissions by an equal amount, say by 1 ton. (This is equivalent to every person being required to return a stated number of bottles per year.) The result will be that instead of 8 tons of CO_2 being released into the atmosphere each year, total amount of pollution will be reduced from 8 tons of CO_2 per year down to 4. Looking at Figure 8.1, for firm A, the cost of pollution reduction is \$100 per ton per year; if firm A, which is currently producing 2 tons of CO_2 annually must reduce pollution by 1 ton per year, that is, from 2 tons to 1 ton, the annual cost to firm A will be \$100; for firm B, which is also polluting 2 tons of CO_2, its annual cost of reducing pollution by 1 ton, from 2 tons to 1 ton, is \$200; for firm C, which is polluting only 1 ton of CO_2, its annual cost of reducing pollution by 1 ton, from 1 ton to 0 tons, is \$400; and, firm D, which is polluting 3 tons of CO_2, its annual cost of reducing pollution by 1 ton, from 3 tons to 2 tons, is \$500.

In sum, total cost to society of removing or cleaning up 4 tons per year (8 – 4 = 4) will be \$100 for A, plus \$200 for B, plus \$400 for C, plus \$500 for D: \$100 + \$200 + \$400 + \$500 = \$1,200 per year in order to reduce CO_2 pollution from 8 tons down to 4 tons per year.

Carbon Tax Approach

Next, consider the total cost of pollution reduction from 8 tons to 4 tons per year via the carbon tax method. If an annual tax of slightly more than \$200 per ton is levied on CO_2 emission and firms are given a choice whether to cease polluting or continue polluting and pay the tax, firms A and B would determine that buying pollution control equipment and totally cleaning up is cheaper than paying the annual tax of \$200 per ton of pollution, and firms C and D would determine that paying the carbon tax is cheaper than eliminating pollution. Why? Assume the annual tax is \$250 per ton; each of the four firms can choose either to eliminate pollution or alternatively to pay the tax. If firm A were to cease polluting completely, that is, reduce pollution from 2 tons to 0 tons, its annual cost of pollution reduction would be 2 tons multiplied by \$100 per ton, which is a total of \$200 per year. Alternatively, if instead of cleaning up, firm A were to pay the annual tax, its 2 tons of pollution multiplied by a tax of \$250 per ton equals \$500 per year. Since \$200 < \$500, ceasing to pollute is cheaper for firm A. For firm B, the annual cost of cleaning up is equal to 2 tons multiplied by \$200 per ton, which is a total of \$400; if instead of cleaning up, firm B pays the annual tax, its 2 tons multiplied by a tax of \$250 per ton equals \$500 per year. Since \$400 < \$500, ceasing pollution is cheaper for firm B.

Firms C and D would determine that it is cheaper for them to pay the tax rather than clean up: for firm C, the annual cost cleaning up its 1 ton of CO_2 is \$400; if instead of cleaning up, firm C pays the tax, it will pay a tax of \$250 per ton multiplied by 1 ton, which equals \$250. Since \$250 < \$400, firm C will pay the tax. For firm D, the annual cost of cleaning up is equal to 3 tons multiplied by \$500 per ton, which is a total of \$1500; if instead of cleaning up, firm D pays the annual tax, its 3 tons multiplied by a tax of \$250 per ton equals \$750. Since \$750 < \$1500, firm D will pay the tax.

In sum, if the annual carbon tax is \$250 per ton, 4 tons of CO_2 would be removed at a cost of (2 × \$100) = \$200 to firm A for cleanup of 2 tons, plus a cost of (2 × \$200) = \$400 to firm B for cleanup of 2 tons, or \$200 + \$400 = \$600 per year, the cost of which is half of the cost of requiring each of the four firms to remove 1 ton, \$1200. Since \$600 < \$1200, a carbon tax is a cheaper way to cleanup pollution than is mandating equal amounts of pollution reduction by each firm.

Revenue Neutral Carbon Tax

Firm C pays \$250 × 1 ton = \$250 in taxes; firm D pays \$250 × 3 tons = \$750 in taxes. Total taxes paid by firms C and D are \$250 + \$750 = \$1,000. What happens to the taxes these firms pay? Assume the legislature that imposes the tax wants it to be *revenue neutral*; that is, the tax will neither increase nor decrease government revenue but will be returned to the industry on a pro rata basis. In other words, the purpose of the tax is to improve economic efficiency, not to raise revenue. Each firm, A, B, C, and D, receives a transfer payment from the government equal to total proceeds from the tax divided by four, \$1,000/4 = \$250.

Alternatively, government could use the proceeds from the tax to offset some of the negative effects of pollution, say by spending more on health care for those whose emphysema is exacerbated by inhalation of smoke from polluting factories; or, proceeds from the tax could become part of general revenue used to fund government operations. If the carbon tax is not revenue neutral, it may have significant redistributive consequences, like most taxes, because revenue from tax payers to government may benefit non-tax paying individuals.

In order to give drivers an incentive to buy fuel-efficient cars, that is, to penalize buyers of gas guzzlers, a revenue neutral gasoline tax has been suggested; that is, all revenue from the gas tax would be remitted to all tax payers in the form of say, tax credits, so that those who bought fuel-efficient vehicles would benefit from the tax and buyers of gas guzzlers would be penalized. Whenever government taxes one party and gives the proceeds of the tax to a party without requiring current services, such payments are called a *transfer* payment. Transfer payments by government, for example, when government uses the taxes collected from one person in order to pay social security benefits to another person, have no effect on resource utilization until they are spent by individuals receiving such payments.

Sale of Pollution Rights

A second approach, the *sale of pollution rights*, where the right to pollute a specified number of tons of CO_2 is sold at auction to highest bidders, is the converse of a carbon tax and has symmetrical results. Referring to Figure 8.1, assume again that government wants to reduce CO_2 emissions from firms by 4 tons annually, from 8 tons to 4 tons. The government can auction off the right to pollute 4 tons, one ton at a time. At these auctions, C and D, who have knowledge of all firms' costs, know that if they bid slightly more than $200 per ton, A and B would not want to buy the right to pollute because A's cleanup cost is $100 per ton and B's cleanup cost is $200 per ton. Thus, at bids of slightly more than $200 per ton, C would buy the right to pollute 1 ton, and D would buy the right to pollute 3 tons. The carbon tax and the auction of the right to pollute are like heads and tails of the same coin. In each case, firms must choose whether it is cheaper to pay cleanup costs and not pollute or pay the government and pollute. In both carbon tax and sale of pollution rights, firms A and B cleanup and firms C and D continue polluting; firms C and D either pay a tax of slightly more than $200/ton or buy pollution rights at slightly more than $200/ton.

Under the carbon tax approach, government determines level of the tax (tax is the independent variable) and waits to see what quantity of pollution reduction occurs at that level of tax (quantity is the dependent variable), whereas under sale of pollution rights approach, government determines quantity of pollution that will be allowed (quantity is the independent variable) and waits to see what price of pollution rights in the auction market will clear the market (price is the dependent variable). A higher tax (or fewer pollution rights) will reduce the quantity of emissions but raise product prices; a lower tax (or more pollution rights) will increases the quantity of emissions and lower product prices. Whether government should increase, decrease, or allow the tax (pollution rights) to remain the same can be determined by marginal cost-benefit analysis, which trades-off health benefits from reducing pollution against costs of pollution reduction in terms of higher product prices.

Government can employ the iterative (one step at a time) method to achieve the efficient level of tax or pollution rights; that is, any time a small (marginal) increase in tax level (or reduction of pollution rights) results in gains as measured by reduced health costs and reduced global warming that exceed monetary losses as measured by the rise in product prices, another small change in the same direction is justified. At some point, when an additional small change in tax level in the same direction causes losses to exceed gains for the first time, the optimal level of tax has been achieved. However, because such analyses are often speculative and subject to large errors, legislatures usually mandate a specific amount of pollution reduction below that which occurred in a chosen base or reference line year.

Cap and Trade (Tradable Emission Permits)

A third method for reducing pollution is for government to issue a fixed number of *tradable emission permits* to existing firms free of charge, called *cap and trade*. For example, if it is desired to reduce pollution from 8 to 4 tons of CO_2 per year, the total amount of pollution allowed is capped at 4 tons of CO_2. Each of the existing four firms, A–D would be given a permit which allows it either to emit 1 ton of CO_2 or sell this permit to other firms. Firms whose pollution is not licensed by a permit must either cease polluting, that is, clean up, or cease production.

There will be a market in tradable pollution rights. The results of cap and trade with 4 tradable emission permits are the same as under the tax or sale of pollution rights approach: firms A and B would each sell their rights to pollute 1 ton of CO_2 to firm D, which values pollution rights more than they do: firm D's right to emit 1 ton, which was allotted to it by government, plus the right to emit 2 tons, which it bought from firms A and B, gives it a right to emit 3 tons; that is, it would not have to reduce pollution. Firms A and B, which have sold their rights to pollute, either have to eliminate pollution or cease production; they will each choose to clean up 2 tons of pollution for a total of 4 tons. And, firm C keeps its allotted right to emit 1 ton of CO_2, so it continues to pollute. In sum, the now familiar result is that firms A and B reduce pollution by a total of 4 tons and firms C and D continue polluting a total of 4 tons. Overall pollution has been reduced from 8 to 4 tons.

How is this result achieved? Recall that in chapter 7, two-person bargaining, one buyer and one seller, was shown to result in a range of prices; price was indeterminate. As the number of buyers and sellers increase, the indeterminate price range is reduced; and, when the number of buyers and sellers is large, which is the definition of a competitive market, the law of one price governs and price is determinate. When only firms A and D are in the permit market, the indeterminate price range per permit is $100–$500; when firms A, B, C, and D are in the permit market, the price range per permit is narrowed to $200–$400. (Can you explain this result by creating a supply and demand table like Table 7.1?) As the number of firms in the cap and trade program increase, the range of prices continues to narrow until the law of one price governs and price is determinate.

Carbon Tax, Sale of Pollution Rights, and Cap and Trade Compared

The carbon tax, sale of the pollution rights, and cap and trade methods all reduce pollution by the same amount, from 8 tons to 4 tons per year; and, under all three methods, firms A and B cease polluting and firms C and D continue to pollute.

Now consider the income redistributive effects of the three methods. Under cap and trade, income transfers stay in the private sector, among private firms, which buy and sell tradable emissions permits; no proceeds from the sale of permits go to government. If carbon tax and sale of pollution rights methods are revenue neutral, that is, proceeds of the carbon tax or auction of pollution rights are distributed pro rata back to firms that pay the tax, these two methods are similar to cap and trade because government revenue is not increased under any of the three methods.

However, if the revenue from a carbon tax or sale of pollution rights enhances government revenue, that is, these methods are not revenue neutral, income distribution effects of the tax and sale of pollution rights methods of pollution reduction are significantly different from cap and trade. Carbon tax and sale of pollution rights, which are not revenue neutral, involve substantial redistribution of income from the private sector to the government.

A unique feature of cap and trade, as opposed to a carbon tax or sale of pollution rights methods, is that government, if it chooses, can treat different firms differently for reasons of equity, that is, allocate different amounts of pollution permits to different firms. For example, firms that have old technology and for whom pollution cleanup costs are expensive can be allotted more pollution permits than firms with newer technology, which can reduce pollution less expensively. In Figure 8.1, firm D, which has antiquated technology and high cost of pollution reduction, could be given the right to pollute 2 tons per year instead of 1 ton per year on grounds of equity.

Cap and trade may create barriers to entry. Since pollution permits are allocated to existing firms without charge based upon historical output, a potential drawback to the cap and trade method is that it may create barriers to entry in industries to which it applies because potential entrants would have to purchase permits from existing permit holders.

ECONOMICS VERSUS GAME THEORY APPROACHES TO POLLUTION EXTERNALITIES

The elegance and simplicity of the carbon tax, sale of pollution rights, and cap and trade methods as solutions for externalities are impressive. In the economic approach, externalities are a mere pricing problem; that is, if price is set at the efficient level, that is, a level that takes into account all social as well as private costs, the externalities problem is solved. However, if externalities are viewed through the lens of game theory, specifically N-Person Prisoners' Dilemma, externalities are very difficult to solve political problems involving a sharp conflict between individual self-interested behavior and social cooperation. Consider the difficulty the countries of the world have had in reaching enforceable agreements on carbon emissions.

PERFORMANCE STANDARDS TO CONTROL EXTERNALITIES

A third method, which can be used to control externalities like smoke pollution (in addition to the property rights or tax approaches), is for government to set *performance standards* for allowable amounts of pollutants via laws or administrative rules. This method internalizes externalities by requiring business to bear the cost of eliminating them, which is passed along to consumers in the form of higher product prices. For example, the federal government has stipulated the maximum amount of pollutants that automobiles can legally emit; this is an emissions standard. Automobile manufacturers have met this standard by installing catalytic converters on cars in order to reduce exhaust emissions.

A potential problem with the performance standards approach is that government could mandate a standard which is so stringent that the cost of eliminating the externality exceeds the social cost of the damage from it. For example, the Delaney Amendment to the US Food and Drug Law[4] established an absolute standard that required the Food and Drug Administration (FDA) to ban any food additive that was proved to cause cancer in man or animals no matter how weak the causative link. This absolute standard led the FDA to ban saccharine, a sugar substitute, even though studies showed that the potential benefits from saccharine in terms of the reduced probability of heart attacks and diabetes brought on by obesity outweighed the minuscule increased probability of death from cancer. Subsequently, Congress, under intense public pressure, overturned the saccharine ban with special legislation. Moreover, the Delaney Amendment required the FDA to ban some *additives* that have less cancer-causing potential than that found in some *natural foods*, such as some species of fish that contain large amounts of mercury and other toxic chemicals. The FDA has since changed the interpretation of the Delaney Amendment from an absolute standard to a *de minimis* standard, by which barely detectable cancer-causing potential is ignored.

Performance standards, like fuel efficiency standards (say, 40 miles/gallon for all vehicles by 2020), which mandate an outcome but not how it is achieved, encourage creativity and innovation by business, which is tasked with finding a technological solution that will satisfy the mandate. Performance standards should be distinguished from design standards, which are inefficient. For example, a design standard, which mandates that all ladders must have rungs one inch thick, would be appropriate for wood ladders but not for aluminum or plastic ladders. Design standards are inefficient because they are inflexible and actually discourage changes in technology. A performance standard for a ladder would be one that required each ladder rung to be able to withstand a weight of say, 500 pounds, and allowed manufacturers to use any materials and structural configuration which conformed to that standard.

Performance Standards Versus Carbon Tax or Cap and Trade

In general, the performance standards are appropriate where the cost of regulation is uniform across sources. For example, uniform standards are set for home fire alarms and carbon monoxide detectors. Performance standards are also appropriate when, as is the case with automobiles, an infrequent or one-time determination whether there has been compliance with a standard is less costly and less subject to evasion than is constant monitoring of the level of exhaust emissions by millions of automobiles. Carbon tax, pollution rights, or cap and trade are appropriate where the cost of eliminating pollution varies greatly among the different sources of pollution, say old versus new factories, or where monitoring and compliance costs are not too high because the number of polluters is small or manageable.

PUBLIC GOODS

A third market failure occurs because some goods are public as opposed to private. A *private* good is one that, when consumed by one person, cannot be consumed by others. Apples, automobiles, houses, haircuts, and accounting services are examples.

Marginal Cost Equals Zero and Non-Excludability

Public goods have one or both of two characteristics: First, once a public good is produced, the marginal cost of supplying it to another individual is zero. Television broadcasting has this characteristic. Once a television program is being broadcast, it costs almost nothing for another person to view the program. The zero marginal cost characteristic is also called the non-rivalrous consumption characteristic.

Second, individuals who do not pay for a public good usually cannot be excluded from consuming or benefiting from it. This is its *non-excludability* characteristic, which is also called the *free rider* problem. Pure examples are national defense, open-air concerts, and displays of fireworks. Excluding people who do not pay for fireworks from viewing them is not possible; exclusion of those who do not pay for national defense is also not practical.

Some other public goods exhibit only one of the two characteristics: television has the zero marginal cost characteristic, but because broadcasters can scramble signals by encoding them or send signals through cables, free riders can be excluded. Although the cost to a library of having one additional book borrowed is almost nothing (up to the library's capacity), libraries could exclude free riders who are not willing to pay a fee for taking out books. Libraries, which educate and entertain their patrons, also produce the positive externalities that result from education. Thus, government subsidies for libraries are justified by two types of market failures, public goods and externalities. An analysis similar to libraries applies to the provision of streets, street lighting, snow removal, public transportation, spraying swamps to kill disease-bearing mosquitoes, and other social infrastructure like police and firefighting.

Public goods that have both characteristics, marginal cost of supplying the good to another person is zero and non-excludability or free rider problem, can be analyzed as N-Person Prisoners' Dilemmas. Consider the problem of providing for national defense through private markets. Purely self-interested individuals will not voluntarily pay for national defense: even if everyone else agrees to pay for national defense, a self-interested person will be a free rider because national defense will be provided whether or not he or she pays; if no one voluntarily pays for national defense, a self-interested person will not to pay so others will not get a free ride at her or his expense. In the context of game theory, there is a dominant strategy—always be a free rider. If all self-interested individuals are free riders, public goods will not be privately produced and society as a whole will be worse off. This is another example in which individual self-interested behavior must be overcome through politically mandated cooperation.

Consider immunization against childhood diseases, such as mumps, measles, and diphtheria, which is one of the most dramatic examples of the non-excludability characteristic or free rider problem. Even if immunizations are provided to the public without charge, some self-interested individuals will not get immunized because they will reason that if everyone else gets immunized, they will be protected from the disease due to herd protection because everyone else is immunized. If everyone thinks this way, no one will be immunized. In order to control the adverse consequences of free riding behavior, some states have passed laws requiring proof of vaccination against childhood diseases as a precondition for registration at school.

However, due to religious and other exemptions from immunization laws, which have been gathering adherents due to unfounded rumors concerning supposed side effects of immunization, childhood diseases, after many years of nearly being wiped out in the United States, are increasing again in many communities; herd protection has been lost due to the growing number of parents who refuse to have their children vaccinated. As a result, some states have begun to interpret exemption criteria narrowly or require parents who refuse to have their children vaccinated to home school them.

Although immunization has the free rider problem, it does not have the other characteristic of public goods, marginal cost equal zero, because the real or opportunity cost of immunizing an additional person, although relatively small, is not zero, whether paid by individuals or government. Only public goods that have both public good characteristics are Prisoners' Dilemmas. Immunization is not a Prisoners' Dilemma because although it is in one's self-interest *not* to get vaccinated if everyone else does get vaccinated, it is in one's self-interest *to be* vaccinated if no one else get vaccinated. In a Prisoners' Dilemma, one's self-interest is always not to take the cooperative action regardless of what others do. In order to be a Prisoners' Dilemma, a public good must have both characteristics, marginal cost equal zero and non-excludability or free rider problem.

How Should Public Goods Be Paid For?

When markets fail due to the existence of public goods, its two characteristics suggest conflicting social policies: The marginal cost equals zero characteristic, once produced it costs nothing to supply the good to another person, suggests that individuals should use such goods free of charge. Why? A fundamental theorem of economics is that the price a consumer pays for any good should equal marginal cost of producing the unit he or she consumes; a price in excess of marginal cost reduces and distorts consumption choices by creating artificial scarcity, which misallocates resources. The zero marginal cost characteristic of public goods suggests provision of public goods by government, public subsidy to private producers, or money prizes to producers (determining the efficient money prize for say, an invention, is problematical). Government production or subsidies to private producers are used to produce national defense, streets, public health measures, and much other social infrastructure.

The other characteristic of public goods, the free rider or non-excludability problem, suggests that, where possible, the institution of private property should be used to solve the excludability problem thereby transforming public goods into private goods so that individuals have incentives to produce such goods. This method is used with respect to patents and copyrights, which encourage the production of useful ideas, writings, and other tangible creations. If creators of tangible ideas received no compensation, there would be less incentive to invent, innovate, or engage in artistic endeavor.

If music, magazine articles, and books are transmitted via computers over the internet, an efficient price may be to charge for use of such information in ways discussed below, thus potentially eliminating the free rider problem. In our age of information, public goods issues with respect to determining the efficient price for intellectual property (trademarks, copyrights, designs, computer software, music, and patents) and the rights to privacy and publicity (e.g., use of celebrity photographs) are becoming increasingly more important.

The current controversy concerning the price that should be paid for music and information that is being downloaded from the internet is an example of a public good, where the efficient price involves weighing the two conflicting policies discussed above, marginal cost equal zero versus incentives for private provision. The maximal position by copyright holders is that copyright holders should have complete control over and get paid for each use at a price determined by them (Pareto); the minimal position espoused by users is that music and information downloaded from the internet should be free. Two intermediate positions are a) blanket licenses, and b) using the *fair use* exemption in US copyright law to protect noncommercial file sharing, both of which are discussed below.

Under the blanket music licensing system, composers and performing artists give a music clearing house non-exclusive authority to license performance rights to users;[5] in turn, a clearing house issues a blanket or a per program license to users that allows use of any or all of compositions in its catalogue for a negotiated fee or for a fixed percentage of total advertising revenue, which percentage does not vary according to the number of compositions used. For example, a radio station may agree to pay say, three per cent, of its advertising revenue to a music clearing house which distributes funds from the blanket license to composers and artists based upon surveys that determine frequency of music performances. Because collective selling of music can easily be abused if all composers and artists belong to a single music clearing house (monopoly), the system has resulted in antitrust litigation.[6]

In the *Sony Corporation of America v. Universal City Studios, Inc.*,[7] the US Supreme Court held that recording copyrighted television programs and movies in their entirety on a video cassette recorder, when done at home for noncommercial purposes, is fair use. The plaintiff copyright owners had asked the Court to declare that such copying is a copyright infringement and asked it to require sellers of blank video cassettes to pay a tax on each cassette, which would be distributed to copyright owners by the US Copyright Royalty Tribunal in a manner similar to the blanket licensing system discussed above.

Whether the method of paying for downloaded information and music will be dominated by copyright owner control over each use, fair use, blanket licenses, or some combination remains to be seen. Currently, blanket licensing of music seems to be dominating: for a nominal monthly (blanket) fee, subscribing members of such services can listen to any music in service's catalogue without a per use charge. Although subscribers do not own music, they can listen to music on any of electronic device they own at any time. Query? Given this option, why pirate music (download music illegally)? On the other hand, some artists claim that royalties currently being received under this system are too low. One remedy is for composers and musicians, who believe they are being underpaid, to opt out of the blanket licensing system, which means that their music would be licensed individually, say on their own websites. As stressed above, the nature of public goods is such that the efficient price for such goods is not easy to determine. Price is greatly influenced by both technology and governmental policy. For example, the switch from selling music on CDs to downloading music over the internet caused a drastic drop in price per music track. The market, the courts, and legislators all have input into resolving the question, what is the efficient price for music and other public good information?

Will the Efficient Amount of Public Goods Be Produced?

The general view among economists is that public goods will be under produced due to the free rider problem; that is, an efficient allocation of resources would include more public goods than are in fact produced. Even the institution of private property in the form of patent and copyright protection may not totally eliminate underproduction of such public goods. For example, *Consumer Reports* is a monthly magazine that evaluates consumer products; many people who do not buy the magazine take advantage of the information in it by word-of-mouth, by borrowing a copy of it from a subscriber, or by reading it in libraries. If the magazine's publishers could require that all those who made use of the magazine pay for its use, the increase in revenue could possibly result in more extensive coverage of products and might increase the number of competitors in the product evaluation field.

It has also been argued that markets may produce too much information in situations where producers of information can obtain profits from speculation and even misinformation.[8] For example, an investor who is the first to find out that two corporations intend to merge can often make large profits by purchasing the target company's stock before others are aware of the impending merger. The duplicative efforts of many people attempting to be first in the race to discover this information causes large amount of resources to be devoted to this activity. The gains to society from having the market price of converge to the equilibrium price faster because this activity may be small compared to the cost of a great deal of duplicative activity.

Moreover, there are gains to be made by producing misinformation as where someone wishes to benefit from selling (unloading) overvalued stock to others. In general, where it is difficult for consumers of information to assess the value of information before buying a good or service, unregulated markets may produce an inefficiently large amount of misinformation; for example, unregulated health claims for food and drugs.

On the other hand, the efficient financial market hypothesis, discussed in chapter 7, assumes that financial information is disseminated so rapidly that self-interested individuals will realize that much duplicative effort is futile and will not engage in it.

Some government production, regulation, or subsidizing of information creation is justified due to its public goods character. In addition, government-mandated, private information creation as, for example, mandated nutritional labeling on food products or calorie count labeling in restaurants, will probably increase efficiency by having more informed consumers.

MARKET FAILURES IN ECONOMICS VERSUS GAME THEORY

Three market failures, commons problems (destruction of resources due to lack of ownership of a resource), externalities (spillover effects from activities), and public goods (collective consumption) are overlapping economic concepts; consider three commonalities:

1. All three market failures are related to *ownership* and *exclusion* (free rider) problems: if a commons cannot be owned, it will be destroyed by overuse; if a public

good cannot be owned in the sense that free riders cannot be prohibited from consuming it, it will not be created; and, if ownership of property does not include the right to forbid others from engaging in activities that impose negative externalities on property, the value of property can be diminished substantially (for example, property owners who suffer smoke damage from factories).

2. Commons problems, public goods, and externalities involve *inefficient pricing*: if a commons is not owned, it is undervalued (if a commons is privatized, its value and price would rise dramatically); if the price of public goods is equal to marginal cost, that is, zero, the incentive to produce public goods is significantly diminished; and, if there are externalities, which represent a discrepancy between private and social cost, products are priced at less than their true cost to society which price should include the lesser of cost of damage or cost of preventing such damage.
3. All three market failures, commons problems, externalities, and public goods are Prisoners' Dilemmas. In each case, a self-interested person has a dominant strategy: do not cooperate. In the context of Prisoners' Dilemmas, market failures are not merely ownership, activity, and pricing problems but are very difficult to solve political problems involving the conflict between individual self-interested behavior and social cooperation. Consider the great difficulty that the US Congress has reaching a consensus on social infrastructure spending involving roads, bridges, public transportation, scientific research, and action against global warming and climate change. In the context of game theory, solving market failures requires overcoming individual self-interest and achieving at least a minimum level of cooperation.

GOVERNMENT FAILURE

A theme of this book is that when individual self-interest works well for society, as for example, competitive markets, government should not interfere, but that when individual self-interest results in market failures and market imperfections, government may have to intervene to ameliorate the negative consequences of self-interested behavior. However, government intervention may exacerbate market failures rather than ameliorate them and may create additional problems collectively known as government failure. Moreover, much government failure is inherent in government.

Voting Paradox as Prisoners' Dilemma

There are public good (free rider) aspects to voting by citizens, who believe that great efforts on their part to become informed and to participate actively will ultimately make very little difference in outcome. As discussed in chapter 7, whether or not to vote is an N-Person, Repeated Prisoners' Dilemma, which may or may not achieve the cooperative solution, that is, everyone voting. If the actions of many individuals implicitly reflect the reasoning of the one-time Prisoners' Dilemma and they do not vote, government that results from such voting will be skewed and not accurately reflect the will of all the people.

Interest Group Politics: Concentrated Benefits and Dispersed Costs

Political interest and pressure groups may distort government goals. A small group of individuals with similar interests may band together and contribute a significant sum of money to the campaign of a candidate who is running for political office, the implicit quid pro quo for which is legislation that will benefit the contributing group. One example is minimum price legislation for milk that benefits dairy farmers. Because the burden on each member of the public from special interest legislation is small, there is likely to be neither a great deal of resistance to special interest legislation nor clamor for its repeal. In addition, it is difficult to organize a broad coalition (large group) to resist special interests because the larger the group, the harder it is to overcome the free rider problem.[9] Political interest groups are successful because the *benefits are concentrated among the few* while the *costs are dispersed among the many*. The cumulative negative effects of a great deal of special interest legislation are significant.

Political interest groups can be modeled as N-Person Prisoners' Dilemma. Even though all individuals will be better off if no one supports special interest legislation, individuals know that if all individuals support special interest legislation favorable to themselves, their self-interest will be served by supporting special interest legislation, and even if no other individuals support special interest legislation favorable to themselves their self-interest will be served by supporting special interest legislation.

Government as Monopoly: Inadequate Feedback and Inefficient Incentives

Government is by nature a monopolist with respect to public functions it performs. Private companies usually receive very quick feedback if they make bad decisions because the bottom line (profit) suffers from their mistakes. However, in government, even if programs are not responsive to citizens' needs, bureaucrats often do not suffer any detrimental consequences because governmental programs often have monopoly status; there is no competition that forces government to be cost efficient or productive. For example, many claim that one of the major reasons why public schools do an inadequate job of educating inner city youth is because of the absence of alternative schools.

In private business, those who contribute most to a firm's profits usually get paid the highest salaries. In government or civil service jobs, everyone at the same grade level receives the same pay regardless of whether they perform well or badly. Moreover, civil service protection against dismissal, which was instituted in order to end the evils of the spoils or patronage system under which someone newly elected to political office would fire all existing employees regardless of whether they were doing a good job and replace them with his cronies, often has the unintended consequence of making it extremely difficult to dismiss even lazy and incompetent workers. The separation of effort and rewards in government service often leads to lack of incentive to do a good job.

The recent trend toward privatizing government services, in order to increase efficiency, may also foster a return of the spoils system through the back door by linking privatization and patronage. Private firms that provide services to government may make political campaign contributions to elected officials who decide which firm to hire, so-called pay to play.

Regulatory Capture and Revolving Door

Often those who staff public agencies expect to leave government and work for the same firms they were formerly regulating. This revolving door, which creates conflicts of interest when public servants gauge the effect of their current regulatory actions on their prospects of future employment, is known as *regulatory capture*. For example, a lawyer for the Securities Exchange Commission (SEC), who is supposed to investigate and seek sanctions against insider trading by executives who illegally take advantage of privileged, private information, may be reluctant to act aggressively if he or she wishes to be employed by the same executives in the future.

Conflicting Objectives

Profit-making firms usually have one object, which is to make profits; they can pursue this object with single mindedness. Government often has conflicting objectives and goals. Private schools can focus exclusively on educating students and can dismiss those who are disruptive of good order; however, public schools cannot dismiss disruptive students easily because they have at least two goals, which are to educate students and to socialize students, so they will be decent citizens; sometimes these goals conflict. If disruptive individuals cannot be dismissed easily, the education of students who want to learn suffers.

PUBLIC SCHOOLS, CHARTER SCHOOLS, AND VOUCHERS

As discussed above, public financing of education (as distinct from public provision of education) is required due to externalities and public goods type market failures. The competitive market cannot supply universal education: given the high cost of education and substantial income inequality in the United States, reliance on the market would result in many parents being unable or unwilling to educate their children. As discussed above, parents who would most likely not educate their children are Rawls's worst-off (poorest) group.

Although market failure theory demonstrates that education must be publicly financed, it does not require that it be publicly provided. Many critics of public schools claim that a major reason why they do an inadequate job of educating inner city youth is due to the absence of alternative schools; that is, they claim government provision of education is an example of government failure due to monopoly. As a remedy, they propose two alternatives to public schools: 1) students should be given *vouchers*, which are monetary stipends that they can use to attend either public schools or private schools, or 2) students should be given a choice between traditional public schools and *charter schools*, which are public schools that are freed from bureaucratic regulation, but whose continued existence is dependent upon student performance.

In order to level the playing field of competition among private schools that accept vouchers, charter schools, and public schools and not disadvantage the worst-off group, three conditions must be satisfied. First, there should be equal financing per pupil. Private schools should not be allowed to charge tuition and fees in addition to vouchers. Otherwise, the voucher program will merely be a subsidy for the rich and a drain on the public treasury. For example, if a student voucher is worth $5,000 and a private school's tuition is $25,000, so that a student is charged a voucher plus $20,000, only the relatively wealthy can afford such a school; the net result of such a voucher program is the draining of public funds for the benefit of the rich, which program surely disadvantages the worst-off group. Moreover, competition between a private school that spends $25,000 per student and a public school that spends $5,000 per student is hardly fair or on a level playing field. The authors of an economic model concluded that a pure voucher program would lead to stratification of students by ability and income.[10]

Second, students who are to attend private schools with vouchers or charter schools should be selected randomly by a lottery from all students (as opposed to being selected from the subset of students who have actively involved parents who choose to enter the lottery); allowing private schools with vouchers or charter schools to cherry pick the best students is neither fair competition with public schools nor a model for how public schools can improve the lot of the worst-off students.

Third, private schools and charter school should not be able to dismiss disruptive students or students whose academic performance is low in order to boost scores on achievement tests, leaving public schools with "poor" students (poor in both performance and income).

If private schools, which accept vouchers, and charter schools are allowed to violate these three rules, the rationale for public funding of education, which is to eliminate negative externalities that occur from not educating the poorest and most disadvantaged children, is undermined.

NOTES

1. See James Buchanan, Robert Tollison, and Gordon Tullock, eds. *Toward a Theory of the Rent-Seeking Society* (College Station: Texas A&M University Press, 1980).

2. "The Tragedy of the Commons," *Science* 162 (1968): 1243–1248.

3. The current dispute over "choice" between public and private education, which is discussed below, is concerned with how education should be provided, not whether education should be publicly funded or subsidized.

4. The Food Additives Amendment of 1958 amended the Food, Drugs, and Cosmetic Act of 1938.

5. The two largest clearing houses in the United States are the American Society of Composers, Authors, and Publishers (ASCAP) and Broadcast Music Incorporated (BMI).

6. With respect to motion pictures (but not radio and television), the blanket licensing system was held to violate the antitrust laws, after which the Department of Justice negotiated a consent decrees with ASCAP and BMI. The blanket licensing arrangement has also been the subject of numerous private antitrust suits.

7. 464 U.S. 417 (1984).

8. Jack Hirschleifer, "The Private and Social Value of Information and the Reward to Innovative Activity," *American Economic Review* 61 (1971): 561–574.

9. Mancur Olson, *The Logic of Collective Action: Public Goods and the Theory of Groups* (Cambridge, MA: Harvard Economic Studies, 1971).

10. Dennis Epple and Richard Romano, "Competition between Private and Public Schools, Vouchers, and Peer-Group Effects," *American Economic Review* 88 (1998): 33–62.

Chapter 9

Five Requirements for Competitive Markets

We next consider the social consequences of self-interested behavior on the assumption that commons problems, public goods, and negative externalities have been eliminated or at least controlled by appropriate laws and institutions; that is, there is a police force and a court system that prevents theft, fraud, and duress; there are environmental protection systems and social infrastructure such as streets, schools, libraries, public health measures, and banking, communication, and transportation systems. Thus constrained, self-interested behavior will be channeled into markets.

In this chapter, the theory of supply and demand in markets is explained; the theory predicts there is a tendency for price to move toward the equilibrium price and quantity, which is the price where the quantity of a good, say bananas, that buyers are willing and able to buy is equal to the quantity that sellers are willing to produce and sell. The tendency of price to move toward equilibrium price will be strong if markets are competitive; it will be weak if there are monopolies or firms in an industry form cartels and set prices collusively. If monopolies or cartels exist, individual self-interest will result in prices that are above the efficient, equilibrium level.

Economists usually specify five requirements that must be fulfilled for markets to be competitive:

1. Many buyers and sellers, so that no one buyer or seller is large enough to affect price.
2. Products are standardized, fungible, or at least good substitutes for each other.
3. Buyers and sellers have good or complete information, that is, they know prices of products, services, labor, and materials and where they are available.
4. Resource mobility; there are no barriers to entry or exit in any industry.
5. Negative externalities are internalized via taxation or regulation.

Absence of any one of the five requirements can cause markets to be inefficient, in which circumstances, government intervention may be required.

If a market is roughly characterized by these five necessary conditions, it is at least workably competitive.

The chapter begins by explaining the theory of supply and demand and determination of equilibrium price and quantity; market surplus is distinguished from consumer surplus. The case against price controls by government is stated. The five requirements of competition are stated and discussed; the reasons monopolies and cartels are inefficient and why they might exist are examined. Then, the requirements, for products to be good substitutes for each other, for consumers to have good information, and for resource mobility, are discussed.

DEMAND CURVE

A demand relation or *demand curve*, which is depicted in Figure 9.1, shows the maximum quantity of a good or service that consumers are willing to buy at different prices over a stated time period. Since price and quantity demanded are usually inversely related, demand curves have negative slopes. This inverse relationship between lower (higher) prices and higher (lower) quantity demanded is called the *law of downward sloping demand* or law of demand for short. There are two reasons why demand curves are downward sloping: income effect and substitution effect. First, consider the income effect. If the price of bread declines, an individual with a given income can afford more bread at the lower price; so, the effect of a price reduction is like an increase in her income. If she previously bought bread, she may now buy more bread without increasing her total expenditure; even if she did not buy bread at the old price, buying bread is now easier at the new, lower price. This is the *income effect* of a price change. A price reduction causes a positive income effect, which is likely to result in some increase in demand for a good; a price rise causes a negative income effect, which is likely to result in some decrease in demand for a good.

Second, consider the *substitution effect*. If the price of bread declines, an individual who used to buy rice, potatoes, or yams at the old price of bread may now buy bread at the lower price instead of substitute goods.[1] A price decline causes a positive substitution effect, which is likely to result in some increase in an individual's demand for a

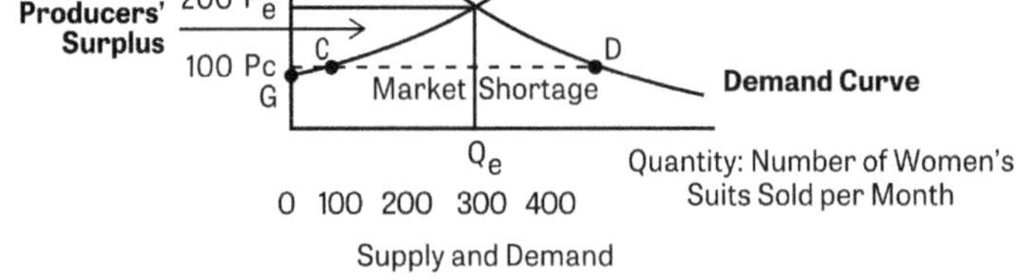

Figure 9.1 Supply, Demand, and Equilibrium Price. *Source*: Created by the author.

good; conversely, a price rise causes a negative substitution effect, which is likely to result in some decrease in an individual's demand for a good as he or she shifts to consumption of close substitutes. Due to income and substitution effects, a price decrease will usually increase quantity demanded and a price increase will usually decrease quantity demanded. In sum, the law of downward sloping demand is explained by income and substitution effects.

The law of demand is not confined to goods with explicit prices. Broadly speaking, the law of demand means that rational individuals respond to incentives. Consider two examples: unpopular teachers sometimes try to increase class enrollments by raising students' grades; other things being equal, hard graders have smaller class enrollments than easy graders. Second, an increase in either the severity of punishment or the likelihood of apprehension and conviction will usually reduce the amount of crime.

SUPPLY CURVE

A supply relation or *supply curve*, which is depicted in Figure 9.1, states the maximum quantity that sellers would be willing to sell at different prices over a given time period. Cost of supplying any quantity of a good or service depends upon cost of production, which is determined by prices of inputs (such as labor, materials, and overhead or fixed costs), technology (whether hand tools or high tech.), and the size of the production units (small business, or mass production on assembly lines). Supply curves usually slope upward from left to right, which implies that there is a positive relationship between cost and quantity supplied; that is, a larger quantity will only be supplied at a higher price per unit.

There are two reasons why per unit cost usually rises as quantity increases: law of diminishing returns and law of increasing costs, both of which were discussed in chapter 1. The law of eventually diminishing marginal returns states that expanding production in an existing factory or farm of fixed size will become increasingly costly. The law of increasing costs states that when resources are transferred from production of one product to the production of a technologically different product, the cost of producing a marginal unit of the product whose output is expanding generally increases because resources tend to be specialized, at least in part, so that when resources are transferred from doing what they are relatively good at doing to what they are relatively bad at doing, some of their productivity is lost.

EQUILIBRIUM PRICE AND QUANTITY

Supply and demand curves for a product, say women's suits at a department store, are depicted together in Figure 9.1. Although demand and supply curves are not visible to the store manager, it is easy to explain why she will be aware of the consequences of supply and demand and the approximate relationship of the curves to each other. Assume that the manager receives a new shipment of suits from a manufacturer and that she decides to sell each suit for a price of $400. The suits, each of which carries a $400 price tag, are put on racks on the sales floor. One month passes and only a few suits are sold. Although the manager cannot see the supply and demand curves, she knows that at a price of $400 per suit, supply exceeds demand because few suits are selling; that is, there is a *market surplus*, which is defined as the excess of quantity supplied over quantity demanded at a given price and is depicted by the line segment A-B in Figure 9.1. The store manager will realize that she must reduce price below $400 in order to sell more suits per month. As price is reduced more consumers buy suits and market surplus will begin to shrink.

What would have happened if the manager had originally priced each suit at $100? Again, although she cannot see supply and demand curves, she would observe that her department would have a hard time keeping up with demand for suits; suits would be "flying off the rack" because at $100 per suit, consumers believe these suits are a good buy. At a price of $100 per suit, there is a *market shortage*, which is defined as the excess of quantity demanded over quantity supplied at a given price and is depicted by line segment C-D. She will realize that price is too low. As price is raised, fewer consumers buy suits.

The theory of supply and demand predicts that there is a *tendency* for price to move toward the *equilibrium* price, P_e, which in this market is $200 per suit and equilibrium quantity, Q_e, which is 300 suits per month, as depicted in Figure 9.1. At this price, the number of suits buyers are willing and able to buy is equal to the number of suits that sellers are willing to produce and sell. A price of $200 per suit and a quantity of 300 suits per month is the equilibrium price and quantity.

In terms of social welfare, the equilibrium market price has three attractive properties: One, at equilibrium price, *markets clear*, that is, the quantity demanded by buyers is equal to the quantity supplied by sellers, so there are neither shortages nor surpluses. Two, at equilibrium price, what buyers pay for a unit of product is equal to the extra (marginal) cost to society of producing one more unit of that product. Three, there is an efficient resource allocation, which means there is no way to rearrange production so as to produce products that consumers desire more cheaply.

One of the major advantages of organizing production through markets, which is a decentralized production system, is that markets economize on the amount of information that any one person, whether buyer or seller, needs to know. For example, in order to be successful in business, a person need only be familiar with her or his particular business and industry; he or she need have little or no knowledge of other industries or how the economy as a whole functions. Moreover, a great deal of information concerning supply and demand is impounded in the price of a product or input: a high price informs buyers that the product's production requires a great deal of society's resources, is desired by many competing buyers, or both; a high price informs sellers that substantial profit is to be had by someone who can produce this product at less than the current market price.

CONSUMERS' SURPLUS DISTINGUISHED FROM MARKET SURPLUS

Recall that market surplus is defined as the excess of quantity supplied over quantity demanded when price is higher than equilibrium price. The concept of market surplus, which is depicted by line segment A-B in Figure 9.1, is not to be confused with the concept of consumers' surplus, which is one of the major benefits of competitive markets and is explained as follows:

A demand curve depicts the quantity consumers would be willing and able to buy at every price; this means that some consumers would be willing to pay a higher than equilibrium price, P_e. In Figure 9.1, quadrangle-like area, $0FEQ_e$, represents the total amount that consumers would be willing and able to pay for quantity Q_e of this good. Rectangular area, $0P_eEQ_e$, represents the actual amount consumers have to pay for this good at the efficient price and quantity (that is, product price multiplied by quantity sold equals total sales revenue of sellers). Subtracting the amount consumers actually pay from the amount they would be willing to pay leaves triangle-like area P_eEF, which is *consumers' surplus*; it represents the amount that consumers would be willing to pay but do not have to pay due to competition among sellers for buyers. Consumers' surplus is a major social welfare benefit from markets. (If sellers were able to engage in perfect price discrimination, that is, charge each buyer the maximum amount he or she would be willing and able to pay for each unit of product, sellers would capture the entire consumers' surplus; sellers' total revenue would be the rectangular-like area $0FEQ_e$.)

Another social benefit from markets is producers' surplus. Triangle-like area P_eEG depicts *producers' surplus*; it equals the difference between what sellers receive for their output, P_e, and the minimum amount that is necessary to cause them to be willing to sell their output, which is depicted by curved line segment, G-E, on the supply curve; that is, producers' surplus is the amount that sellers receive over and above the minimum price necessary to call forth their supply.

Competition among sellers causes price to decline to P_e, which benefits those consumers who are willing to pay more than the equilibrium price; and, competition among buyers causes price to rise to P_e, which benefits sellers who are willing to supply goods or services for less than the equilibrium price. The sum of consumers' surplus and producers' surplus, that is, the sum of the two triangle-like areas, $P_eEF + P_eEG = GFE$, is a measure of the social welfare benefit from markets.

The theory of supply and demand does not predict that price and quantity will actually be the efficient levels, only that there is a tendency toward the efficient price and quantity. There are a number of reasons why the tendency toward the efficient price may not occur; that is, there are reasons (discussed below) why the tendency to efficiency inherent in markets may be thwarted. It is of great importance for social policy to be able to predict whether unregulated markets will in fact tend toward the efficient price. If the tendency toward the efficient price and quantity is strong, the invisible hand theorem will be true, and a government policy of laissez-faire (which is defined as no government intervention in the economy beyond that necessary to eliminate market failures and create social infrastructure) is justified; but, if the tendency toward the efficient price is significantly impeded by factors discussed below, further government restrictions on self-interested behavior are warranted. In other words, without further assumptions about how markets work, it is still not possible to state that the invisible hand theorem, which implies that government should not intervene in markets, is true. These further assumptions concern the number of market participants, product characteristics, information, and resource mobility. Before discussing these factors, the case against price controls by government is stated next.

CASE AGAINST GOVERNMENT PRICE CONTROLS

Government intervention in the form of legislation that sets maximum or minimum prices causes shortages, surpluses, black markets, and often requires rationing and queuing. For example, assume government enacts a law that requires suits to be sold at a price that is above the equilibrium market price; this minimum price is a price floor, P_f, in Figure 9.1. Because price is above the equilibrium level, a surplus will result; the surplus is represented by line segment A-B. Sellers of suits who cannot find buyers for the surplus suits at the legal price may become desperate for money and violate the law by offering to sell suits for less than the legal minimum price, unless the government agrees to be a buyer-of-last-resort. Thus, minimum price legislation not only causes surpluses, it can be enforced only if government supports such legislation as buyer-of-last-resort with funds from the treasury. This means that citizens as consumers must pay a higher than equilibrium price for suits, and as taxpayers, they must also finance the government's buying and storage of surplus suits.

Now, assume government enacts a law that sets a maximum price, say a maximum price on suits, requiring that suits be sold at a price that is significantly below the equilibrium market price; this maximum price is a price ceiling, P_c, in Figure 9.1. Because price is below the equilibrium level, a shortage of suits will result; the shortage is represented by line segment C-D in Figure 9.1. As a result of the shortage, queues develop in which customers must wait in order to be served on a first-come, first-served basis. When products are rationed by queues as well as money, a product has a double price, its monetary price plus its price in terms of the time a consumer must wait before being allowed to buy the product at the legal price.

It is likely that many potential buyers of suits, who do not want to wait in queues, may be willing to violate the maximum price law by offering to buy suits at a higher than legal price. When buyers offer to pay more than the legal price or suppliers charge a premium over the legal price, a secondary or "black market" develops. In order to prevent the development of black markets, a regulatory agency or police force must be established to ensure that the maximum price law is not violated.

In order to prevent black markets and to deal fairly with the hardship from shortages due to price ceilings, a system of rationing, which allocates goods by a non-price mechanism, may also have to be instituted. Under rationing each buyer must have a coupon as well as money to buy the good—in effect there are two kinds of money or two prices. Since different individuals have different views of equity, establishing fair criteria for the distribution of coupons is often very difficult. Should suit rationing coupons be apportioned according to family size, occupational need, individual or family income, or some other more or less equitable criterion?

The case against government intervention in markets to set prices is that price controls impair or destroy the three attractive properties of competitive markets: one, markets do not clear, that is, there are shortages or surpluses because quantity demanded by buyers is not equal to quantity supplied by sellers; second, consumers pay a price that is either greater than or less than marginal cost of production; third, resources are inefficiently allocated, which means that too little or too much of the price controlled good is produced in comparison to what would be produced by competitive markets.

FIVE REQUIREMENTS FOR COMPETITIVE MARKETS

The tendency toward efficient price and quantity will be strong if markets are competitive. Economists usually specify five requirements that must be fulfilled for markets to be competitive:

1. Existence of many buyers and sellers, so that no one buyer or seller is large enough to affect price. One of the main reasons why competition is beneficial for society is that no one person or firm has power over another. If a seller tries to charge a price that is exorbitant, the buyer can say, "You're not the only seller of widgets," and can buy widgets from one of the myriad other sellers who are willing to sell at the market clearing, efficient price. So, too, if a buyer attempts to coerce a seller into selling a product or service at an exceedingly low price, the seller has myriad other buyers to whom he or she can sell the product at the efficient market price. If condition 1) is not satisfied, that is, there are one or few sellers or one or few buyers, price will usually be above or below the efficient market equilibrium price.
2. Products are standardized, fungible, or at least good substitutes for each other. If products are good substitutes for each other in terms of function and quality, then the only factor that consumers consider is price, in which case, if condition 1), existence of many buyers and sellers, is satisfied, price will be at the efficient level. On the other hand, if products are *differentiated* because consumers have brand loyalty, sellers of differentiated products have, to a greater or lesser extent, some power to raise price above the competitive level. For example, if consumers prefer Coca Cola or Pepsi Cola to Brand X, Y, or Z Colas, the former two sellers can charge a price above the competitive level.
3. Good or complete information means that all buyers and sellers know all relevant prices and where products, services, labor, and materials are available. In competitive markets, fraud and mistake are less likely due to the complete information assumption. Recall that in game theory, complete information means that both players know entire game matrix so bluffing or fraud is impossible. However, if information is incomplete, especially if information is asymmetric so that some players have private information not known to others (discussed in chapter 16), there can be serious market imperfections, inefficiencies, and even market failure.
4. Easy entry or resource mobility means that there are no barriers to entry in any industry. If sellers are making large profits, the resource mobility condition along with complete information condition guarantees that new sellers will enter the industry in order to seize the opportunity to earn high profits; new entrants will continue to enter until prices are competed down to the efficient level and profit is reduced to the normal rate of return on capital (opportunity cost).
5. Either there are no negative externalities or externalities are internalized via taxation or regulation.

If a market is roughly characterized by these five necessary conditions, it is workably competitive. Assuming market failures are eliminated by legal institutions so that appropriate social infrastructure exists, the theory of supply and demand plus the five requirements discussed in the previous paragraph are known as the model of competition. Competitive markets have five socially attractive properties, the first three of which were discussed above: one, markets clear because supply equals demand; two, product prices equal marginal cost of producing a unit of product; three, there is efficient resource allocation; four, buyers and sellers benefit from consumers' surplus and producers' surplus; and five, competitive markets have the attractive political or democratic characteristic that no person or firm has power over any other person or firm; competition decentralizes both economic and political power.

Absence of any one of the five conditions for competition can cause markets to be inefficient, in which circumstances, government intervention may be required. Each of the necessary conditions will be examined below, starting with the first condition, the number of buyers and sellers; the effect of one or a few sellers on economic welfare will be emphasized.

MONOPOLY, MONOPSONY, OLIGOPOLY, AND CARTELS

The tendency of market price toward the efficient level will not occur if there is only one seller of a product. A single seller of a product is a *monopolist*, whose power to raise price above the efficient level depends upon its ability to reduce supply so as to create an artificial shortage. The ability of a firm to sustain price above the efficient level is called *monopoly power*. The greater a seller's ability to raise price above efficient level, the greater the monopoly power. A monopolist's power to raise price depends upon closeness of substitute products; fewer substitutes means greater monopoly power. For example, a monopolist of trucks would have less power to raise price than a monopolist of motor vehicles because there are more close substitutes for trucks, say cars, than for motor vehicles, which is a more general product market. The broader a monopolist's market, the more power it has to raise price.

Monopolies are bad for consumers in three ways: First, a monopolist restricts output below the efficient level by creating an artificial shortage, which allows it to raise price above the equilibrium price. Second, a monopoly misallocates resources; too little of the monopolized good is produced; and, the excess resources not used by the monopolist are employed in competitive industries, which means that too much is produced in competitive industries. Restriction of output by monopoly misallocates resources. Third, the monopoly price, which is higher than efficient price, allows a monopolist to transfer some consumers' surplus to itself; that is, there is a redistribution of income from consumers to the monopolist. In addition, the rise in price and restriction in output by a monopolist causes some consumers' surplus to be destroyed, which is called *dead weight loss*. Empirical studies on the welfare costs of monopoly, dead weight loss, in the United States are inconclusive: the results vary considerably from less than one percent to more than 13 percent of GDP. However, even a relatively small percent of GDP represents a considerable amount of resources.[2] In addition, the transfer of consumer surplus to monopoly profit can amount to a significant redistribution of income.[3]

Just as a monopolist or single seller is able to raise price by restricting supply so as to create an artificial shortage, a *monopsonist* or single buyer is able to lower price by restricting demand, so as to create an artificial surplus. To understand the possible magnitude of the income redistribution effects of monopoly and monopsony, consider a famous example of monopsony from professional sports. Before the 1970s, professional baseball players were bound by a "reserve" clause in their contracts, which stipulated that after a player signed his first professional contract with a team he could not play for any other professional team unless that team sold his contract to another team. Since there was only a single (monopsony) buyer for an athlete's services, the reserve clause caused athletes' salaries to be relatively low. After the reserve clause was abrogated in the 1970s, professional baseball players, who become free agents after an initial period of six years, could sell their services to the highest bidder in a competitive market. Salaries of free agents in professional sports have soared even though the competitive nature and performance in professional sports leagues has hardly changed (see chapter 11). That is, the abrogation of the reserve clause, which reduced the monopsony power of teams (power to depress wages), has had little impact on efficiency but has had a tremendous impact on the distribution of sports income; that is, there has been a substantial transfer of producers' surplus from monopsonist team owners to players. The redistribution of income due to transfer of consumers' surplus to monopolists is also probably significant and substantial.

Why do Monopolies Exist?

Why do monopolies exist? Why are there no alternative sellers in some markets? There are several reasons: an individual may be the owner of a scarce resource, such as a mineral, like iron or gold, or a talent, like unusual athletic or managerial ability. A second reason for the existence of monopolies has to do with the relationship between the most efficient size firm, which depends upon both technological considerations (production by hand or by mass production) and the extent of the market (neighborhood, city, regional, national, or international market).

Whether a particular activity will be a monopoly depends upon the size or extent of the market in relation to a product or service's technology. The retailing of groceries is a simple example. A single grocery store in a desert community of a hundred inhabitants is surely a natural monopoly. Only one grocery store can exist in such a small community because the extent of the market is not big enough for two grocery stores to operate profitably. The monopoly store will have some power to raise price above the efficient level. The limit on such power to raise price is determined by the cost of buying and transporting groceries from an alternative source to the desert community. On the other hand, a single grocery store in a large metropolis is not a natural monopoly and has no power to raise price above the efficient level because it has many competitors due to the large market.

The power to raise price, monopoly power, is not all-or-nothing, but exists on a continuum. Although usually not classified as a monopoly, a convenience grocery store located several miles from the nearest supermarket may have some weak monopoly power, that is, the power to raise price slightly. Customers are willing to pay slightly higher prices for the convenience of not having to travel several miles to a supermarket. The store is not called a "convenience" store for nothing! Thus, a firm need not literally be a monopoly in the single seller sense to have some or even a great deal of monopoly power, which is the power to raise price above the competitive price. For example, in the 1960s, International Business Machines (IBM), although not a monopoly in the single seller sense, sold approximately 75 percent of the main frame computers and was thought to have had substantial monopoly power, power to raise price.

Economies of Scale and Natural Monopoly

Monopoly often exists because there are vast economies of scale in an industry. Economies of scale in an industry means that a large firm can produce a unit of output more cheaply than a small firm can. If a large firm can produce a unit of output more cheaply, it can underprice small firms and drive them out of business; at the limit, one large firm may be able to supply an entire market more cheaply than two of more firms. Such a firm is called a "natural monopoly."

Consider the following example of economies of scale, which is called the *container principle* and depends on two geometric definitions learned in grade school: circumference of a circle, $C = \pi D$, and area of a circle, $A = \pi r^2$. Query? Is it more efficient to build two oil pipelines to transport oil, each of which is one foot in diameter, or one oil pipeline, which is two feet in diameter? The cost of building a pipeline is a function of the amount of steel in the pipeline, that is, its circumference, which is given by the formula, $C = \pi D$ times its length. Thus, the circumference of a pipeline with a one foot diameter, 1′, is $C = \pi\,(1') = 1'\pi$ times length; the combined circumference of two 1′ pipelines is $2(1'\pi) = 2'\pi$ times length. Similarly, the circumference of a pipeline with a 2′ diameter is also $2'\pi$ times length. Thus, cost of building two pipelines with 1′ diameters is roughly equal to cost of building one pipeline with a 2′ diameter.

However, the amount of oil that can be pumped through a pipeline is a function of the area of a circle, which is given by the formula, $A = \pi r^2$. If a pipeline's diameter is 1′, its radius is ½′ and area of a cross section is, $A = \pi(\,½')^2 = (1/4')\pi$. Thus, two 1′ pipelines have a combined area of $2(1/4')\pi = (½')\pi$. A pipeline with 2′ diameter has a 1′ radius; area of a cross section is, $A = \pi(1')^2 = 1'\pi$. Thus, combined cost of constructing two pipelines, each with a 1′ diameter is roughly the same as the cost of constructing one pipeline with a 2′ diameter, but twice as much oil can be pumped through a 2′ pipeline as can be pumped through two 1′ pipelines; alternatively, 2′ pipeline can pump four times as much oil as 1′ pipeline.[4] Thus, the larger the pipeline, the cheaper is the cost to transport oil. Oil pipelines are natural monopolies. This means that having competition in shipment of oil by pipeline is neither efficient nor feasible.

The container principle illustrates the concept of economies of scale, which is defined as a production process in which a proportionate increase in all input costs (e.g., double input costs) results in a *less* than proportionate increase in total output cost (less than double total output costs); economies of scale translates into lower costs per unit of output. Increases in scale are usually accompanied by changes in technology: that is, changes from hand tools to assembly lines to assembly lines with robots. If there are economies of scale such that one firm can supply an entire market at less cost than two or more firms, that market is a natural monopoly.[5]

Government regulation of natural monopoly is often required to prevent consumers from being exploited by self-interested natural monopolists, who, if unconstrained, would charge consumers the monopoly price.

Oligopoly

Oligopoly is defined as competition among a few sellers. Oligopolies exist for two reasons: First, economies of scale may be large enough so that large firms have a significant cost advantage over small firms, but not so large that one firm can supply the entire industry efficiently, that is, economies of scale are not so great as to result in natural monopoly. Second, it has been argued that oligopoly, that is, competition among a few firms, is an unnatural market structure, resulting from the US antitrust (anti-competition) laws; without antitrust laws, which prohibit mergers that result in monopolies, industries with a few competitors would probably either merge together to become a monopoly or one firm would eventually drive others out and become a monopoly.[6] Examples of industries with an oligopoly market structure are computers, automobiles, airlines, "big" oil, beer, and warehouse grocery stores.

There are many models and some empirical evidence that suggests the market structure of oligopoly, a few firms, results in prices that are higher than the efficient price but lower than monopoly price. For example, oligopoly pricing can be modeled as a Repeated Prisoners' Dilemma, which suggests that when there are few competitors, all realize that aggressive price cutting (non-cooperative strategy) is not advantageous because any price cut by one firm will be quickly matched by other firms due to *mutual interdependence recognized* (retaliation for non-cooperative behavior will be swift). On the other hand, as in any Repeated Prisoners' Dilemma model, cooperation can easily break down, so occasional price wars will occur.

However, if there is a great deal of technological change in an industry characterized by oligopoly, as, for example, in the computer industry, the conclusion that oligopoly will result in higher than efficient level prices may not be true because tacit collusion among firms engaged in technological competition is unlikely to be sustained; if technological change is rapid, firms realize that falling behind technologically is fatal. Price is less important than being first with the next innovation.

There is also some empirical evidence that an industry with few sellers, that is, oligopoly, is more conducive to technological change than is an industry with many small sellers, that is, competition, because each of many small sellers many not be able to fund research and development, which costs are often huge; oligopoly may also foster more technological change than monopoly because a monopolist may be too lazy or have too little incentive to do research and development because it does not feel threatened by competitors.

Cartels as N-Person, Repeated Prisoners' Dilemmas

Explicit agreements by two or more competitors to restrict output and raise price, as distinct from an *implicit* agreement or tacit understanding inherent in oligopoly, are called *cartels*, price fixing, or collusion. Firms enter into cartels in an attempt to mimic the behavior of monopolists by banding together in order to restrict output so as to raise price and increase profits above the equilibrium level. Cartel members agree to charge a higher than the efficient price, which agreements violate federal antitrust law and are therefore illegal and unenforceable.

As discussed in chapter 7, competitive markets are N-Person, Prisoners' Dilemmas. Cartel behavior is an example of N-Person, Prisoners' Dilemma in which the competitors are trying to achieve the cooperative solution, which also happens to be illegal. For the cartel to be successful, all members must fulfill their pledges to restrict output, so that an artificially created shortage can be sustained, causing the market price to stay above the efficient price level. However, each self-interested member of a cartel will be tempted to cheat on the agreement, realizing that if all other members of the cartel restrict output as agreed, the single firm that cheats on the agreement and does not restrict its output can make a great deal of profit; if all members cheat on the agreement, believing erroneously that all other members will honor the agreement, a cartel breaks down and price sinks back to the efficient level. As the number of cartel members increases, the likelihood that at least one firm will cheat on the agreement and increase output beyond that allowed by the cartel, which will cause price to drop and create a tendency for the cartel to break down, becomes increasingly likely.

On the other hand, because each day the cartel agreement is sustained can be thought of as another game, or repeat play, a cartel can also be modeled as an N-Person, Repeated Prisoners' Dilemma, in which trust, due to consistent collusion, may be built up over time. Whether the tendency for cartels in a one-shot Prisoners' Dilemma to break down or the possible tendency for cartels in a Repeated Prisoners' Dilemma to be sustained will dominate is difficult to predict. Unlike the Prisoners' Dilemma model of externalities, where the cooperative solution is socially desirable, the cooperative solution to Prisoners' Dilemma model of cartels is socially undesirable. In other words, just as we hope that there is no honor among thieves, we also hope that cartels will break down. The most famous cartel is the Organization of Petroleum Exporting Countries (OPEC); its recent inability to control the supply of oil due to increasing world production of oil is well known.

STANDARDIZED PRODUCTS OR GOOD SUBSTITUTES

To the extent that consumers regard products produced by different manufactures in the same industry as good substitutes for each other, the limiting case of which is where all manufacturers produce standardized, homogeneous, or fungible products, consumer choice among products will be based solely on price and competition will tend to drive market price toward the efficient level.[7] This tendency toward the efficient price will be impeded in industries in which consumers regard products of some manufacturers as being differentiated, that is, having no good substitutes. The fewer and less satisfactory are the substitutes for a product of a particular manufacturer, the greater is its monopoly power. Thus, market price may not tend to the efficient price in industries where products are not good substitutes for each other. For example, if consumers regard Apple computers as being substantially different and preferred to Brand X computers, the former will have some power to raise price above the efficient level, which power may be significant.

WELL-INFORMED CONSUMERS

The question of whether products in an industry are approximately standardized, that is, have good substitutes, is closely related to the question of whether consumers have enough information to compare and evaluate products. The tendency toward efficient level will be impeded if buyers are not well-informed about what products are available, where they can be bought, at what prices, and whether good substitutes exist. In addition, self-interested sellers of products and services often have an incentive to misinform buyers.

For example, consider nutritional labeling on food products prior to standardization of information being mandated by government; consumers faced a bewildering array of information, misinformation, and outright fraud. By altering portion size, food manufacturers could manipulate information on calories per serving so as to mislead and confuse consumers. Food manufacturers also often made unsubstantiated health claims. Both consumer interest groups and honest food manufacturers applauded the federal government's promulgation of mandatory nutritional information and food labeling standards. If consumers are well-informed, manufacturers compete on a level playing field because misinformation will not be rewarded.

"[I]n the presence of imperfect information or incomplete markets, the economy will not be Pareto efficient; in other words, there will always be some intervention by which the government can make everyone better off."[8] As examples of government intervention that improves information, consider the Textile Fiber Products Identification Act of 1958, under which a label must identify each fiber used to manufacture a clothing product and the percentage of total weight accounted for by each fiber must be stated; the Fair Packaging and Labeling Act of 1966, under which the net quantity in a product must be stated in a uniform and prominent location on the package and expressed in a unit of measure appropriate to the product; the Consumer Protection Act of 1968, under which creditors are required to inform debtors of the annual percentage rate (APR) of interest as calculated on the outstanding balance as opposed to misleading debtors by calculating interest on the original balance, which later method understates the true interest rate by approximately 50 percent; the Magnuson-Moss Warranty Act of 1975, under which the conditions of a written warranty must be stated "in simple and readily understood language" and manufacturers of products that include warranties must specify the length of the warranty, the period being not less than a year, and whether the warranty is *full*, which means that charges for repair and replacement during the warranty period are minimal or nil, or the warranty is *limited* for stated product failures; Dodd-Frank Wall Street Reform and Consumer Protection Act (2010) under which Consumer Financial Protection Bureau is tasked with promoting fairness and transparency in mortgages, credit cards, and other consumer financial products and services; the Uniform Commercial Code, under which sales of goods automatically include implied warranties of merchantability (average quality) and fitness for particular purpose (goods chosen based upon seller's skill and judgment) unless such warranties are disclaimed in conspicuous language;[9] state statutes or insurance-department regulations which stipulate provisions in standard form fire or health insurance policies; and, the "plain language" statutes of many states.

Similar considerations justify judicial scrutiny and government efforts to monitor standardized contracts between consumers and large corporations so as to promote disclosure and to regulate terms of product warranties, insurance contracts, rental agreements, and installment sales contracts.

If product quality cannot be observed by buyers prior to purchase, markets may not achieve the efficient price due to the problem of asymmetric information—the seller knows whether the product is of good quality or defective, but the buyer does not find out until after he or she buys and uses the product. In markets with repeat purchases, sometimes called experience goods, say groceries and toiletries, this problem can be overcome because sellers can establish and maintain reputations for quality products.

However, if products are bought infrequently, sellers' reputations are unknown, warranties are non-existent, and product quality cannot be observed by buyers prior to purchase, defective products, called "lemons," may make up a large part of the market.[10] Government regulations, which require mandatory disclosure of defects in used cars, that product warranties be clearly written, and that warranty disclaimers be clear and conspicuous, are justified in markets where there are significant imperfections due to asymmetric information.

Lack of adequate information also justifies government safety standards for consumer products; standards of purity, safety, and efficacy for food and drugs; and standards for health and safety in the work place. Information inadequacies experienced by workers who are often uninformed about job-related risks (such as job-related risks in asbestos factories), justify the existence of government efforts to improve work place information.

RESOURCE MOBILITY: EASY ENTRY AND EXIT

The tendency of market price toward the efficient level will be impeded to the extent that resources of land, labor, and capital cannot move freely from one industry to another and from one region to another; that is, the tendency toward the efficient price requires resource mobility or free entry and exit from industries. When resources are mobile, individuals allocate their resources in markets and regions where they will get the highest rate of return. If price and profits are substantially above the efficient level in some industries or geographic regions, entrepreneurs will shift resources to those markets so that price will tend to drop to the efficient level and profit above opportunity cost will be competed away. The rapid shift of resources from one industry to another means that there is easy entry and exit. If there are barriers to entry, discussed below, which prevent resources from moving to markets where the rate of return is highest, resources will not be mobile and price will tend to stay above the efficient level.

The easy exit aspect of resource mobility also means that there are no *sunk costs*, that is, assets that have no other uses than the current one. For example, compare starting an airline with starting a railroad. The major cost of an airline is its airplanes. If the airline is unsuccessful, airplanes can be sold to some other airline, so that it may be possible to recoup much of the initial investment. However, if a railroad is unsuccessful, although railroad engines and cars can be sold, its investment in right of way, roadbeds, tracks, and depots represent sunk costs that have few or no profitable alternative uses. If exit from an industry is not easy due to sunk costs, such costs make investors reluctant to enter and so are barriers to entry.

There are several other barriers to entry that impede resource mobility: First, unique resources or skills may be required for production in a particular industry, such as an essential mineral resource; those who do not have access to such mineral encounter a barrier to entry that keeps them out of that industry. However, over time, the advance of knowledge and technological change often reduce barriers to entry due to a scarce resource. For example, although the lack of access to iron ore was at one time a formidable barrier to entry into steel manufacturing, the advent of steel alloys, recycling, and substitutes like aluminum, and plastics has reduced the effect of this barrier.

Second, if an entire market can be supplied by one efficient size firm, then there are technological or natural monopoly barriers to entry. Over time, the natural monopoly barrier to entry also may tend to erode due to technological change. For example, the communications industry, which was once a natural monopoly (telephone), is fast becoming an intensely competitive industry due to microwave transmission of telephone calls, fiber optics, computer routing and storing of information, and satellite communications.

A third and possibly the most persistent barrier to entry is government-created monopoly. Much tariff protection and quotas on foreign importation of goods are explained as special interest barriers to entry rather than national revenue-raising measures. Government barriers to entry are examples of government failure.

Consider the relationship between lack of consumer information and government-created barriers to entry in professional licensing requirements. On the one hand, the lay public cannot easily judge the qualifications of physicians, lawyers, pharmacists, and the like; moreover, incompetent practitioners in these and other professions can create severe externalities, as, for example, a physician who misdiagnoses an epidemic disease. Thus, government may need to determine some quality of service characteristics, for example, by setting licensing standards, which is justified due to externalities and the public good aspects of information. On the other hand, governments usually allow licensing boards to be dominated by members of the profession being regulated rather than require that a majority of a board's members be from outside the profession. The result is that under the guise of raising standards in the profession, many professional licensing boards use their power to restrict entry (supply) in order to raise the profession's income to the detriment of consumers.[11] The principle that should be observed with respect to government regulation of markets, especially state and local regulation, which tend to be more easily captured by special interests, is that such regulation should be the absolute minimum scope necessary to rectify market failures.

CONCLUSION

As indicated above, the theory of supply and demand is too general to predict whether the tendency toward the efficient price will actually be realized without some government intervention. What additional requirements in addition to the control of market failures and provision of infrastructure are necessary so that the invisible hand theorem will be true and a government policy of laissez-faire is justified? Economists usually specify five requirements (including control of externalities):

1. Existence of many buyers and sellers;
2. All products have good substitutes;
3. All buyers and sellers are well-informed about all relevant markets;
4. Possibility of resource mobility or easy entry and exit;
5. Either presence of no negative externalities or externalities are internalized through taxes or regulation.

The theory of supply and demand plus the five requirements are known as the model of competition. Competitive markets have four very attractive properties: First, markets clear; that is, there are neither shortages nor surpluses because the quantity demanded by buyers equals the quantity supplied by sellers. Second, price per unit consumers pay is equal to the marginal cost of producing a unit in all markets. Third, there is efficient resource allocation, which means there is no way to rearrange production so as to produce products that consumers desire more cheaply. Fourth, the sum of consumers' surplus and producers' surplus, which is a measure of the increase in social welfare due to competitive markets, is maximized.

If any of the five necessary requirements for competition are substantially unfulfilled, markets may not be self-correcting, and some government intervention may be required to improve social welfare. However, if the five requirements of competition are satisfied, with or without government intervention, it is almost possible to say that individual self-interest as expressed through markets will maximize social welfare. There are two caveats: the need for government regulation of insurance markets, which is discussed in chapters 15 and 16, and the fact that short-lived, self-interested individuals often make decisions that have serious irreversible, long-term negative consequences, which is the subject of the next chapter.

NOTES

1. Goods that are "substitutes" are contrasted with "complements," which are goods that are usually consumed together like coffee and cream, automobiles and gasoline, spaghetti and sauce, and so on. If two goods are complements, a reduction in the price of one good causes an increase in the quantity demanded for the other, and a rise in price of one good causes a decrease in the quantity demanded for the other. Independent goods, such as popcorn and yachts, are unrelated goods such that a change in the price of one has no effect on the quantity demanded of the other good.

2. Don Waldman and Elizabeth Jensen, *Industrial Organization: Theory and Practice* (New York: Routledge, 1997), 39–40.

3. National Association of Attorneys General, *Horizontal Merger Guidelines of the National Association of Attorneys General* (Washington, DC: Bureau of National Affairs, 1987), §2.

4. The container principle also applies to warehouses and buildings in general, which are essentially cubes. Doubling the length and width of a side of a warehouse or building causes

the cost of the area of a building to rise by a factor of 4, but its cubic capacity rises by a factor of 8.

5. W. Kip Viscusi, John Vernon, and Joseph Harrington, Jr., *Economics of Regulation and Antitrust*, 3rd ed. (Cambridge, MA: MIT Press, 2000), 339–344.

6. Donald Dewey, *The Theory of Imperfect Competition: A Radical Reconstruction* (New York: Columbia University Press, 1969).

7. Paradoxically, product standardization also may facilitate collusion; if all firms sell identical products, cartel members, who do not have to determine which firm will be allowed to sell its product for a higher price, have one less issue that could cause disagreement among them.

8. Joseph Stiglitz, "The Private Uses of Public Interests: Incentives and Institutions," *Journal of Economic Perspectives* 12 (1998): 3.

9. Uniform Commercial Code, § 2–316(2).

10. George Akerlof, "The Market for Lemons: Uncertainty and the Market Mechanism," *Quarterly Journal of Economics* 84 (1970): 488–500.

11. H. Craig Petersen, *Business and Government*, 4th ed. (New York: HarperCollins, 1993), 434.

Chapter 10

In the Long Run We Are All Dead

This chapter is concerned with whether decisions made by short-lived, self-interested individuals are consistent with long-run (intergenerational) social welfare. Assume a man and woman are in conversation and one says, "I'm lucky. I have no parents, so financing Medicare isn't a problem, and I have no children, so the climate change isn't a problem."

Consider two examples of current decisions made by self-interested persons that involve intergenerational equity: why should a self-interested person living today be concerned with whether the ozone layer, which protects the earth from the sun's ultraviolet rays, will exist a century from now? Who cares whether the Brazilian rain forest is depleted in the distant future when we are all dead? Decisions, some of which are irreversible, which are made by short-lived, self-interested individuals, are often not in the interest of future generations.

The concept of present discounted value or present value for short is explained; economists and business persons use this method to discount income that an investment or asset produces in the future; the sum of discounted future annual returns is an asset's present value. As shown below, if a high, market interest rate is used to discount future income by the present value method, income received more than thirty years in the future is nearly worthless, so the future disappears. When current decisions have intergenerational ramifications, say those involving wildlife, virgin forests, global warming and climate change, the present discounted value approach is contrasted with a sustainability approach. Although "in the long we are all dead,"[1] nations live in the long run, so society ought to take a longer view than that inherent in private markets where the value of future income and future costs are determined by decisions of short-lived, self-interested individuals.

MALTHUS VERSUS MARKET INDUCED SUBSTITUTES FOR DEPLETABLE RESOURCES

In 1798, Thomas Malthus published his *Essay on the Principle of Population*[2] in which he argued that population tends to grow faster than food supply: by analogy, he said that food supply grows by an arithmetic progression (1, 3, 5, 7, . . .) while population grows by a geometric progression (2, 4, 8, 16, . . .). The reason he thought that population would tend to outstrip food supply was that the law of diminishing returns applies to increases in food, especially on the relatively small island of England where Malthus lived, whereas population growth is subject to no such check. Since population growth cannot outstrip food supply for long, Malthus argued that the two must necessarily be kept in balance by the ravages of war, famine, and pestilence; he did not put much credence in birth control or abstinence from sex as a means of controlling population growth because he believed that the sex drive is so strong as to be ungovernable for the majority of mankind. No wonder that after reading Malthus's *Essay*, Thomas Carlyle called economics "the dismal science."

Time has proven Malthus's prediction wrong, at least in the western world. It is ironic that Malthus, who lived in the midst of the Industrial Revolution which began around 1750, did not foresee that technological change in the form of mechanized farm machinery, fertilizer, and hybrid seeds would eventually revolutionize agricultural methods, increasing its output manyfold. (In the United States, the percent of the population engaged in agriculture went from above 90 percent before the American Revolution to considerably less than 3 percent currently.)

Economists derisively refer to any argument that predicts crippling long-run shortages or depletion of resources as Malthusian. The standard view is that if a particular resource begins to be depleted, its price will rise, giving strong incentives for entrepreneurs to discover ways to economize on its use and to find substitutes for the scarce resource. With respect to energy resources, the shift over the past several centuries from wood to coal to oil, along with the current search for a cheap gas, wind, solar, or nuclear energy as a substitute for oil, seems to fit this pattern. Query: Will the change to renewable energy occur quickly enough to avert disastrous climate change and global warming? Will someone invent a technology that will mimic the way trees remove carbon dioxide from the atmosphere?

PRESENT DISCOUNTED VALUE LAYS WASTE TO THE FUTURE

Just as Carlyle reproached nineteenth-century economists for their over pessimistic or dismal view of the future, Garrett Hardin, in "Dr. Pangloss meets Cassandra,"[3] chided twentieth-century economists for their over-optimistic view of the future. He said that economists'

> standard technique of "discounting" the future . . . lays waste to the real future. . . . What counts most is what income the land can bring in right now. High interest has the effect of virtually destroying the future—in the economist's calculations. . . . A society that listens only to economists ratchets its way to destruction. . . .
>
> Economic libertarians and doctrinaire free-market economists who concede no limits to the simple-minded method of discounting the future are today's providentialists. Pure economics will, in their view create the best of all possible worlds. We need another Voltaire to write a new *Candide*.[4]

Hardin's argument has two steps: one, the higher the interest rate that is used to discount the future, the shorter the private market's view of the future; two, if, based on the market's short view of the future, irrevocable decisions are made with respect to resources that in hindsight become more valuable in other uses, the market will have made the wrong decision. Consider the potential long-term damage from destruction of rain forests as well as ocean acidification, global warming, and climate change that results from burning fossil fuels. If such destruction occurs as a result of the decisions of millions of short-lived, self-interested individuals, one solution is enact regulations or tax economic activities that have huge negative externalities that are only manifest after a number of generations (see chapter 8). Society should, as a matter of intergenerational equity, take a longer view than that inherent in the future discounted by a high market interest rate.

In order to explain Hardin's argument more clearly, how economists and business persons discount future annual income that an investment asset or resource produces will be explained; the sum of such discounted annual returns is an asset's present discounted value or present value of the asset. Then, the effect of different interest rates on present value and social welfare are considered.

PRESENT DISCOUNTED VALUE OF A FUTURE INCOME STREAM

As an example of the present discounted value of a future income stream, consider the choice one has after winning a lottery. Assume someone wins a \$2 million lottery, the payoff from which is \$100,000 a year for 20 years; the winner may be given an option of foregoing the annual payments and taking a current lump sum of \$1,250,000. The current lump sum is the present discounted value of the 20-year income stream. How is present discounted value determined?

Assume someone offers to sell you a business that earns a profit of \$1 per year forever starting today or sell you an annuity that pays \$1 per year forever starting today (\$1 is merely illustrative, it could be \$10,000 or \$1 million per year). How much should you pay to buy this business or this annuity now; that is, how much should pay now for the right to receive an annual income stream starting today of \$1, \$1, \$1, \$1, \$1, . . . to infinity? Answer: In order determine the current value of this infinite constant income stream, we must compare each payment to its *opportunity cost* which is the income stream that would be generated if each annual payment of \$1 after today's payment earned interest in a bank up to the year in which it was received. Starting today, the infinite income stream begins with \$1. Next, if \$1 is deposited in a bank for one year at 5% interest, it would be worth \$1.05 a year from now. The calculation is: \$1 + (.05)(\$1) = \$1.05. If \$1 is deposited in a bank for two years at 5% interest, it would be worth \$1.05 + (.05)(\$1.05) = \$1.103 two years from now; if \$1 is deposited in a bank for three years at a 5% interest, it would be worth \$1.103 + (0.5)(\$1.103) = \$1.158 three years from now; and so on.

Compare a) the constant annual income stream of \$1 per year starting today with b) the income stream where each annual payment after the first earned interest in a bank at 5% interest until withdrawn by dividing a) by b): \$1/\$1, \$1/(\$105), \$1/(\$1.103),

\$1/(\$1.158), ... The result of division is the infinite series 1, .95, .91, .86, ... In other words, \$1 today is the same in both income streams. But \$1 received a year from now is worth only 95% of the sum one would get if the money was left in a bank at .05 interest for a year; \$1 received two years from now is worth only 91% of the sum one would get if \$1 was left in a bank at .05 interest for two years; \$1 three years from now is worth only 86% of the sum one would get if \$1 was left in a bank at .05 interest for three years; and so on.

This infinite series 1, .95, .91, .86, . . . can be written as 1, .95, $(.95)^2$, $(.95)^3$. The *sum* of this infinite mathematical series $1 + .95 + (.95)^2 + (.95)^3 + \ldots$ approaches a surprisingly simple limit: 1/.05 = 20 (where .05 is the interest rate).

Thus, when compared to an income stream that would be generated if each annual payment earned interest at .05 in a bank until withdrawn, the present value of a constant income stream of \$1 per year forever has a value of only \$1/.05 = \$20. This formula is true for any annual amount of money at .05 interest; for example, the present value of \$100 a year forever at .05 interest equals \$100/.05 = \$2,000.

The present value formula for an annual income stream of any amount of money at any interest rate is $PV = A/r$, where PV stands for present discounted value, A stands for constant annual payment, and r stands for the interest rate. For example, the present value of annual payments of \$100 forever at .10 interest is, \$100/.10 = \$1,000.

The higher interest rate, the larger a sum of money would be if each annual payment earned interest in a bank until withdrawn, which means that the higher the interest rate, the less a constant annual income stream is worth because more interest is forgone. Conversely, the lower the interest rate, the more a constant annual income stream is worth because less interest is forgone. We conclude that is there is an inverse relationship between the interest rate and the present discounted value of an annual income stream to infinity. Some examples follow.

Effect of Market Interest Rates on Present Discounted Value

Consider the present value of an income stream of \$1 per year to infinity for interest rates of 2%, 5%, 10%, and 15%.

$$PV = A/r = \$1/.02 = \$50$$
$$\$1/.05 = \$20$$
$$\$1/.10 = \$10$$
$$\$1/.15 = \$6.7$$

Note the inverse relationship between the interest rate and the present value of an annual income stream to infinity. If a low interest rate, 2%, is used, a future income stream is worth 7 ½ times as much as a future income stream discounted with a high interest rate, 15% (\$50 vs. \$6.7)

As a rough rule of thumb, many businesses use 15% as the minimum rate of return (before taxes) that is necessary to justify an investment. At 15% interest, the value of a \$1 annual payment in year 34 is worth less than .01 of the value of \$1 deposited in a bank at .05 for 34 years; that is, at a .15 interest rate, the future at 34 years and beyond is nearly totally worthless and disappears. At 10%, it takes 50 years before the present value of $1 per year is worth less than .01 of the value of $1 deposited in a bank at .10 for 34 years; at 5%, it takes approximately 95 years before the future is totally discounted; and, at 2%, it takes approximately 233 years before the future is totally discounted. In other words, the higher the interest rate that is used to discount the future, the less a self-interested decision maker would consider the long-run consequences of her decisions.

Why do many private businesses discount future income at the relatively high interest rate of 15%, which totally disregards consequences beyond 34 years? Business investment is inherently risky: new products and technologies make the old obsolete, consumer tastes change, and predicting business conditions more than 34 years from the present is totally speculative. Moreover, if a business has not recouped its investment in 30 years, it probably never will. Discounting the future with a high interest rate by business persons is short-term rational and efficient.

INTERGENERATIONAL EQUITY: DISCOUNTING THE FUTURE VERSUS SUSTAINABLE HUMAN WELFARE

When intergenerational matters are under consideration, such as the long-run effects of global warming, ocean acidification, and endangered species, the way in which the private market discounts the future, which is based on the self-interest of short-lived individuals, is likely to be detrimental to the welfare of future inhabitants on planet earth. In this section, the present discounted value approach to intergenerational equity is contrasted with the sustainability approach, in which discounting of the future is eschewed.

Utilitarianism, which dates back at least to Jeremy Bentham and John Stuart Mill, among others, attempts to maximize a social good. Pareto efficiency, Kaldor-Hicks, and marginal cost-benefit efficiency are utilitarian concepts; they attempt to maximize total income of currently living individuals. One of utilitarianism's biggest drawbacks is its total indifference to inequality, whether such inequality is intragenerational or intergenerational. Chapter 4 was concerned with Rawls's attack on utilitarianism's intragenerational inequality; he argued that Pareto, Kaldor-Hicks and marginal cost-benefit efficiency result in a society in which income is so unequal as to be neither fair nor just.

In the context of intergenerational equity, utilitarianism's indifference to inequality comes into even sharper focus because, depending upon assumptions, it could require *entire* early generations, not just some classes within a generation, to sacrifice their wellbeing for later generations, or vice versa. The next two sections present a comparison of the present discounted value and sustainability approaches to intergenerational equity.

Maximizing Total Income, Economic Growth, and Discounting the Future

Assume that the intergenerational social goal is utilitarian, that is, the goal is to maximize the total income (more abstractly to maximize total utility) of mankind over time, starting from now (because the past cannot be changed) to an infinite number of generations into the future, without regard to intergenerational inequality. Assume further that technological change proceeds at a moderate but steady pace, so that if society does not consume all its current production, but saves and invests some of its resources, economic growth will occur over time. Given the goal of maximizing total income over all generations, early generations should save as much as possible, that is, significantly reduce their standard of living so as to increase the standard of living in future generations; such a program would maximize total income or utility over all generations. Incidentally, income of the last generation would also be maximized. A fine intergenerational inequality indeed!

Although a majority of US citizens generally accept utilitarian goals (Pareto, Kaldor-Hicks, and marginal cost-benefit efficiency), which are indifferent to current intragenerational inequality, few would accept a utilitarian goal that requires a major sacrifice in wellbeing from *entire* current generations for the wellbeing of latter generations who will not exist until far in the future. How can this scenario be prevented?

In order to prevent this scenario, in which early generations sacrifice a great deal of their welfare for latter generations from occurring, economists have traditionally discounted future income with the present value method. However, as shown in the previous section, if a high or market rate of interest is used to discount future income, maximizing total income over generations becomes very present oriented, that is, favors early generations over later ones. If the future is discounted with a high interest rate, so that future income beyond thirty of fifty years is worthless according to present discounted value, the welfare of future generations can be totally disregarded. As Harden said, discounting the future based upon a market determined interest rate lays waste to the future. Utilitarianism's indifference to inequality allows it to sanction a sacrifice of either current or future generations' welfare.

Now add greenhouse gases into the analysis, so that there are three main variables in the utilitarian maximization problem: potential economic growth, discounting future income, and greenhouse gases that result in global warming and climate change. In a model constructed by William Nordhaus in which future income and costs of greenhouse gases are discounted with a rate of interest that is market determined, doing little about greenhouse gases until the middle of the twenty-first century when costs of global warming and climate change, even if discounted, rise considerably, is justified.[5] Delayed action on climate policy is also supported by the assumption that technological progress will, in addition to economic growth, produce green technology even if there is no specific investment in it, so that if we wait until the future, global warming will be cheaper to combat.[6]

If mankind waits until the middle of this century to combat global warming as suggested by a model in which the future is substantially discounted, whether there will be time enough to slow the rise in temperature or whether a technological fix will be found so that little need be done, no one can know. However, continual increases in global temperature and the proliferation extreme weather events should cause reasonably prudent persons to be less sanguine about a future in which nothing is done about global warming until the middle of the twenty-first century.

In another study on the intergenerational economics of climate change by Nicholas Stern, discounting the future was justified on the probability that our species, like many species before us, will become extinct in the future.[7] Discounting the future based upon the probability of human species extinction has two specific problems: First, no one knows what the probability of extinction is or how to measure it. Second, if our actions significantly influence the probability of extinction, a catastrophic event or a tipping point could occur; as may be the case with global warming. As the probability of one of these events increases due to global warming, ever stronger attempts to combat global warming become economically justified.[8] Since the rate of interest used by Stern to discount future income and costs is low, he concluded that benefits of prompt and strong action on climate outweigh costs.

If the rate of interest used to discount the future is small enough, whether the future is discounted or not will be immaterial; early action will be justified. The strong policy differences by Nordhaus and Stern are largely accounted for by the difference in interest rates with which the future is discounted.[9] To repeat, if market determined interest rates are used to discount the future, delay or inaction on climate change is recommended; if low rates of interest are used to discount the future, the future is highly valued and global warming and climate change will be combated much sooner and with urgency.

Sustainable Human Welfare

Although it is prudent and moral for living individuals to discount their future income (in the long run we are all dead), do we, the living, have a moral right to discount the future of unborn generations? Non-economists, like Hardin, have long argued that discounting of the future is both bad economics and immoral with respect to our descendants; this normative argument has begun to be accepted by some mainstream economists who eschew both the goal of maximizing total income or utility over time and discounting the future; they choose instead the goal of sustainable human welfare.

The goal of sustainability is related to Rawls' maximin principle, in which a change is positive if it benefits the worst-off group. The sustainability goal can be interpreted in two ways: unidirectional, that is, forward looking only, in which every future generation has the right to be as at least as well off as its predecessors,[10] or bidirectional, in which current generations also have the right to be as well off as later generations.[11] However, the problem with the bidirectional sustainability goal is that if current generations are required to be as well off as later generations, economic growth would be ruled out (even if population grows, per capita income would remain constant).

There are two ways out of the no-growth constraint that limits bidirectional sustainability. One is to assume that current generations favor human development and so do not enforce their right to be as well off as future generations; they agree to reduce their current welfare to permit future growth. In other words, current generations agree to give the "gift of growth." Two, current generations agree to interpret sustainable human welfare unidirectionally.

Rawls was concerned with both *intra*generational and *inter*generational consequences of saving for the future. Rawls did not agree to a rate of saving for the future that will maximize the welfare of the least well-off generation, unless the consequences of saving on current intragenerational equity is also considered; that is, in deciding how much to save for the benefit of future generations, the effect of saving on the least-advantaged group within current generations must also be considered. Should a society choose to increase its savings rate so as to benefit future generations by increasing current income inequality, that is, by making the lot of the current least-advantaged group even worse? Rawls would be unlikely to agree to cut food stamps for the poor in order use those resources to save and invest for future growth. He combined the just saving principle, due regard for future generations, with his maximin principle, due regard for the least-advantaged groups within any generation, by "supposing that this [savings] principle is defined from the standpoint of the least advantaged in each generation."[12]

Lavador, Roemer, and Silvestre (LRS) constructed a model in which sustainable, balanced economic growth is possible over generations, without discounting, because of productivity in the education sector.[13] Investment in education, which increases over time, supports balanced growth in two ways: First, education propels technical change (including green technology) and material economic growth. Second, education is an end in itself, a positive good, which people both want and have the capacity to consume more of as generations advance, so that later generations simultaneously increase their consumption of education (in all forms, including the arts) and (substitute) reduce their want for material goods. Given a finite planet earth, if individuals' appetite for an ever increasing amounts of material goods is not curbed, the law of diminishing returns would prohibit long-term, balanced growth (earth is of fixed size) and predicts possible disaster. In a sense, increasing demand for education as a consumption good and decreasing consumption of material goods obviates the need for crude discounting or dismissal of the future.

LRS then incorporate global warming and climate change in their model. They accept the scientific consensus that the nations of the world should attempt to limit the rise in global temperature to less than a 2°C from current average temperature and limit concentration of CO_2 in the atmosphere to less than 450 ppm (parts per million of air).[14] In order to meet those climate goals, their model shows that nations of the world will have to invest in green technology and reduce their rate of economic growth, which, of course, will not be without social costs.

NOTES

1. "Long run is a misleading guide to current affairs. In the long run we are all dead." John M. Keynes, *A Tract on Monetary Reform* (Cambridge: Cambridge University Press, 1923), ch. 3.
2. The full title of which is *An Essay on the Principle of Population as It Affects the Future Improvement of Society* (Cambridge: Cambridge University Press, 2016).
3. *The New Republic* (October 28, 1981), 31–34.
4. *Ibid.*, 34.
5. William D. Nordhaus, *A Question of Balance: Weighing the Options on Global Warming Policies* (New Haven, CT: Yale University Press, 2008), 14; Humberto Llavador, John E. Roemer, and Joaquim Silvestre, *Sustainability for a Warming Planet* (Cambridge, MA: Harvard University Press, 2015), 189.
6. Llavador, Roemer, and Silvestre, note 5, 191.

7. Nicholas Stern, *The Economics of Climate Change: The Stern Review* (Cambridge: Cambridge University Press, 2011).

8. When the value of a variable in an economic model is determined by and within the model, it is an "endogenous" variable. If the value of a variable in an economic model is taken from outside the model, say the limit that earth temperature should not exceed, it is an "exogenous" variable.

9. Thomas Piketty, *Capital in the Twenty-First Century* (Cambridge, MA: Belknap, 2014), 568.

10. Robert M. Solow, "Intergenerational Equity and Exhaustible Resources," *Review of Economic Studies* 41 (1974): 29, 30.

11. Llavador, Roemer, and Silvestre, note 5, 264.

12. John Rawls, *A Theory of Justice* (Cambridge, MA: Belknap, 1971), 257–258.

13. Llavador, Roemer, and Silvestre, note 5, 87–88.

14. *Ibid.*, 133–137.

Part III

GENERAL CASE VERSUS HARD CASE METHODOLOGIES

Rottenberg, Coase, Calabresi, and Posner

Part III contains four chapters. In chapter 11, property rights and liability rules are defined; these rights and rules are used in legal irrelevance theorems, which state the conditions under which an efficient outcome is independent of legal rights or rules; that is, conditions under which more than one property right or liability rule will achieve an efficient outcome. The truth or falsity of legal irrelevance theorems or any economic theory is evaluated differently by the general case methodology of economics and the hard case methodology in law, which methodologies are often inconsistent.

Then, Rottenberg's theorem on the effect of a change in property rights on free markets is explained, after which its corollary on the effect of a change in liability rules on free markets is explained.

In chapter 12, the two Coase theorems (his actual theorem and the one attributed to him) on the effect of property rights or liability rules on the coordination of mutually interfering activities (negative externalities) is discussed.

In chapter 13, Judge Calabresi's four criteria for allocating accident costs (negative externalities) common to several activities are discussed.

Chapter 14 explains Judge Posner's economic analysis of common (judge-made); he showed that the law of contracts, property, and torts has an underlying economic unity that cuts across these doctrinally defined fields.

Chapter 11

Rottenberg's Theorem

Effect of a Change in Property Rights on Free Markets

In this chapter, property rights and liability rules are defined; these rights and rules are used in legal irrelevance theorems, which state the conditions under which an efficient outcome is independent of legal rights or rules; that is, conditions under which more than one property right or liability rule will achieve an efficient outcome. The truth or falsity of legal irrelevance theorems or any economic theory is evaluated differently by the general case methodology of economics and the hard case methodology in law, which methodologies are often inconsistent.

Next, Rottenberg's theorem on the effect of a change in property rights on free markets is explained. Rottenberg compared a labor market for baseball players governed by the "reserve clause," which gave baseball team owners monopsony (single buyer) power over players, with a labor market in which players are free agents; that is, under free agency, players have a property right in their own labor services and can contract with any of many teams who are in competition with each other for their services. He concluded that regardless of whether baseball players are bound by the reserve clause or are free agents, the best players would end up on rich, big city teams that most value them; that is, an efficient outcome is independent of property right regimes. He also concluded that the distribution of income is substantially different with a change in property rights because the disfavored party will incur a large loss of income: under reserve clause, players incur a large income loss; under a shift to free agency, team owners incur a large income loss.

Rottenberg's legal irrelevance theorem on the effect of a change in property rights in free markets is shown to be consistent with both the general case methodology of economics and the hard case methodology of law. By contrast, in the next chapter, the two Coase theorems (his actual theorem and the one attributed to him) are shown to be "true" in the context of economics' general case methodology because they are insightful but false in law's hard case methodology because they are not useful to judges in specific cases. Moreover, for clarity of thought, Rottenberg's theorem concerning free markets and Coase's theorems concerning mutually interfering activities (negative externalities), which are often conflated, should be distinguished clearly. Conflating Rottenberg's and Coase's theorems results in an economic theorem that is insightful due to its broad generality but false in law because it is not useful in actual cases.

Rottenberg's article also contains a theory or model of the optimal distribution of talent among teams in a professional baseball league, or any professional sports league. This model is of interest because it is a Prisoners' Dilemma, in which teams both compete and cooperate. The model is consistent with a major theme of this book, which is that outside of free markets, unconstrained individual self-interested behavior is often less efficient than joint self-interested behavior or cooperative behavior.

Then, a corollary to Rottenberg's legal irrelevance theorem concerning the effect of a change in liability rules (rather than property rights) on free markets is stated and three examples, which involve *default* liability rules from the UCC Article 2, Sales (of Goods), are discussed: one, liability rule for goods that are lost or damaged in transit from seller to buyer; two, liability rule for the time at which payment of goods is incurred; and three, liability rule for defective goods, known as the perfect tender rule. In each of these cases, the UCC's Article 2 default liability rules that favor sellers are compared to alternative legal liability rules that favor buyers. In these examples, the party who bears the risk of loss will, as a result of contracts made in the context of free markets, bargain for a compensating price differential equal to any benefit conferred or any cost saved by the party who is favored by the legal liability rule; that is, the efficient outcome is independent of legal liability rule and does not result in any change in the distribution of income.

Two exceptions to the corollary on the effect of a change in liability rules, and the reasons why, are discussed below and in subsequent chapters. One exception involves the shift from negligence-based liability for employee, on-the-job injuries to workers' compensation; the other involves the shift from negligence-based liability for defectively manufactured products to strict liability. Because there is a principled way to distinguish specific cases for which the corollary is true, that is, an accurate guide to public policy, from the two exceptions for which it is false, the corollary is both insightful and useful in specific cases.

In sum, in both Rottenberg's theorem for property rights in free markets and its corollary for legal liability rules, different legal regimes result in efficient outcomes; however, the difference is that when property rights are changed, the party who is favored by the property right gains income while the disfavored party loses income, whereas when legal liability rules are changed, the distribution of income is not altered.

PROPERTY RIGHTS AND LIABILITY RULES DEFINED (PLUS INALIENABILITY RULES)

Calabresi and Melamed showed that in deciding any civil (private) lawsuit, a court must make two decisions: First, it must decide which party is entitled to win; this decision is one of determining *entitlement*. Second, the court has to decide whether to protect the winning party's entitlement by a *property right*, often enforceable by injunction that forbids others from interfering with one's property, or by a *liability rule*, enforceable by damages.[1]

For example, if someone possesses an entitlement to an object, such as an automobile, that entitlement is usually protected by a property right. Another person who wants to transfer the legal entitlement to herself can do so only if she pays a price satisfactory to the person who currently owns the entitlement to the property. In other words, an entitlement protected by a property right can be transferred legally only via a voluntary transaction, that is, by a Pareto improvement.

The price at which the property right is sold includes the monetary equivalent of the entitled party's subjective or sentimental value of the object. If the market value for someone's automobile is $10,000, but the owner would not voluntarily part with it for less than $20,000 due to his sentimental attachment to it, another person who wanted that specific automobile would have to agree to the owner's price in order to get legal entitlement to it. One of the property rights of an owner of real property or tangible personal property (chattels) is the right to prevent another from interfering with or trespassing on the property. Courts often enforce property rights with injunctions; the trespassing party is enjoined or forbidden to interfere with the other party's property right unless he or she pays a price satisfactory to the other party. The tort (civil wrong) of trespass against property is also enforced by both injunction and assessment of damages against the trespasser.

On the other hand, if a party has an entitlement that is protected only by a liability rule, another party may destroy or take that entitlement and pay compensation at the market price for equivalent property; if the parties cannot agree on the market value, price will be determined by a court. For example, a party who destroys another's automobile in a collision due to her or his negligence must pay for the automobile at the market price of equivalent property; and, if the parties cannot agree on the market price, the price will be determined by a court.

Consider these examples: 1) If the automobile collision, discussed in the previous paragraph, also resulted in personal injury and lost income, a liability rule would, in addition to property damage, cover hospital bills, lost wages, and loss of life. 2) If a buyer of a product, which is dangerously defective, is injured or her property is damaged or destroyed due to the product's defect, as where an automobile's brakes fail and it crashes into a garage, the owner is entitled to hospital bills, lost wages, and market value of property damaged. 3) If goods, which are being transported by seller to buyer, are destroyed due to seller's negligent packing, seller may be legally liable for lost profits (consequential damages) incurred by buyer due to delay in receiving substitute goods. 4) If a worker is injured on the job, he or she is protected by a workers' compensation liability rule that entitles him to a percentage of lost wages. 5) If government uses its power of eminent domain to take property for public use, it must compensate the owner at the market price for similar property; and, if the parties fail to agree on a price, the amount will be determined by a court using a liability rule.

An entitlement protected by a liability rule does not include the entitled party's personal or subjective valuation of the object. Continuing the automobile example discussed in the previous paragraphs, if the owner's automobile was destroyed in a collision as a result of another's negligent driving, a court would require the responsible party to compensate the owner in the amount of the market value, $10,000 (not $20,000 which includes the owner's subjective value). An entitlement protected by a liability rule is usually worth significantly less in monetary terms than is an entitlement that is protected by a property right because it does not include the plaintiff's subjective value.

In sum, if property is protected by a property right, it can be legally transferred only by a voluntary (Pareto) transaction. A person whose property is protected by a property right has a great deal of bargaining power; he or she can demand a price that is satisfactory to her or him. If property is protected by a liability rule, it can be destroyed or taken for market value as determined by a court. A person whose property is destroyed or taken and is protected by a mere liability rule has little or no bargaining power; a court will attempt to determine the fair market value of the property without regard for any subjective value the entitled party may have.

Calabresi and Melamed also said that entitlements may be protected by *inalienability rules*. An entitlement is inalienable to the extent that its transfer is not permitted between a willing buyer and a willing seller. For example, a person is not allowed to sell himself into slavery, sell a body part, or to enforce contracts with minors or others who are legally incompetent to make transactions.[2] Inalienability rules are not relevant to Rottenberg's theorem, the Coase theorems, or Calabresi's analysis of the efficient allocation of risk common to several activities, which are discussed in the next two chapters.[3]

LEGAL IRRELEVANCE THEOREMS

In this book, there are a number of theories, theorems, and corollaries that are defined in terms of property rights or liability rules,[4] the best known of which are the Coase theorems (Coase's actual theorem and the one attributed to him are both discussed in chapter 12), which state circumstances under which two or more property rights or liability rules, with seemingly different consequences, all result in economically efficient outcomes. In other words, these theorems conclude that an efficient outcome is independent of property rights or liability rules. They are called *legal irrelevance* theorems. This chapter is concerned with Rottenberg's irrelevance theorem on the effect of a change in property rights on free markets and its corollary on the effect of a change in liability rules on free markets.

Legal irrelevance theorems are extremely sensitive to assumptions and circumstances, such as whether a party's entitlement is protected by a property right or liability rule, whether parties act simultaneously or sequentially, and whether parties have complete or incomplete information. Legal irrelevance theorems are true under some assumptions but false under other assumptions. In addition, due to the different methodologies of economics and of law, the general case methodology of economics and the hard case methodology or law, each discipline answers the question of whether these theorems are true or false differently.

ECONOMICS' GENERAL CASE VERSUS LAW'S HARD CASE METHODOLOGIES

A law professor said: "The trouble with law and economics is that it's insightful but often not useful." The resolution of this paradox is provided by understanding the different, sometimes conflicting, methodologies of economics and of law.

Economics' General Case Methodology

In economics, if a theory, theorem, or empirical proposition explains, say, 80 percent of cases to which it is applicable, it is extremely insightful; it can also be said to be roughly "true." For example, the Head Start pre-school program was developed to help at-risk children from disadvantaged homes do well in school and later life. Although not all children have been able to overcome the disadvantages of their background as a result of this program, economic analyses tend to show that the benefits from the program outweigh the costs[5] (Kaldor-Hicks or marginal cost-benefit efficiency). In other words, as an empirical economic proposition, enough children are helped by Head Start so that the program is efficient and worthwhile. However, it is not possible to predict which specific children will be helped significantly and which will not be helped. In economics, if a particular instance or example conflicts with a theory, theorem, or empirical proposition, it is often said to be an outlier or anomaly, which can usually be ignored because it does not affect whether the theory or empirical proposition is correct in many or even the overwhelming majority of other cases.

Law's Hard Case Methodology

I repurpose Ronald Dworkin's theory of constitutional interpretation and use it as the methodological principle for testing the validity of economic propositions, theories, or theorems in law. Dworkin argued that in order to be a valid theory of interpretation of the US Constitution, a theory must pass the test of being able to explain and justify why a precedent, such as *Plessy v. Ferguson,*[6] which held that state laws mandating racial segregation of public facilities (separate railway cars) was constitutionally permissible under the doctrine of "separate but equal," could be overruled legitimately by *Brown v. Board of Education,*[7] which held that racial segregation in public schools was unconstitutional on grounds that "separate education facilities are inherently unequal." Dworkin argued that the theory of constitutional interpretation known as "historicism" or "originalism" fails this test because it cannot explain and justify why *Brown* legitimately overruled *Plessy*; originalism fails as a valid theory of constitutional interpretation because it cannot explain change in law.[8]

In this context, *Brown* is what Dworkin called a *hard* case (chapter 5), which is a case that rises to the top of the appellate court pyramid because it involves conflicting principles or is interstitial with respect to those principles.[9] These cases rise to top appellate courts because they are difficult to decide. Hard cases are outliers or anomalies. Hard cases are distinguished from *easy* cases which are defined as cases that fall clearly within settled and uncontroversial rules of law. In short, economics is concerned with the general case whereas appellate courts are often concerned with hard cases, anomalies, and outliers.

In law, the truth or falsity of an economic theory or theorem will be tested by whether it is consistent with and can explain decisions in hard cases or anomalies. If an economic theory or theorem conflicts with the holding in a hard case that the legal community generally agrees to have been decided correctly, the theory or theorem cannot be used to decide (is not dispositive of) specific cases unless there is a principled way to distinguish specific cases in which it is true, that is, an accurate guide to public policy, from cases in which it is false. If there is no principled way of distinguishing cases for which the theory or theorem is true from those in which it is false, the economic theorem may be insightful in identifying a general tendency but it is not useful to judges in specific cases. As shown in this and the next chapter, Rottenberg's theorem and corollary are true in both the methodologies of economics and of law, but Coase's two theorems are true in the general case methodology of economics because they are insightful but are false in the hard case methodology of law because they are not useful in specific cases.

THE BASEBALL PLAYERS' LABOR MARKET

Rottenberg's article, "The Baseball Players' Labor Market,"[10] which was published in 1956, four years before Coase's article (discussed in the next chapter), concerns an example which is easily generalized into a theorem concerning the effect of a change in property rights on free markets. In addition to the theorem, his article also contains a theory or model of the optimal distribution of talent among teams in a professional baseball league, or any professional sports league, which can be modeled as a Prisoners' Dilemma.

Baseball's Reserve Clause Versus Free Agency

When Rottenberg wrote his article, a baseball player's right to play for any team ended when he signed his first professional contract; the reserve clause in this and subsequent contracts bound him to play for his current team or a team to which his contract was sold. The reserve clause, by creating a monopsony for player services, significantly reduced the compensation of a major league baseball player below what it would be in a free market.

Major League Baseball argued that the reserve clause's purpose was to equalize talent among teams. However, Rottenberg determined that from 1920 to 1951, the New York Yankees were first in the American League eighteen times; the Chicago White Sox were never first. So, the reserve clause did not distribute talent evenly.

In 1975, after several court decisions, arbitration awards, and collective bargaining between players and management, the reserve clause was abrogated in baseball (subsequently in other professional sports as well); currently, after six years, baseball players become free agents, which means they are legally allowed to negotiate with any team for their services.

EFFECT OF A CHANGE IN PROPERTY RIGHTS ON FREE MARKETS

Rottenberg arrived at two conclusions: one, whether the labor market for players' services included the reserve clause (each player can only negotiate with a single team for his services) or players are free agents in a competitive market for their services is irrelevant to the efficient outcome; in either case, players will end up with the team for which the player's services produce the most value. Since teams in big cities have larger markets and greater revenue than teams in small cities, the rich, big city teams can buy the most talented players. In other words, twenty years before free agency became a reality, Rottenberg predicted that regardless of whether baseball players were bound by the reserve clause or they have property rights in their own services, the best players would end up on rich, big city teams.

Two, the distribution of income is significantly different under the two alternative property right regimes of reserve clause and free agency: "The difference is only that in a market subject to the reserve rule, part of value of the player's services is paid to the team that sells his contract, and part of the value is kept by the team that holds his contract; in a free market the player gets his full value."[11] So, under the reserve clause, teams own a share of a player's services from his labor; under free agency, a player owns, that is, has a property right in his own services.

Consider Alex Rodriguez (A-Rod), one of baseball's biggest stars: he started his career with the Seattle Mariners, became a free agent, signed a ten-year contract with the Texas Rangers for $252 million in 2000, was traded to the New York Yankees, and in 2007, exercised his option to get out of the remainder of his contract, and signed a second ten-year contract with the Yankees for $275 million. As Rottenberg predicted, A-Rod began his career in a small baseball market, went to a larger more lucrative market, and finally ended up with the Yankees, who have one of the biggest and most valuable markets. In other words, A-Rod ended up where his services are most valuable, on the Yankees, a big-market team. The difference is that under the reserve clause, the baseball team owners would have split the major share of the two ¼ billion dollar contracts among themselves while under free agency A-Rod owned the property right to his own services and received his full value.

Rottenberg's theorem is stated formally as follows:

1. *If* parties transfer goods or services in free markets,
2. law specifies which party is favored by a property right (e.g., players or teams),
3. *then* bargains will the achieve efficient use of resources; that is, the efficient outcome is independent of property rights, and
4. *the disfavored party will incur a loss equal to the benefit conferred or cost saved by the favored party.*

Effect of a Change in Property Rights on Copyright Markets

The scope of copyright protection for ideas expressed in tangible form extends only to the actual expression as opposed to the ideas underlying the expression, which are not protected. For example, two or more authors of say, accounting textbooks, usually explicate the same ideas, only the presentation differs; these textbooks are in competition with each other. Notwithstanding copyright protection, the market for similar or the same ideas expressed in tangible form is very competitive.

The US Copyright Act of 1790 granted copyright for a term of 14 years, with a right of renewal for another 14 years if the author was still alive at the end of the first term. Copyright protection was continually extended over the years: to periods of 28 and 14 years respectively, to two periods of 28 years, and then to life of the author plus 50 years. In 1976, copyright protection was extended to life of the author plus 75 years. In 1998, the Sony Bono Copyright Term Extension Act (CTEA) extended copyright protection for works published before 1978 and still under copyright on October 27, 1998, to life of the author plus 95 years.

The numerous extensions of the term of copyright protection in the United States are an example which fits assumptions of the theorem for the effect of a change in property rights on free markets: the extensions on the term of copyrights involve changes in property rights that probably have some effect on the distribution of income because they increase revenue on copyright works. In recognition of its purpose and effect on income from copyrights, CTEA is derisively called the Mickey Mouse Protection Act.

Do the extensions of copyrights impair efficiency? We live in the age of information, which was discussed as a public good in chapter 8. The marginal cost equals zero characteristic of information (non-rivalrous consumption), which can be digitalized and transmitted over the internet, suggests that a low price, close to zero, is efficient. However, copyright law is an attempt to overcome the difficulty of excluding free riders; if creativity is to be encouraged, copyright law suggests a price for individual works that is significantly greater than zero. In between these extremes, there are myriad intermediate positions depending on the scope of fair use, whether songs and information are paid for per use or by blanket license whereby users have unlimited use of information for a monthly fee, and the longevity of copyright protection. For example, assume that single songs sell for $1 per song and that an Apple iPod has a capacity to hold 5,000 songs. Is it efficient or distributionally fair for a person to pay $5,000 in order to have a personal library of songs? Markets for information are continually experimenting and seeking efficient ways to distribute information. Copyright extensions and/or collective bargaining between copyright owners and music distributors or internet providers may affect income distribution without greatly impairing markets for efficient information distribution.

Patent Rights Are Not Exchanged on Free Markets

The scope of patent protection is much broader than copyright protection. Patents, which protect any "new" and "useful" process, machine, matter, or improvement, give the holder government-protected monopoly right to the useful ideas underlying the patent; abstract ideas, like the general theory of relativity, are not protected. The term of protection, which is for twenty years, is much shorter than copyright protection due to the much greater potential impact of patent protection on efficiency. Extension of the term of patents beyond twenty years would have a potentially large negative effect on efficiency by restricting access to new and advanced technology as well as a large effect on income distribution. Patent protection is not within Rottenberg's theorem because patent monopolies are not traded in free markets where buyers have good substitutes.

Rottenberg's Theorem Is Consistent with Economic and Legal Methodologies

Even though some persons own scarce resources (monopoly), like the non-replicable skills of some baseball players, Rottenberg's theorem is true in free markets because persons or firms in free markets have alternative persons or firms with which to deal. If free markets are at least workably competitive, hard cases, which conflict with the theorem, are unlikely. In other words, Rottenberg's theorem is true in both the methodologies of economics and of law. As shown in the next chapter, Coase's theorems (his actual theorem and the one attributed to him) are "true" in the context of economics' general case methodology but are false in law's hard case methodology.

TEAM VERSUS LEAGUE REVENUE MAXIMIZATION IN PROFESSIONAL SPORTS

Rottenberg also formulated a theory of the optimal distribution of talent in a professional baseball league or any professional sports league, which theory can be modeled as a Prisoners' Dilemma because teams both compete and cooperate. His theoretical league consists of two teams, a rich team and a poor team, which are owned by one firm (joint management); their total revenue depends upon the number of customers in the two baseball stadiums.

Maximizing League Income Requires Roughly Equal Talent Distribution

Rottenberg assumes the cooperative goal for the league, maximization of combined league income (cooperative solution to the Prisoners' Dilemma); this goal requires that its rich team wins a few more games than its poor team, but that outcome of games be uncertain and talent be approximately evenly distributed, albeit slightly in favor of the big city team. Assume that management of the jointly owned teams allows the rich team to add star players while the players on the poor team remain constant. What will happen to the combined total revenue of both teams? Total revenue will have the usual mound shape: as the rich team begins to buy star players, combined revenue will at first rise, reach its apex, and then decline because it will become less and less uncertain as to which team will win any given game. In other words, the law of diminishing returns will set in at some point so that the addition of another star player to the rich team does not maximize league income. Also, a first star player for the poor team, by increasing uncertainty as to which team will win, will increase combined total revenue of both teams more than say, the third or fourth star for the rich team will add to combined total revenue. (Note that whether players are bound by a reserve clause or have a property in their own services is irrelevant to the optimal distribution of talent in a league.)

Maximizing Individual Team Income Concentrates Talent (Prisoners' Dilemma)

Now, instead of the league owning both teams, assume that a league has two *independently* owned teams, one rich and one poor, whose competition and cooperation is modeled by a Prisoners' Dilemma. When teams are owned independently, there is a conflict between individual self-interested teams and cooperative league goals; teams can either adopt a self-interested, non-cooperative strategy that maximizes individual team revenue or a cooperative strategy that maximizes combined league revenue.

If both adopted cooperative strategies, each would try to maximize combined league income as in Rottenberg's optimal talent distribution model above. However, consider possible non-cooperative strategies for rich team or poor team in which they attempt to maximize their own success rather than league success; also consider league strategies that counter uncooperative strategies.

Rich Team's Non-cooperative Strategies

Assume a rich team, let's call it the New York Yankees, adopts a non-cooperative strategy: it could either 1) maximize team revenue or 2) maximize the number of team wins (some recent research suggests that a team's core "fans" are more interested in their team winning than uncertainty of outcome[12]), which latter goal is even less consistent with Rottenberg's assumptions that outcomes of games be uncertain, differences in the quality of play among teams be not great, and talent be more or less equally distributed.

If teams maximize combined income, poor team's interests are *internal* to rich team's calculations; however, if rich team maximizes its own income, the negative effect of its increased revenue on poor team's revenue is *external* to its calculations. Rich team (with its larger market), whose goal is to maximize its income, will add additional star players until *its* marginal revenue from another star player is less than the first star player adds to the revenue of poor team with its smaller market. The law of diminishing returns, which makes the addition of another star player reduce increases in income, will set in at a later point when rich team maximizes its own income rather than maximizes combined total revenue of both teams, so there will be a tendency for rich team to have significantly more star players than poor team. If rich team adopts the goal of maximizing the number of wins, it will add star players even if they would be more valuable on poor team so that talent will be even more concentrated than if it attempted to maximize its income. Question: If all teams adopt the extreme uncooperative strategy of maximizing the number of wins rather than income, can any team make a profit? Will players be overpaid in the sense that they are paid more than their contribution to team income? Is this extremely uncooperative strategy equivalent to "rent-seeking" behavior, such as when 100 individuals each spend $1000 dollars searching for a sunken treasure worth $50,000, or they compete for a government granted monopoly, or millions of individuals all attempt to outperform the stock market?

What rule or rules do professional sports leagues adopt to prevent or at least reduce the effectiveness of non-cooperative strategies that tend to concentrate talent on the rich team? Answer: 1) Leagues institute player drafts in which the team with the worst record gets to choose the best available college or minor league players and is the only team who is allowed to hire them for a period of years before they become free agents. 2) Compensation or salary caps can be instituted that limit total team labor compensation or salary that can be paid to an individual player. 3) Teams can share television revenue without regard to market size. (What are the effects on competition of equal revenue sharing from television by professional football teams versus each team keeping most of its own television revenue as in professional baseball?) 4) Teams with the best winning records can be scheduled to play against each other more frequently.

Poor Team's Non-cooperative Strategy

Assume that poor team adopts a non-cooperative strategy: it could maximize individual team income by selling promising players to rich teams and employing only players who agree to low salaries. A poor team can generate revenue parasitically from other teams' expenditure on star players because it will have large attendance when teams with highly paid star players come to its city, even though their fans would not expect the team to win. This strategy maximizes revenue for a poor team because it maximizes the difference between team revenue and cost of players' salaries. What rule can a league adopt to prevent or at least reduce the effectiveness of a non-cooperative strategy by poor team? Answer: The league can require both minimum compensation per player and minimum total labor compensation.

Professional Athletes as Financial Assets

Recently, some professional sports teams have been managed as if a team's players are a collection of financial assets (regardless of the team's market size). Such teams buy players who are undervalued; that is, players whose market price is less than their value to the team as determined by individual performance statistics; they sell players if another team offers more than they are worth. The goal is to produce the greatest gross financial margin; improving team performance is an incidental byproduct.[13] For example, a team that managed its players as if they are financial assets sells overvalued players to a rich team that wants to maximize the number of wins. Question: Can a poor team, which adopts the financial model, that is, searches for undervalued (assets) players, compete with rich team, which has the goal of maximizing wins? Alternatively, will rich team that overpays its players necessarily achieve the most wins?

ROTTENBERG'S COROLLARY: EFFECT OF A CHANGE IN LIABILITY RULES ON FREE MARKETS

Rottenberg's theorem, which is couched in terms of a change in property rights in free markets, suggests a corollary for liability rules. (In chapter 13, Calabresi suggests that this corollary was known since the early 1900s.) Although both the theorem and its corollary are equivalent in terms of an efficient result, the property right and liability rule versions of the theorem have significantly different distributional consequences.

The theorem on the effect of a change in liability rules on free markets will be stated, after which three examples will be discussed from the UCC, Article 2, Sales (of Goods): one, liability rule for goods that are lost or damaged during transit from seller to buyer; two, liability rule for when payment for goods is incurred; and three, liability rule for defective goods, known as the perfect tender rule. In each of these cases, the UCC's Article 2 default legal liability rules that favor sellers are compared to alternative liability rules that favor buyers. Each example illustrates the corollary on the effect of a change in liability rules on free markets.

Rottenberg's corollary is:

1. *If* parties transfer goods or services in free markets,
2. law specifies which party is favored by a *liability rule* (e.g., buyer vs. seller,),
3. *then* contracts allocating liability in free markets will be efficient regardless of which party is favored by the liability rule; that is, the efficient outcome is independent of the liability rule, and
4. *the disfavored party will receive a compensating price differential equal to any benefit conferred or any cost saved by the favored party.*

UCC'S Article 2, Sales (of Goods), Default Liability Rules

The UCC, which governs commercial law in forty-nine states[14] is remarkable for at least three reasons: First, the UCC is a drastic change from the primacy of the judge-made (common) law to a codified or statutory system of commercial law.[15]

Second, the UCC states that excepting obligations of good faith, diligence, reasonableness, and care, which obligations may not be disclaimed, "the effect of provisions of this Act may be varied by agreement."[16] In other words, the UCC's rules are *default* rules; that is, all its rules are subject to change by agreement of parties to a contract. (By analogy, a default font on a computer is the font, for example, "Times New Roman," in which the computer is programed to type. A user may change this font to Arial or any other of the myriad available fonts.)

Third, after centuries of playing the game of title, title, who's got the title, under UCC Article 2, Sales (of goods), judges and lawyers now play the game of risk, risk, who's got the risk in order to make commercial law conform to business practices and economic theory rather than hoary legal tradition.

UCC Article 2, Sales, governs contracts and sales of *goods*, which are defined as things that are movable at the time of identification to the contract;[17] that is, tangible, personal property, like automobiles, doughnuts, and stereo sets. Other contracts, say those involving services, for example, haircuts, real estate, and employment, are governed by the judge-made (common) law, that is, contract rules developed by judges over centuries in Anglo-American contract cases.

Place for Delivery of Goods

UCC §2–308, Absence of Specified Place for Delivery, states: "Unless otherwise agreed . . . the place for delivery of goods is the seller's place of business." In UCC parlance, this is called a *shipment contract.* Under a shipment contract, buyer pays the cost of transporting the goods and insurance. Once the seller transfers the goods to a carrier, buyer bears the risk of loss; that is, if the goods are lost in transit due to carrier's negligence, buyer has to pay seller and must sue carrier for compensation. Under a shipment contract, the liability rule favors seller. Seller's price for goods will not include the cost of transportation plus insurance; that is, the price that buyer, who is the disfavored party, must pay to seller will be reduced by a compensating price reduction.

Assume that buyer prefers to have the goods under contract delivered to buyer's place of business even though the UCC default rule states that delivery is to be at seller's place of business. In UCC parlance, buyer wants to negotiate *delivery contract.* Under a delivery contract, seller pays for transportation and retains risk of loss until the goods are delivered by carrier to buyer; that is, if the goods are lost in transit due to carrier's negligence, seller will not be paid for the goods by buyer and seller will have to sue carrier. Assuming seller retains risk of loss until delivery is completed, how much can seller charge buyer to deliver goods to buyer's place of business? In a free market, seller can charge no more than the cost of delivering goods plus insurance because both parties have many alternative buyers or sellers with whom they can deal.

Under the corollary on the effect of a change liability rules in free markets, if the law clearly specifies which party a liability rule favors (seller), free markets will achieve the efficient use of resources; that is, the efficient outcome is independent of the liability rule because disfavored party (buyer) will receive a compensating price differential equal to the cost saved by the favored party (price reduction equal to the transportation and insurance cost saved by seller).

Time for Payment of Goods

UCC §2–310(a), Open Time for Payment or Running of Credit states, "Unless otherwise agreed . . . payment is due at the time and place at which the buyer is to receive the goods even though the place of shipment is the place of delivery." Delivery of goods and payment are to be concurrent. In circumstances where buyer wants say, 30 days credit before payment, the default rule, under which payment occurs on delivery, can be changed by agreement. Assuming competitive credit markets, which is likely in business-to-business transactions, the price of credit will be the interest rate on the unpaid balance until payment occurs. The efficient outcome is independent of the liability rule because buyer will pay a compensating price differential equal to the benefit conferred, which is not having to pay for the goods at the time of delivery, that is, for the right to buy on credit.

Breach of Contract of Goods

UCC §2–601, Improper Delivery, states: "[U]nless otherwise agreed . . . if the goods or the tender of delivery fail in any respect to conform to the contract, the buyer may (a) reject the whole; or (b) accept the whole; or (c) accept any commercial unit or units and reject the rest." This section is known as the UCC's *perfect tender rule*, which means that if the goods fail to conform to the contract in any respect, the buyer can totally reject the goods. For example, a manufacturer sells and sends 100 pairs of hiking boots to buyer, who is a retail seller of hiking equipment. After receiving the boots, seller notices that they are excellent boots except that they have one small defect: each right boot of the 100 pairs is missing one eyelet on the top-right side of the laces. The defect may be small but the boots are unsalable by retailer. Retailer has the right to reject the entire shipment of boots. Under the UCC, there are provisions that soften this stark, all-or-nothing rule: if the time for performance has not expired, manufacturer can resend goods that conform to the contract; or, if seller has a reasonable understanding that defects can be *cured* by adjustments or repair of goods without great inconvenience to buyer, extra time is allowed.[18] For example, if a new automobile has some minor defect in operation or cosmetic damage, seller usually has the right to repair such defects rather than replace the automobile with another.

In addition, parties can by agreement change the perfect tender default rule back to the judge-made (common law) rule for breach of contract, which governs most contracts other than those involving goods, such as contracts involving real estate, services, and labor contracts.

Under common law, when there is a breach of contract due to improper performance, a judge determines whether the breaching party *substantially performed* the contract or there was a *material breach*. This distinction is easily explained with an example: Suppose a contractor, who agrees to build a house for buyer, does a workman-like job in building the house, but paints two rooms in the house a hideous green instead of sky-blue as required by the contract. The buyer must accept the house because the contractor substantially performed the contract, but buyer is entitled to a reduction in price (compensatory damages) equal to the cost of repainting rooms in conformity to the contract. If buyer does not accept the house when there is substantial performance with small defects, the buyer, not seller, is in material breach.

Now change the facts: suppose before taking possession of a newly built house, buyer notices that there is a large crack running the entire length of the foundation, the roof leaks, there are water stains on the ceilings, floors are uneven, and in several rooms, corner joints do not meet so that insulation is exposed. This would be a material breach entitling the buyer to refuse to accept the house and demand return of all payments made.

The UCC's perfect tender, all-or-nothing, liability rule for breach of contract involving goods is a major departure from the common law distinction between substantial performance and material breach. Why would some sellers and buyers agree to change from the UCC's perfect tender liability rule back to the common law liability rules? For example, the common law breach of contract rules may be more efficient for sellers and buyers of perishable goods. Assume a fruit cooperative sells large amounts of fruit to a supermarket chain and that the parties have a long-standing relationship. The parties may agree that if fruit has small imperfections, rather than being totally rejected, a mutually agreed upon amount will be deducted, equal to the diminution in value of the fruit. This is another example of the corollary for liability rules. Since the wholesale market for fruit is most probably free and competitive, that is, both buyers and sellers have many alternative parties with whom to deal, the reduction in price will be equal to actual diminution in value due to imperfection of produce. The legal rule will be independent of the efficient result.

In order to understand the potential magnitude of a change from the common law rule on breach of contract to the UCC's perfect tender rule, consider *Pittsley v. Houser*,[19] in which seller sold a carpet plus installation to buyer. After installation, buyer complained that carpet seams were visible, gaps appeared, it did not lay flat in some areas, and it failed to reach the wall in some areas. After several attempts to fix the installation, buyer was unsatisfied and filed a lawsuit. The issue was whether the dispute involved goods, so that UCC Article 2 governed the case or whether it involved services governed by the common law of contract. The trial court ruled that the common law governed the case and that revocation of the contract was not available because there was substantial performance. The court held that the breach was not *material* because damages for improper installation were only $250 plus $150 expended in moving furniture prior to seller's attempt to repair. The court determined that buyer must pay $902 remaining on the contract, minus $400 in damages.

The Court of Appeals said the only question on appeal was whether the UCC Article 2 was applicable to the case. Using the *predominant factor* test, it held that because the installation (services) cost was only $700 of the total cost of $4,319 for carpet (goods) plus installation, "the sale of the carpet was the predominant factor in the contract with the installation being merely incidental to the purchase." The trial court's ruling was reversed; on remand, the trial court was required to apply UCC Article 2 to this transaction. Since the UCC applies to this case, the UCC's perfect tender rule is applicable, which means that seller must either fix the installation to buyer's reasonable satisfaction or give buyer a total refund. A different outcome indeed!

Two Exceptions to the Corollary

Two major examples in which a change in liability rules does not result in complete compensating price or wage differentials, and the specific reasons why, are discussed below and in subsequent chapters: A shift from negligence-based liability for employee on-the-job injuries to workers' compensation probably created additional benefits for employees because it reduced employers' monopsony power; the change also internalized the cost of industrial accidents to employers. The corollary to Rottenberg's theorem for a change in liability rules was well known by economists at the beginning of the twentieth century but its assumptions were inaccurate with respect to the shift to workers' compensation. Chapter 18, on torts, shows that a shift from negligence-based liability for defectively manufactured products to strict liability enhances efficiency due to asymmetric information which favors manufacturers over consumers; the shift also internalizes the cost of dangerously defective products to manufacturers. However, because there is a principled way to distinguish specific cases in which the corollary is true from cases in which it is false, it is an accurate guide to public policy; that is, the corollary is both insightful and useful in specific cases.

NOTES

1. Guido Calabresi and Douglas Melamed, "Property Rules, Liability Rules, and Inalienability: One View of the Cathedral," *Harvard Law Review* 85 (1972): 1089–1093, 1106–1111.
2. *Ibid.*, 1092–1093, 1111–15.
3. Inalienability rules are relevant to the theoretical argument made by Landes and Posner in favor of a market for babies. Elizabeth Landes and Richard Posner, "The Economics of the Baby Shortage," *Journal of Legal Studies* 7 (1978): 323–348.
4. Rottenberg's theorem (chapter 11) and Coase's two theorems (chapter 12); in torts (chapter 18), the theorem that all-or-nothing negligence rules and comparative negligence rules are equally efficient; in contracts (chapter 19), the theorems that in perfect competition, all direct damage rules for breach of contract are equally efficient and the lost-volume seller rule, which determines whether a buyer who breaches a contract is legally liable for seller's lost profits, is irrelevant to efficiency.
5. Office of Planning, Research, and Evaluation, US Department of Health and Human Services, *Head Start Impact Study and Follow-up, 2000–2015.*
6. 163 U.S. 537 (1896).
7. 347 U.S. 483 (1954).
8. Ronald Dworkin, *Law's Empire* (Cambridge, MA: Belknap, 1986), 106, 359–365, 379–392.
9. *Ibid.*, 106.
10. Simon Rottenberg, "The Baseball Players' Labor Market," *Journal of Political Economy* 64 (1956): 242–258.
11. *Ibid.*, 256.
12. Simon Kuper and Stefan Szymanski, *Soccernomics* (New York: HarperCollins, 2016).
13. *Ibid.*; Michael Lewis, *Moneyball* (New York: W. W. Norton, 2013).
14. The UCC has been adopted all the states except Louisiana, which is influenced by French law.
15. Subjects in UCC are Sales (of Goods), Leases (of Goods), Commercial Paper (Negotiable Instruments), Bank Deposits and Collections, Funds Transfers, Letters of Credit, Warehouse Receipts, Bills of Lading, Documents of Title, Investment Security, and Secured Transactions.

16. UCC §1–102(3) Variation by Agreement.
17. UCC § 2–105(1).
18. UCC § 2–508(1–2).
19. 875 P.2d 232 (Id. Ct. App. 1994).

Chapter 12

Coase's Theorems

Property Rights, Liability Rules, and Mutually Interfering Activities

In his classic article, "The Problem of Social Cost,"[1] Ronald Coase attacked the view accepted by most economists then as well as now that the appropriate remedies for smoke pollution (public bad type externalities) should be: a) hold the factory liable for smoke damage, b) tax the emission of smoke equivalent to the monetary damage done, or c) exclude the factory from residential districts. Coase contended that these courses of action are inappropriate because "they lead to results which are not necessarily, or even usually, desirable."

Coase attacked these generally accepted public policies and stated a theorem positing certain conditions under which mutually interfering activities (externalities) will be coordinated efficiently via private bargaining, without the need of taxation or government regulation. Coase's actual theorem and the one attributed to him will be stated and critiqued in this chapter.

TWO COASE THEOREMS

Coase constructed several numerical examples and analyzed a number of common law cases, some of which will be discussed in this chapter, from which he concluded:

1. *If* there are "actions of business firms which have harmful effects on others,"
2. the law clearly specifies "whether the damaging business is liable or not for damage caused," and
3. there are "costless market transactions," which transfer legal liability from those who have the law specified in their favor to those who do not,

   then "the ultimate result (which maximizes the value of production) is independent of the legal position."[2]

Transaction costs are the cost of making contracts; such costs include finding persons with whom to deal, exchanging information, negotiating an agreement, and expressing it in writing. "These operations are often extremely costly, sufficiently costly at any rate to prevent many transactions that would be carried out in a world in which the pricing system worked without cost." The condition that transaction costs be zero or at least not prohibitive confines the theorem to disputes involving a small number of parties as opposed to externalities that affect large numbers of persons in regions, countries, and continents (public bads). When large numbers of parties are involved, say externalities like global warming, transaction costs of the billions of necessary bargains are prohibitive and insurmountable (see chapter 8).

Coase discusses examples and cases in which he compared a legal regime of strict liability for damages, which is a liability rule, with no liability, which is a property right. (Recall Calabresi and Melamed's distinction between a property right and a liability rule, discussed in chapter 11.)

Although *strict liability* and *no liability* look and sound alike because both concepts contain the word liability, legally, they are very different and have substantially different economic consequences. Specifically, strict liability for damage done to another's property is a compensatory principle in the law of torts (civil wrongs), where courts determine the level of damages; there is no bargaining between parties (courts attempt to determine what the price of damaged property would be in a competitive market), and neither party has any power over the other.

No liability concerns the law of property, not torts; no liability refers to a right attached to property ownership. No liability for pollution means that in addition to owning the factory's physical land and buildings, there is an additional right attached to this property, which is right to pollute the air. If a factory has no liability for pollution, that is, it has the absolute right to pollute, a person or persons who want the factory to reduce or stop polluting must pay a price satisfactory to the factory owner, which price is not subject to control by courts. The price that the owner, the single seller (monopolist) of the right to pollute, can charge is only limited by its self-interest.

Although, as Coase's theorem suggests, an efficient cooperative bargain between polluter and persons injured by pollution is possible, non-cooperative possibilities and inefficient results are also possible and likely when a party is favored by a property right. Bargaining may not be successful if the party which has a right to inflict uncompensated damage on another attempts to extract most or all of the cooperative surplus from the efficient bargain; or the party favored by the no liability right threatens to exacerbate damage in order to maximize payment from the persons injured by its actions (extortion). As the discussion of *Carpenter v. Double R Cattle Co.*[3] in Chapter 1 showed, courts are aware of the possible negative consequences that may result from no liability rules. Due to the possibility of uncooperative bargaining, Coase's actual theorem is often not useful in specific cases where a property right of no liability is at issue (e.g., *Spur Industries, Inc. v. Del Webb Development Co.*, discussed below). In sum, in Coase's actual theorem, the pairing of a liability rule with a property right has created a great deal of confusion.

The theorem attributed to Coase is:

1. *If* the law clearly specifies *the assignment of property rights*, and
2. transaction costs are zero, *then* private bargaining will result in the efficient outcome, independent of the assignment of property rights.[4]

The theorem attributed to Coase creates additional problems in two ways: one, it is stated in terms of two property rights, which eliminate all judicial restraint on strategic bargaining by a party who is favored by a property right. The converse of a property right to pollute is a property right to clean air, enforceable by an injunction, which forbids anyone from polluting the air unless that person pays a sum satisfactory to a person possessing the property right to clean air. In other words, a property right to clean air protected by an injunction gives any and all individuals with such right the power to prohibit the polluter from operating; that is, the victims of pollution have the ability to shut down the factory. Courts, which are aware of the possible negative consequences that may result from property rights that give a party such *hold-up* power will refuse to grant such power to injured plaintiffs (e.g., *Boomer v. Atlantic Cement Co.* discussed below).

Two, rather than being confined to mutually interfering activities (externalities) as is Coase's actual theorem, the theorem attributed to Coase is conflated with Rottenberg's theorem on the effect of a change in property rights in free markets (chapter 11) and with bargaining in bilateral monopoly, which is discussed below. The elimination of the restraining effect of liability rules on bargaining power and conflation of different types of markets substantially increases the generality of the theorem attributed to Coase, but makes it not useful to courts in specific cases.

COASE'S CATTLE-CROPS EXAMPLE

Coase's initial explanation of his theorem was couched in terms of the now famous cattle-crops example. Coase assumes that a cattle ranch is located next to a farm and that straying cattle cause crop damage. Coase further assumed that the market price of the farmer's crop is $1/ton and that the annual cost of a fence, which would prevent crop damage, is $9. Coase considered two legal regimes, one under which there is strict liability, a liability rule, for crop damage, the other under which there is no liability, a property right, for crop damage. He generalized the cattle-crops example into his theorem, the conclusion of which is that the efficient combination of cattle and crops is *invariant* under both legal regimes; that is, the legal regime "would not affect the allocation of resources but would merely alter the distribution of income and wealth as between the cattle rancher and the farmer."

Strict Liability Versus No Liability for Crop Damage

Consider the rancher's decision process under either strict liability or no liability. Under strict liability, if the rancher currently has no cattle and is deciding whether to enter the cattle business, how valuable must the steer 1 be in order for the rancher to earn a profit? Since the rancher is strictly liable for the 1 ton or $1 of crop damage done by steer 1, the rancher will not maintain steer 1 unless the profit from steer 1 exceeds the $1 of crop damage that it causes. If the profit from the steer 1 exceeds $1, the rancher determines that steer 1 is profitable.

Under no liability, if the rancher currently has no steers and is deciding whether it is profitable to enter the cattle business, how much profit must be earned from steer 1 in order for the rancher to maintain steer 1? Since steer 1 will cause \$1 in crop damage, if the profit from steer 1 is less than \$1, then a bargain can be struck between rancher and the farmer so that the rancher will not go into the cattle business. For example, assume that the rancher would earn a profit of 50¢ from steer 1; both the rancher and the farmer will benefit from a bargain in which the farmer pays the rancher a sum between 50¢ and \$1 not to go into the cattle business. Assume the farmer pays the rancher 75¢ not to add steer 1. The rancher is better off by 25¢ and the farmer is also better off by 25¢ because the farmer pays the rancher 75¢ rather than suffer \$1 crop damage from steer 1. But, if the rancher earns a profit of more than \$1 from steer 1, no bargain can be made because the rancher's profit from steer 1 exceeds the farmer's crop loss of \$1 from steer 1. Thus, under either the legal regime, whether there is strict liability or no liability, the rancher will not add steer 1 unless the profit from steer 1 exceeds \$1.

(Notice, however, if instead of comparing no liability with strict liability, the comparison is between no liability and a property right that the farmer has to forbid the rancher's steer from trespassing on farmer's property, enforceable by injunction, the amount of payment farmer can demand is not limited to \$1, but is up to the price of building a fence which is \$9. In the latter legal regime the efficient solution is much less likely.)

Under what condition will the rancher add enough steers so that under no liability the farmer will build a fence and under strict liability the rancher would build a fence? Under no liability, if the damages the farmer would incur from cattle are greater than \$9, the farmer will build a fence. Under strict liability, if the optimal size herd does more than \$9 in damages, the cost of a fence, and the rancher earns a profit of more than \$9 from the herd, then the rancher will build a fence rather than pay damages.

Fencing-In Versus Fencing-Out in the United States

As a general rule, in the eastern part of the United States, states hold owners of livestock legally liable for damage they do on another's property; that is, owners of livestock are required to *fence-in* their livestock. However, many western states follow the open range principle; they reverse the fence-in duty and require landowners to *fence-out* livestock from their property. If, in the east, there is more land devoted to farming than to livestock, and the reverse is true in the west, both rules are efficient in that both rules will minimize the total amount of fencing that is required.

INCOME REDISTRIBUTION AND THE INVARIANCE HYPOTHESIS

Although the allocation of resources is invariant with respect to legal regime in the cattle-crops example which compares no liability to strict liability, a shift from one legal regime to the other causes a significant shift in income distribution. For example, if maintaining a large herd of cattle is efficient, the rancher pays for the fence under strict liability, whereas under no liability, the farmer pays for the fence.

The possibility of large swings in income distribution under alternative legal regimes has two significant consequences: First, it is generally agreed that due shifts in income distribution under different legal regimes, the invariance hypothesis, that is, the claim the resource allocation will be identical under different legal regimes, is false. The reason is that unless farmers and ranchers spend their incomes on the same goods and services, the allocation of resources will be different in a regime in which the income distribution favors ranchers than in a regime in which the income distribution favors farmers. The fallback position is to argue that although the allocation of resources is not invariant, it is efficient; that is, even in the presence of mutually interfering activities such as ranching and farming, bargaining between the parties will result in the efficient coordination of their activities. Second, the possibility of large swings in income distribution under different legal regimes points to the fatal flaw in the Coase Theorem: the party who is favored by the law can often use the monopoly power inherent in the legal entitlement, especially if protected by a property right, to extort money from the other party, with the result that bargaining may break down.

THE LEGAL EXTORTION (HOLD-UP) PROBLEM

Consider Coase's own example of extortion. He extends the cattle-crops example to consider a situation in which it would be unprofitable for the farmer to grow crops on a strip of land because the value of the crop on the market is $10 but the cost of cultivating the crop is $11. In the absence of the rancher, the land would not be cultivated because the farmer would lose $1. However, given the rancher as a neighbor and a legal regime of strict liability for crop damage, if the strip of land were cultivated, the entire crop may be destroyed by the cattle and the rancher would be forced to pay $10 to the farmer. Although the farmer would lose $1, the rancher could lose $10. According to Coase there would be bargaining: "The farmer would not be able to obtain a payment greater than the cost of fencing off this piece of land [between $0 and $9] nor so high as to lead the cattle-raiser to abandon the use of the neighbouring property."

On the contrary, isn't it more likely that the rancher would refuse to accede to such legal extortion and would reply: "Nine dollars for a fence, but not one cent for tribute." If the western movies depicting the epic battles between ranchers (cattlemen) and farmers (sodbusters) are to be believed, the rancher in this scenario is more likely to shoot or strike the farmer rather than make a Coasian bargain. Consider another example under no liability for crop damage (the rancher has a property right to allow his cattle to range freely). How would the farmer respond to the rancher who says, "3,000 of my cattle are going to stray (stampede) onto your land and destroy your entire crop and anything in their path unless you either build an extremely strong fence on your property at cost of $20,000 or pay me $19,999"?

Consider a third example which involves two property rights rather than, as above, the contrast between a property right and a liability rule: contrast a regime with a right to quiet enjoyment of one's property versus a legal regime with no right to quiet enjoyment. Under a legal regime in which there is a right to quiet enjoyment, a land owner can get an injunction against excessive or unreasonable noise. On the other hand, under a legal regime in which there is no liability for making noise, that is, no right to quiet enjoyment of "real" (land) property, your neighbor says to you that she intends to point her stereo speakers at your house and play music at top volume twenty-four hours a day, every day, unless you pay her $5,000. Is your response likely to be, "Let's make a deal?" In circumstances, where one legal regime gives one party monopoly (hold-up) power over another and the converse legal regime gives the other party monopoly power, both the market structure of bilateral monopoly and game theory (discussed below) are analytically superior to the Coase Theorem because they more accurately model how shifts in legal institutions may have substantial effects on bargaining power, income distribution, and efficiency.

BARGAINING VERSION OF THE COASE THEOREM REQUIRES COOPERATION

An attempt to salvage the Coase Theorem is known as the bargaining version: *If* 1) the law clearly specifies property rights, 2) transaction costs are zero, and 3) all possible bargains are consummated, *then* the allocation of resources will be efficient regardless of legal assignment of liability.[5] Bargaining can coordinate mutually interfering activities because it is possible for both parties to gain from a division of the economic surplus produced by such coordination. Recall that under a regime of no liability in the cattle-crops example, if the addition of steer 1 yielded less than $1 in profit to the rancher, say 50¢, it was possible to make bargain that would benefit both rancher and farmer, who stands to suffer a $1 loss from crop damage, say, where the farmer paid the rancher 75¢ not to maintain steer 1. However, if the rancher demanded 99¢ as the price of not adding steer 1 and the farmer refused to pay more than 51¢, no deal would be possible. In that case, the rancher would add steer 1 even though the result would be an inefficient allocation of resources. Bargains may fail to materialize because self-interested parties may be unable to agree on how to share the 50¢ economic surplus that results from the efficient coordination of ranching and farming. Robert Cooter wrote:

> The costs of communicating, writing a contract, and enforcing its terms are the transaction costs. . . . These costs sometimes constitute an obstacle to exchange. However, there is another obstacle of an entirely different kind, namely the absence of a competitive price. The parties must haggle over the price until they can agree upon how to distribute the gains from trade. There is no guarantee that the rational pursuit of self-interest will permit agreement. If we interpret zero transaction costs to mean there is no dispute over price, then we have dissolved the bargaining game. . . .
>
> The error in the bargaining version of the Coase Theorem is to suppose that the obstacle to cooperation is the cost of communication, rather than the strategic nature of the situation.[6]

Cooter's point is that outside of the context of competitive markets, efficiency is often "contingent upon agreement about distribution. . . . To eliminate the possibility of noncooperation, we would have to eliminate the problem of distribution." In order for the bargaining version of the Coase Theorem to be true, it must be assumed that the parties bargain cooperatively rather self-interestedly. However, this assumption violates the usual assumption in economics that individuals are self-interested. The assumption that people are usually cooperative in situations where there are mutually interfering activities is unrealistic in a litigious society such as ours in which the ethical cannon that lawyers should represent the interests of their clients zealously[7] is often interpreted to mean that a lawyer ought to press every legal advantage to the fullest extent.

BILATERAL MONOPOLY

In a bilateral monopoly, a monopolist, single seller, must negotiate with a monopsonist, single buyer. The most-often-used bilateral monopoly paradigm is a contract negotiation between the management of the only factory in town and the only union in town, which union represents all employees. The union of all employees is a monopolist because it represents the only seller of labor. The factory's management is a monopsonist because it is the only buyer of labor. The monopolist must negotiate a contract with the monopsonist. Bilateral monopoly is a negotiation whereby two parties have no alternative but to deal with each other, yet each party attempts to take advantage of the other. The union is trying to bring management to its knees so that it can insist on higher than competitive wages and featherbedding work rules; management is trying to break the union and turn the factory into a sweat shop with lower than competitive wages. If the two parties cannot agree, that is, reach an impasse in negotiations, the parties may grant a third party coercive power and allow it to impose a settlement, such as binding arbitration.

Consider the following three properties of bilateral monopoly: First, in bilateral monopoly, unlike competition where parties have many alternative persons with which to deal, two parties must deal with each other; they have "locked horns." Both parties attempt to behave opportunistically; they try to exert monopoly power over the other. Second, the results in terms of price and quantity are indeterminate: whether a bilateral monopoly market results in the competitive price and efficient quantity of output or tends to monopoly or monopsony price and inefficient output depends upon the bargaining power of the parties, which is influenced by who has the deepest pocket.

Third, the results in terms of price and quantity are greatly influenced by whether legal institutions and government intervention favor one party or are neutral. If legal institutions and government intervention favor neither party and the parties have similar economic strength, the results from bilateral monopoly negotiations, in terms of price and quantity of output, will tend to be the same as in competitive markets. However, if legal rules are biased in favor of one party or the other, the favored party can exert monopoly or monopsony power and act opportunistically over the other.

Bias occurs if, at one extreme (as in the nineteenth century), a union could be enjoined as an illegal conspiracy, or at the other extreme, union members have the legal right to blockade all entrances to the factory during a strike. In general, if one party is favored by legal institutions and government intervention, the results in terms of price and output will tend toward either monopoly or monopsony and will therefore be different from the results of competition; the transaction that results can, in general, be characterized as both inefficient and distributionally unfair.

The bilateral monopoly model shows that legal rules matter even when transactions costs are low. The bilateral monopoly model suggests that institutionally sensitive courts should be wary of protecting entitlements with property rights as opposed to liability rules when the property rights would result in inefficient outcomes and large shifts in income distribution; in these situations courts should use the threat implicit in their equity powers (which are similar to compulsory arbitration) to suggest ways in which parties can coordinate mutually interfering activities. (Bilateral monopoly figures prominently in Judge Posner's economic analysis of judge-made law, which is the subject of chapter 14.)

An optimal solution which coordinates two antagonistic interests, such as labor and management, requires that the two interests be merged. Such optimal coordination could occur if the union and management were one as when the union owns all of the company's common stock through its pension fund. In that case, the bilateral monopoly problem would disappear because the worker-owners would have a unity of interests.

UNIFIED MANAGEMENT OF MUTUALLY INTERFERING ACTIVITIES

In another well-known article by Coase, "The Nature of the Firm,"[8] he set out to explain why coordination sometimes occurs by the price mechanism and sometimes by firms, the distinguishing mark of which is "the supersession of the price mechanism." In other words, if markets are such an efficient means of coordinating production, why do firms exist at all? "The main reason why it is profitable to establish a firm would seem to be that there is a cost of using the price mechanism." Negotiating and making separate contracts for every transaction that occurs in markets is costly.

Coase said that there are other disadvantages of using markets, chief of which involves the difficulty of forecasting long-term needs. The longer the term of contract, whether for services or supplies, the less possible it is for the buyer to specify exactly what is needed from the other party. This is known as the problem of "incomplete contracts." Firms emerge in response to the inefficiencies arising from incomplete contracts because a firm's owner can simply direct employees to do the right (efficient) thing. A firm comes into existence whenever it is cheaper to organize production administratively than through market transactions.

Coase then discussed the question of what limits the size of the firm. His answer was that "a firm will tend to expand until the costs of organising an extra transaction within the firm become equal to the costs of carrying out the same transaction by means of an exchange on the open market or the costs of organising transactions in another firm."[9]

Within the firm, production is coordinated through hierarchical, administrative decisions without need for bargains. In the cattle-crops example, a landowner whose land encompassed both the ranch and the farm would maximize the net return on the entire tract by taking account of the mutual interference between cattle and crop activities. Such consolidations may be especially useful in the presence of mutually interfering activities where contracts specifying what the parties have agreed to do or not do with respect to noise, odors, the size of herds, and so on may require long and detailed contracts that may be difficult to enforce. On the other hand, joint ownership incurs difficulties because of the transaction costs in combining many parcels of land, lack of expertise in unrelated markets, and the problem of loss of control that results when there are many subordinate managers in a large organization.

CASES INVOLVING TWO PARTIES

Bryant v. Lefever

In *Bryant v. Lefever*,[10] plaintiff complained that defendant, having put timber to a considerable height on his roof thereby preventing wind from blowing over plaintiff's house as it had for twenty years, caused his chimneys to smoke and created a nuisance. The jury, which found for plaintiff, awarded £40 damages. The Court of Appeals reversed the jury. Bramwell, L. J. said:

> But it is said, and the jury have found, that the defendants have done that which has caused a nuisance to the plaintiff's house. We think there is no evidence of this. . . . It is the plaintiff who causes the nuisance. . . . Let the plaintiff cease to light his fire; let him move his chimney; let him carry it higher, and there would be no nuisance. . . . If the plaintiff has not the right to the passage of air, except subject to the defendants' right to build or put timber on their house, then his right is subject to their right, and though a nuisance follows from the exercise of their right, they are not liable.

Cotton, L. J. said:

> Cases to prevent, or to claim damages for, interference with ancient lights, are frequently spoken of as cases of light and air, and the right relied on as a right to the access of light and air. But this is inaccurate. The cases, as a rule, relate solely to the interference with the access of light, and in no case has any injunction been granted to restrain interference with the access of air.

Discussion

1. Who caused the nuisance? The jury verdict for plaintiff is consistent with the maxim, "First in time, first in right" and causation by defendant. The appellate

court said that lighting the fire in the chimney by the plaintiff, first mover, "caused" the nuisance. Scientifically, both activities are necessary causes.

2. Although moving the timber is probably cheaper than extending the chimney in this specific case, the appellate court, which views this case as one among many cases, probably did not want to create a general right to light or air because future land development would be obstructed.

In Figure 12.1, assume the goal is to minimize the cost of mutually interfering activities. Assume the cost of moving defendant's timber is –£5, the cost of raising plaintiff's chimney is –£10, and that if neither cost is incurred plaintiff will suffer –£40 (jury verdict) in damages due to the backing up of smoke in his house.

		Defendant	
		Don't Move Timber	Move Timber
Plaintiff	Don't Raise Chimney	-£40, 0	0, -£5
	Raise Chimney	-£10, 0	-£10, -£5

Figure 12.1 Timber and Chimney Game.

If both parties can choose their strategies independent of the legal rule, Nash equilibrium is (–£10, 0); efficient (least cost) solution is (0, –£5), so Nash equilibrium is not the least cost solution.

Property Rights: Under a no liability rule, the solution is (–£10, 0). Limits of Coasian bargain to achieve least cost, efficient solution, are min.–max., £5–£10. Bargaining will fail if either party demands too much of the £5 cooperative surplus. Assume that the neighbors in Bryant dislike each other so much that cooperative bargaining is impossible. Does this suggest that in cases where bargaining breaks down, the property right should in general be assigned to the party who values it most, or in this case, against the party who can most cheaply resolve the mutual interference?

If plaintiff is entitled to win protected by an injunction, the solution is (0, –£5), which is the efficient solution. No bargains away from the efficient solution are feasible.

Liability Rules: If plaintiff is entitled to win and is protected by liability rule, the efficient solution would be achieved because moving the timber would cost defendant £5, which is cheaper than paying damages, £40, or paying plaintiff to raise the chimney, £10; no Coasian bargain is possible.

If defendant is entitled to win and is protected by a liability rule, plaintiff would pay defendant £5 to take away that entitlement because it costs plaintiff less than raising his chimney.

If one party owned both houses, the joint ownership solution is to move the timber, which is least cost solution.

Sturges v. Bridgman

In *Sturges v. Bridgman,*[11] the defendant, a confectioner, operated a kitchen in the rear of his house where there were two large mortars in which meat and other materials were pounded between the hours of 10 a.m. and 1 p.m. Plaintiff, a physician, occupied a house that had a garden at the rear, which abutted defendant's premises. After plaintiff built a consulting room on the site of the garden, he alleged that the noise from the mortars "prevented him from examining his patients by auscultation for diseases of the chest. He also found it impossible to engage with effect in any occupation which required thought and attention." The defendant contended that he had acquired the right to continue by uninterrupted use for more than twenty years. The court disagreed on the ground that a man cannot be said to consent to an easement of which he had no knowledge, actual or constructive notice.

Defendant, who attempts to buttress his argument by two hypothetical examples, argued that if the rule of this case were generalized, it would create "serious practical inconveniences." One, if a man built a house in the midst of tanneries of Bermondsey, he could put a stop to manufacturing altogether. The court replied that whether an activity is a nuisance depends upon circumstances, and what would be a nuisance in Belgrave Square would not necessarily be so in Bermondsey. Two, "a blacksmith's forge built away from all habitations, but to which, in course of time, habitations approach." The court said that the use of adjoining land should not, for all time and under all circumstances, be restricted by reason of an activity that was first in time. Moreover, the court said that the blacksmith hypothetical was *idem per item* [the same for the same] case with the present case.

Discussion

1. Consider the court's first hypothetical example of a man who builds a house in the midst of tanneries of Bermondsey. Why no nuisance? Answer: The economic gains from tanneries greatly exceed losses to single homeowner due to odors (Kaldor-Hicks).

   Consider the court's second hypothetical example: a blacksmith's forge to which, in course of time, habitations approach. Why is there a nuisance? Answer: The gains to economic growth from the tanneries greatly exceed losses to single homeowner due to odors (Kaldor-Hicks).
2. The court says, "As regards the blacksmith's forge, that is really an *idem per item* [the same for the same] case with the present." In legal rationality, *Sturges* ⊃ blacksmith and blacksmith ⊃ *Sturges* (chapters 5, 6), which means that these cases are mutual precedents. In both, an activity that was first in time does not have the right to prevent subsequent development. However, in economic rationality, the

cases maybe dissimilar; although economic gains from housing expansion greatly exceed blacksmith's losses from having to move, whether the gains to the doctor are greater than the losses to the confectioner is much less certain.

3. The Court said that the "[black]smith in the case supposed might protect himself by taking a sufficient curtilage [enclosed land surrounding a dwelling] to ensure what he does from being at any time an annoyance to his neighbour." Is this a realistic solution? Even if the blacksmith bought surrounding land to protect his right to operate, as development approached, the land would be worth more in housing and he would probably sell his land.
4. In Figure 12.2 below, the rationale behind the numbers is as follows: The relative values of the doctor and the confectioner's activities are stipulated *arbitrarily* in favor of the doctor (in the sense that either activity could have been made the more valuable one without misrepresenting the facts) in £ (pounds). The confectioner earns £10 if he uses his machinery in the morning; the doctor earns £12 if he sees patients in the morning while the confectioner is not working. If the confectioner is working, noise from his machinery makes it impossible for the doctor to examine his patients. So, if both activities occur at the same time, the doctor's payoff is £0. Also, both the doctor and the confectioner both prefer to work in the morning: assume seeing patients in the morning and making hospital rounds in the afternoon is more efficient for the doctor; as for the confectioner, the court said that he had been working generally from 10 a.m. to 1 p.m. for years, so I assume he prefers to work in the a.m. (maybe he is a morning person). The payoffs to both parties if they work at the less favorable time when the other works at the most favorable time are £8 for the doctor and £9 for the confectioner. Lastly, if the confectioner has to move, he earns £0 profits from his present location. This is a complete information, simultaneous game in which the first number in each cell is the doctor's payoff and the second number is the confectioner's payoff.

In Figure 12.2 below, if both parties can choose their strategies independent of the legal rule, Nash equilibrium is (£8, £10); efficient solution is (£12, £9), because it has the highest joint profit.

		Confectioner		
		Make Confections a.m.	Make Confections p.m.	Stop Production (Move)
Doctor	Consult Patients a.m.	0, 10	12, 9	12, 0
	Consult Patients p.m.	8, 10	0, 9	8, 0

Figure 12.2 Confectioner and Doctor Game. *Source*: Created by the author.

Property Rights: Assuming that plaintiff, doctor, wins on a theory of nuisance and his entitlement is protected by a property right, Injunction (forbidding production by the confectioner). Injunctive solution is (£12, £0); efficient solution is (£12, £9), so a Coasian bargain with a min.–max., £1–£9, is possible. A property right-injunction, gives the doctor great bargaining power. If one or the other party demands more than half of cooperative surplus and the other party refuses to accede, bargaining will fail.

Assume defendant, confectioner, wins and his entitlement is protected by a property right, no liability; this solution, (8, £10), is not efficient. To get to efficient solution, a Coasian bargain with a min.–max, £1–£2, is possible.

Liability Rules: Assume the defendant, confectioner, wins and his entitlement is protected by a liability rule; that is, if the doctor pays damages, he can take away the confectioner's entitlement. Doctor pays damages of £1 to take away confectioner's entitlement to work in the a.m. and makes a gain of £4 – £1 = £3. The efficient solution is achieved.

Assume the plaintiff, doctor, wins and his entitlement is protected by a liability rule. The efficient solution is achieved because the confectioner will not pay damages of £2 to take away plaintiff's entitlement to get a gain of £1.

5. If one party owned both businesses, joint ownership solution is, £12, £8, the efficient result. Should the court consider an equitable remedy? That is, should the court dictate the efficient or joint ownership solution? A court may not have enough information to determine the efficient result but could suggest that the parties coordinate their schedules.

Prah v. Maretti

In *Prah v. Maretti*,[12] defendant began constructing a house, the location of which would block sunlight from reaching plaintiff's solar panels. Plaintiff alleged that he is entitled to "unrestricted use of the sun and its solar power" and demanded injunctive relief and damages.

The court said that in the nineteenth and early twentieth centuries, lack of protection for a landowners' access to sunlight was based upon three policy considerations: First, landowners' right to use their property as they wished was jealously guarded. Second, because artificial light could be used for illumination, sunlight was valued only for aesthetic enjoyment. Third, a right to sunlight would hinder property development.

These three policies are now obsolete. First, landowners' use of land is increasingly regulated for the general welfare. Second, access to sunlight as an alternative energy source is now important both to the landowner who invests in solar collectors and to society. Third, the need for easy and rapid development is not as great today as it once was, while the value of sunlight as a source of energy has increased significantly.

The court held that the plaintiff in this case has stated a claim under which relief can be granted and remanded the case for trial. The "law of private nuisance is better suited to resolve landowners' disputes about property development in the 1980's than is a rigid rule which does not recognize a landowner's interest in access to sunlight."

Depending upon the circumstances, obstruction of access to sunlight may or may not constitute a nuisance. The test in each case is whether the conduct complained of is reasonable.

Discussion

1. Why was the court in *Prah* and the US Congress, which encouraged solar energy with tax subsides, concerned with sunlight as energy in the 1980s? After two "oil shocks," the Arab Oil embargo, 1973–1974, and the Iranian revolution, 1978–1979, which caused large rises in oil prices, alternative sources of energy became more valued.
2. The court considered three policy considerations that governed nineteenth- and early-twentieth-century development and said that these policies had become obsolete by the late twentieth century. Do these policies and their obsolescence over time affect a Kaldor-Hicks analysis? As economic costs and benefits from different activities change over time due to changing economic circumstances, Kaldor-Hicks analyses reflect these changes, which in turn are reflected in changed public policy.
3. Apply Kaldor-Hicks to *Prah*. Which activity, residential view or solar heating, is more economically valuable? Do we need more facts? The value of a residential view is the difference between property value with and without the view; the value of solar energy is the net present discounted value of the electric energy generated from the solar panels. Whether view or solar energy is more valuable is difficult to determine; however, since the solar panels can be placed in several places, possibly the view is more valuable.
4. The court rejected the rigid rule that does not recognize a landowner's access to sunlight in favor of the law of private nuisance. This decision is inconsistent with the Coase Theorem, which requires a rigid property right or liability rule whereas the law of nuisance is couched in terms of what is reasonable under the circumstances.
5. One of the plaintiff's theories was based on an analogy with water law's doctrine of prior appropriation in the western United States under which the first user to appropriate the resource has the right of continued use to the exclusion of others. As discussed in *Bryant* and *Sturges*, courts usually reject first-in-time arguments when they conflict with Kaldor-Hicks.
6. Are the parties in *Prah* more likely to bargain after the court's decision? Since the court overturned the bright-line property right, that is, ruled that there is no right to light, and held that the case could proceed to trial under the law of nuisance, the test of which is reasonableness under the circumstances, neither side could be sure of a win after this decision, so both may have thought that bargaining would be prudent.
7. What would be the efficient result if both properties were under joint ownership? Depending upon valuations of the two activities, possibly solar panels could be moved, defendant could pay plaintiff's electric bills, or new home could be reduced in height.

CASES INVOLVING PARTIES WHO BARGAIN AS A GROUP

Spur Industries, Inc. v. Del Webb Development Co.

In *Spur Industries, Inc. v. Del Webb Development Co.*,[13] plaintiff, a development company, bought rural land on which to build a retirement community; the land's price was considerably less than if it were near Phoenix. As the retirement community expanded, it came within 500 feet of defendant's cattle feedlot, at which time 1,300 residential lots were unfit for development because of odors and flies from the feedlot. According to testimony, "a feedlot produces 35 to 40 pounds of wet manure per day, per head, or over a million pounds of wet manure per day for 30,000 head of cattle."

The court held that the feedlot was both a public and private nuisance and enjoined the feedlot's operation. However, it said that courts of equity are also concerned with protecting a lawful, albeit noxious, business from encroachment by others (e.g., tanneries of Bermondsey). The land developer is entitled to a permanent injunction, not because it is without blame, but because of damage to people in the retirement community. The developer must indemnify the feedlot for the cost of moving or shutting down as a result of the permanent injunction.

Discussion

1. Apply Kaldor-Hicks to *Spur*. Which activity, residential housing or cattle feedlot, is more economically valuable? Clearly, a large residential housing development is more valuable than a cattle feedlot (Kaldor-Hicks).
2. If the court *Spur* denied an injunction on grounds that Dell Webb came to the nuisance, would a Coasian (cooperative) bargain occur? Why or why not? What is the most that Spur could demand? At the limit, Spur could demand all Del Webb's profits from all land that would be diminished in value by proximity to its feedlot. Due to hold-up problem, courts rarely give a much lesser valued activity a property right over a higher-valued activity.
3. If the court had determined that Del Webb had an entitlement protected by a property right, which would allow Del Webb to enjoin operation of the feedlot without compensating Spur, would an injunction be efficient? Would it be equitable? An injunction against Spur is efficient under Kaldor-Hicks, which requires that Spur move, but is inequitable. Under Pareto efficiency, which requires that the cost of negative externalities be internalized to the activities that cause them, an injunction without compensation is neither efficient nor equitable. Recall the discussion of *Double R Cattle Co.* (chapter 1), where one of shortcomings of Kaldor-Hicks, when it is used to compare total gains to total losses (rather than marginal gains and losses), was discussed; Kaldor-Hicks can sanction doing nothing about even large negative externalities if the activity is very valuable.
4. In choosing a liability rule, does the court have any freedom to vary damage awards based on distributional considerations, that is, could the court grant Spur

monetary relief for noneconomic damage such as inconvenience without affecting efficiency? A court has some discretion as to damages as long as a court does not make the project economically unfeasible. A court might also want to discourage a high-valued activity from ignoring its effect on surrounding activity when it chooses where to locate.

5. If one corporation owned both the feedlot and the development company, how would it deal with these mutual interfering activities? Joint ownership would move the lesser valued activity, feedlot, and would internalize its moving costs.

Boomer v. Atlantic Cement Co.

In *Boomer v. Atlantic Cement Co.,*[14] land owners who filed a nuisance action against a cement plant alleged injury to their property from dirt, smoke, and vibration emanating from plant (externalities), and asked for injunction and damages. The two mutually interfering activities were landowners' right to quiet enjoyment of their property versus the company's right to produce cement regardless of its effect on surrounding homeowners. The New York Court of Appeals refused to grant landowners a property right enforceable by an injunction, which would prohibit the factory from operating. The court wrote that "total damage to plaintiffs' properties is, however, relatively small [$185,000] in comparison with the value of defendant's operation [in excess of $45,000,000 investment in a plant where over 300 people were employed] and with the consequences of the injunction which plaintiffs seek." The court refused to grant an injunction, which is a legal vehicle that is used to enforce property rights, and instead awarded damages as determined by a court, which is a liability rule that totally eliminates private bargaining.

> The ground for the denial of an injunction, notwithstanding the finding both that there is a nuisance and that plaintiffs have been damaged substantially, is the large disparity in economic consequences of the nuisance and of the injunction. This theory cannot, however, be sustained without overruling a doctrine which has been consistently reaffirmed in several leading cases in this court and which has never been disavowed here, namely that where a nuisance has been found and where there has been any substantial damage shown by the party complaining an injunction will be granted. . . . [T]o follow the rule literally . . . would be to close down the plant at once. . . . The parties could settle this private litigation at any time if defendant paid enough money and the imminent threat of closing the plant would build up the pressure on defendant.[15]

The court, which overruled its long-standing rule that an injunction will be granted where substantial damage due to a nuisance is found, must have believed that private bargaining would not result in an outcome that was efficient or distributionally fair. With injunctive power, which would grant monopoly power to landowners over the cement company, landowners could, at the limit, extort the company's entire profits as the price of removing the injunction. The cement company, rather than paying an extortionate price, may have shut down or curtailed its operations, which would clearly destroy value and not be efficient. The court refused to grant a property right, protected by an injunction, and sensibly limited the remedy to court-determined damages, a liability rule by which the court controlled self-interested behavior of the victorious plaintiffs by totally eliminating any bargaining rather than allowing plaintiffs to have extortionate power.

Discussion

1. If plaintiffs have an entitlement to clean air, which is protected by a property right (injunction), one person can veto a potential bargain (the "hold-up" problem); on the other hand, if the defendant has an entitlement to produce smoke, which is protected by a property right (no liability), one person and by extension all persons can refuse to contribute to compensate the factory on the assumption that everyone else will pay the factory to reduce pollution (free rider problem).
2. Do game theory analysis and the bilateral monopoly model fit *Boomer* better than the Coase Theorem? Even if homeowners bargain as a group, so that there are only two parties, homeowners' association and Atlantic Cement, if homeowners were granted an injunction, they could at the limit demand all of Atlantic's profit as the price of lifting the injunction. The court in *Boomer* was aware of the extortion possibility and refused to grant an injunction. Courts rarely grant a property right to a much lesser valued activity because they rightly assume that there will be exploitive bargaining as predicted by bilateral monopoly and game theory rather than cooperative bargaining predicted by Coase theorems.
3. How would a unified management, which owned both the cement company and the land on which the residences are located, coordinate these activities so as to minimize the negative effect of each activity on the other? Assuming, as stated in the case, that management of the cement company took all currently known precautions to reduce negative externalities, joint management of the two activities would insulate homes to reduce vibration and install air filtration and air condition systems, so as to minimize reduction in rental value of the homes from smoke and dirt.

FOUR LEGAL REGIMES FOR CASES INVOLVING MUTUALLY INTERFERING ACTIVITIES

Before Calabresi and Melamed wrote their article[16] which distinguished property rights from liability rules (discussed in chapter 11), the conventional view was that there were three ways to resolve nuisance claims involving externalities: 1) eliminate the nuisance by granting the plaintiff a property right and injunctive relief; 2) let the activity continue if the defendant pays damages; or, 3) let the activity continue by denying all relief, that is, a no liability property right (the converse of the first alternative). However, Calabresi and Melamed concluded that the statement of three rules in the context of their framework, where there are two possible winners in a suit, and where the winner's entitlement may be protected by either a property right or a liability rule, suggests that there must be a $(2 \times 2 = 4)$ fourth way to resolve claims (the converse of the second alternative): eliminate the activity if the plaintiff pays damages.

Four of the cases discussed above involving mutually interfering activities will be stated in the context of Calabresi and Melamed's 2 × 2 = 4 rules as depicted in Figure 12.3, below.

1. Plaintiff has an entitlement protected by a property right entitling it to an injunction (top-left cell). In *Sturges v. Bridgman*, plaintiff, physician, gets an injunction prohibiting defendant, confectioner, from operating his machinery. The injunction cannot be lifted unless defendant pays compensation in an amount satisfactory to plaintiff; plaintiff has monopoly power over defendant.
2. Defendant has an entitlement protected by a property right, no liability (bottom-left cell). Defendant in *Byran v. Lefever* does not have to remove the timber unless plaintiff pays compensation in an amount satisfactory to defendants; defendant has monopoly power over plaintiff.
3. Plaintiff has an entitlement protected by a liability rule (top-right cell). In *Boomer v. Atlantic Cement Co.*, defendant can take away plaintiffs' entitlement and continue to inflict negative externalities on plaintiffs (dirt, smoke, and vibration) if it pays damages as determined by a court.
4. Defendant has an entitlement protected by a liability rule (bottom right cell). In *Spur Industries, Inc. v. Del Webb Development Co.*, plaintiff can take away defendant's entitlement to inflict odor and flies (negative externalities) on plaintiff if it pays damages as determined by a court. This is Calabresi and Melamed's fourth rule.

When two mutually interfering activities have greatly disparate economic values, courts usually do not grant the much less valued activity a property right due to the hold-up problem, *Boomer*; courts also should not grant a much more valued activity a property right based upon a Kaldor-Hicks analysis, which compares total gains to total losses, because to do so would provide a rationale for inaction against even substantial negative externalities, *Double R Cattle Co.* (chapter 1). Calabresi and Melamed said: "The choice of a liability rule is often made because it facilitates a combination of efficiency and distributive results which would be difficult to achieve under a property rule."[17]

	Entitlement Protected by Property Right	Entitlement Protected by Liability Rule
Plaintiff Entitlement to Win	Injunction Prohibiting Defendant's Activity *Sturges v. Bridgman*	Defendant (more valuable activity), Pays Damages to Continue Pollution or Abates Pollution, Whichever Cheaper *Boomer v. Atlantic Cement Co.*
Defendant Entitlement to Win	No Liability *Bryant v. Lefever*	Plaintiff (more valuable activity) pays Damages to Abate Pollution *Spur Industries v. Dell Webb*

Figure 12.3 Four Property Rights and Liability Rules. *Source*: Created by the author.

Both game theory and bilateral monopoly predict that property right outcomes in bilateral monopoly cases will often be neither efficient nor distributionally fair. In these cases, efficiency and distributional equity are interdependent (rather than separable and independent as in the context of competition).

MUTUALLY INTERFERING ACTIVITIES WHERE HIGH TRANSACTION COSTS PROHIBIT BARGAINS

Where, because of the large numbers involved in mutually interfering activities (externalities), transaction costs are too high to permit bargains between those who have legal entitlements and those who do not, Coase concluded that the legal rule does matter in the sense that one legal entitlement is more consistent with efficient resource allocation than is its converse. In high transaction cost settings, he said that the law should assign the legal entitlement to the side that values it most according to Kaldor-Hicks.[18] This conclusion, with appropriate qualifications related to equitable income distribution, discussed below, is correct. Second, Coase maintained that in high transaction cost settings, a tax adjusted to equal the damage done may not bring about optimal conditions and is not necessarily superior to strict liability or no liability.[19] On the contrary, William Baumol showed that where the externality is of the public bad variety, which affects vast numbers of people (like the usual pollution problem), taxing the generator of the externality is optimal; moreover, compensation that varies with the amount of damage done (e.g., strict liability) is inefficient because those who are damaged by the externality are discouraged from taking precautions to minimize that damage.[20] Baumol's analysis is discussed below.

NO LIABILITY, STRICT LIABILITY, AND OPTIMAL TAXATION

The conclusion that legal rules matter in cases of high transaction costs was based on Coase's now well-known sparks-crops example. In the nineteenth century, steam locomotives on railways emitted sparks that often ignited crops and woods adjacent to the tracks. Given the large number of landowners adjacent to many miles of tracks, high transaction costs (hold-out and free rider problems) prevent the making of bargains that will efficiently coordinate the mutually interfering activities of railroad and farming. Coase says that the question is whether the railway should be liable in conditions in which it is too expensive for such bargains to be made. Once again, he analyzes the problem by asking which of two legal regimes, *strict liability* for crop damage or *no liability* for crop damage results in a more efficient resource allocation. Had he analyzed this problem using two property rights, no liability and trespass, the latter of which would allow any one farmer to prohibit the railroad from operating, the result would be both totally inefficient and untenable.

Although Coase employed a numerical example to illustrate the effects of no liability and strict liability regimes on the mutually interfering activities of railroad (sparks) and farming (crops), the essence of the example is depicted by Figure

12.4. In this figure, the number of trains per day that the railroad chooses to run is measured on the horizontal axis and the railroad's profit is measured on the vertical axis.

No Liability: Farmers' Crop Damage is External to Railroad

Under a regime of no liability for crop damage, farmers will not plant crops on land adjacent to the railroad tracks because cultivation of this land is unprofitable due to the large amount of uncompensated crop damage. Farmers will cultivate land that is far enough away from the tracks so that a profit can be made even if sparks from the railroad cause some crop damage on this land. Both the line that separates land that is profitable to cultivate from that which is not and the amount of crop damage done on profitable land will depend on how many trains the railroad runs per day. Because farmers reduce the amount of land cultivated less than in proportion to the increase in the number of trains, the extra or marginal crop damage is positively related to the number of trains per day. Under a regime of no liability, both the cost of crop damage and the cost of leaving land adjacent to the tracks uncultivated are external costs to the railroad and have no effect on the railroad's decisions.

In Figure 12.4, the curve labeled Railroad's Total Profit Under No Liability is mound shaped, which indicates that increasing the number of trains per day increases total profit up to a point, after which increasing the number of trains per day decreases total profit. The rationale for the shape of the profit curve is as follows: initially, the total number of passengers increases faster than the number of trains per day because more frequent trains increase convenience of train travel up to a point after which the market begins to be saturated and the total number of passengers increases more slowly than the number of trains per day. Under a regime of no liability, the railroad will maximize total profit by running 0A trains per day, which is greater than the efficient number of trains per day. Why?

Under a regime of no liability, resources will be misallocated because there will be too many trains per day, too much crop damage, and too few crops. The reason for the resource misallocation is that the costs of crop damage and lost profit on uncultivated land are internal to farming but external to the railroad.

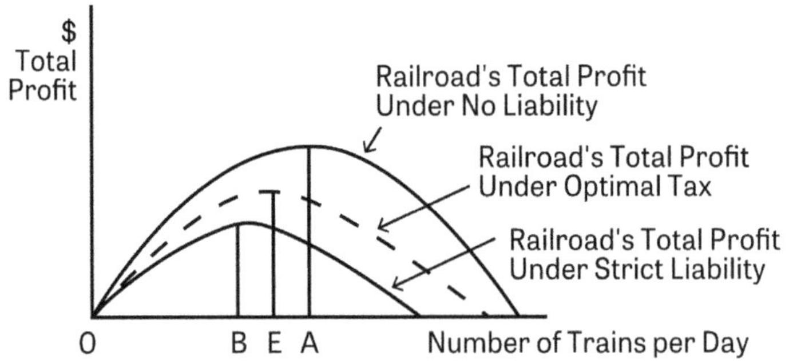

Figure 12.4 No Liability, Strict Liability, and the Efficient Tax. *Source*: Created by the author.

Strict Liability: Railroad's Lost Profit Is External to Farming

Under a regime of strict liability, the amount of crop damage done to farmers in relation to the same number of trains per day is greater than under a regime of no liability because farmers know that they will be fully compensated for crop damage and will plant crops on land adjacent to the tracks, which land was uncultivated under a regime of no liability. Because total profit is reduced when the railroad must compensate farmers for crop damage, the curve labeled Railroad's Total Profit under Strict Liability is beneath the curve labeled Railroad's Total Profit Under No Liability. Under strict liability, the railroad would run 0B trains per day, which is less than the efficient number of trains per day. Why? Because the profit foregone by the railroad from not running trains has no effect on farmer's decisions. Under a regime of strict liability, resources will be misallocated because there will be too few trains per day and too many crops. The reason for the resource misallocation is that the lost profit from not running trains is internal to the railroad but external to farming.

NO LIABILITY VERSUS STRICT LIABILITY

If the issue is limited to whether or not farmers should be compensated for crop damage done by sparks emitted from the railroad (whether there should be no liability or strict liability), Coase concluded that decision should depend upon whether the total value of production would be greater under no liability or strict liability (Kaldor-Hicks). In Coase's numerical example, the total value from railroad and farms under no liability exceeded the total value of production from railroad and farms under strict liability; therefore, he concluded that for the circumstances in his example, no liability was more efficient than strict liability. In fact, the law in England was that "if an engine is constructed with the precautions which science suggests against fire and is used without negligence, they are not responsible at common law for any damage which may be done by sparks."[21] Coase emphasized an important point; courts are rarely willing to shut down the more valuable of two mutually interfering activities as determined by Kaldor-Hicks efficiency. In other words, the Coase Theorem suggests that when transactions costs are high, entitlements should be assigned to the party who values them most, so that the costly process of exchanging property rights is unnecessary.

WHEN BILATERAL CARE IS REQUIRED, BOTH NO LIABILITY AND STRICT LIABILITY ARE INEFFICIENT

Coase's conclusion, that when activities on property interfere with each other and the efficient outcome requires mutual accommodation, both no liability and a strict liability are inefficient because the former results in too few crops and too many trains per day, no liability results too many crops and too few trains per day, has been generalized by economists.

In the law of torts (discussed in chapter 18), economists have shown that where the efficient solution requires that both parties take due care as in collisions between automobiles and pedestrians, both no liability and strict liability are not efficient.[22] No liability results in too little incentive for driver-injurers to take due care; strict liability results in too little incentive for pedestrian-victims to take due care.

In the law of contracts, liability for breach of contract is strict. However, both no liability and strict liability can be inefficient. No liability for breach of contract is obviously inefficient because potential contract breakers would only consider their own gains from breach, as when a buyer, who has a contract to buy wheat for $2 per bushel, has an opportunity to buy the same amount of wheat for $1.5 per bushel. If there were no liability for breach of contract, a buyer would only consider her or his gain from breach, $0.5 per bushel; seller's loss of $0.5 per bushel would be external to buyer's decision. Liability for breach of contract makes buyer internalize seller's potential loss (chapter 19).

Liability for breach of contract, which is strict, may be inefficient in many cases where information is incomplete because it causes the non-breaching party to over rely on the other party and not take into account that it may breach a contract. For example, in the financial meltdown of 2007–2008, buyers of securitized mortgages and credit default swaps so over relied on sellers' ability to satisfy their obligations on these instruments that the US financial system nearly collapsed. (Where information is complete, over reliance does not occur. Chapter 19.)

OPTIMAL TAXES FOR PUBLIC "BAD" TYPE EXTERNALITIES

William Baumol demonstrated that where an externality is of the public good variety like the usual pollution problem, which affects vast numbers of people over large geographic areas, 1) taxing the generator of the externality is optimal and 2) compensation to those who are affected by the externality that varies with the amount of damage done (strict liability) is counterproductive because it discourages victims from taking precautions to minimize that damage.

An appropriately chosen tax, levied only on the factory (without payment of compensation to local residents) is what is needed for optimal resource allocation under pure competition.

> When only smoke emission is taxed, with the tax level based on . . . [the size of the nearby population], what will prevent too many . . . [people] from moving near the smoky factory? The answer is that, when the tax on the externality producer is set properly, the externalities themselves keep down the size of the nearby population. Moreover, the level of the tax will control both the magnitude of the smoke and thereby (indirectly), the size of the nearby population. A high tax rate will discourage smoke and hence encourage migration into the neighborhood. A low tax rate will encourage smoke and, hence, drive residents away. A tax on smoke alone is all that is needed to control the magnitudes of *both* variables.[23]

As stated above, economists now generally agree that a regime of no liability will result in too much smoke and too few residents because the cost of smoke would be external to the factory but internal to the neighbors, and that strict liability will result in too many residents in the neighborhood and too little smoke because the cost of smoke reduction on product prices would be internal to the factory but external to its neighbors. The effluent tax as applied to the usual pollution problem does not encounter these perverse incentive problems.

Determining the Efficient (Optimal) Tax

Even though the efficient level of tax may not be known in advance, it is possible to approach the efficient level by trial and error: after a tax is imposed, the value of the combined output from farming and the railroad is readily observed; then, an iterative process which continues in the direction of any change in tax level that produces an increase in the total social product will converge on the maximum social value.[24]

The efficient tax *maximizes* the combined value of farming and the railroad by *minimizing* the negative effect of each activity on the other. At the efficient tax level, the reduction in railroad profit is equal to the reduction in rental value of farmers' land below the competitive level for land of similar quality. At any higher tax, the reduction of annual railroad profit is greater than the reduction in annual rental value of farmers' land below the competitive value; at any lower tax, the reduction in rental value of farmers' land is greater than the reduction of railroad profit. The Railroad Total Profit Under Optimal Tax curve depicts the effect of the optimal tax on the railroad; under the optimal tax, the railroad would run 0E of trains per day, which is the efficient number.

NOTES

1. Ronald Coase, "The Problem of Social Cost," *Journal of Law and Economics* 3 (1960): 1–44.
2. *Ibid.*, 1, 8, 15.
3. 105 Idaho 320 (1983), *rev'd.* 108 Idaho 602 (1985).
4. Robert Cooter and Thomas Ulen, *Law & Economics*, 6th ed. (New York: Addison-Wesley, 2012), 88.
5. Guido Calabresi, "Transaction Costs, Resource Allocation, and Liability Rules—A Comment," *Journal of Law and Economics* 11 (1969): 67, 68.
6. Robert Cooter, "The Cost of Coase," *Journal of Legal Studies* 11 (1982): 1, 17, 23 (the last or summary sentence of the quotation appears six pages after the other part).
7. Although the word "zealous" has been taken out of the cannons of ethics, it remains in the commentary.
8. Ronald Coase, "The Nature of the Firm," *Economica*, New Series 4 (1937): 386–405.
9. *Ibid.*, 395.
10. (1879) 4 CPD 172.
11. 11 C.D. 852 (1879).
12. 108 Wis.2d 223 (1982).
13. 108 Ariz. 178 (1972).
14. 26 N.Y.2d 219 (1970).
15. 26 N.Y.2d 223, 225.

16. Guido Calabresi and Douglas Melamed, “Property Rules, Liability Rules, and Inalienability: One View of the Cathedral,” *Harvard Law Review* 85 (1972): 1089–1128.
17. *Ibid.*, 1110.
18. Coase, note 1, 34.
19. *Ibid.*, 41–42.
20. William Baumol, “On Taxation and Control of Externalities,” *American Economic Review* 62 (1972): 307–322.
21. Coase, note 1, 30. To this general rule there was an exception for fires not exceeding £200. Coase speculates that “this exception was probably designed to help the small [land] holder.” *Ibid.*
22. John Brown, “Toward an Economic Theory of Liability,” *Journal of Legal Studies* 2 (1973): 323–349.
23. Baumol, note 19, 309, 312.
24. *Ibid.*, 319.

Chapter 13

Calabresi's Criteria for Allocating Accident Costs Common to Several Activities

Judge Guido Calabresi made seminal contributions to law and economics in three areas: he co-authored the article explaining the crucial distinction between property rights and liability rules, which was discussed in chapter 11; he made contributions to tort theory, which will be noted in chapter 8; and he wrote several important articles on risk distribution and a book, *The Cost of Accidents*, which are the subjects of this chapter.[1]

Coase and Calabresi's work have significant similarities and differences: First, Calabresi's initial article on accident-causing activities was published within six months of Coase's article on mutually interfering activities. Second, both theorists were concerned with efficient resource allocation in the context of externalities: Coase was concerned with determining which legal rules will efficiently internalize the costs of mutually interfering activities (externalities), for example, factories' right to pollute versus property owners' right to clean air; Calabresi was concerned with rules which internalize (allocate) the cost of accidents (externalities) common to several activities, for example, allocating costs from injuries in backyard pools among pool manufacturers, homeowners, and negligent victims.

Third, a significant difference is that Coase was concerned with the efficient solution to externalities in which the parties have the potential to bargain with each other (mutually interfering activities) whereas Calabresi was concerned with an efficient solution to externalities in which parties usually do not bargain with each other (accidents). He constructed four criteria for the allocation of accident costs common to several activities: cheapest cost avoider, cheapest insurer, non-externalization of costs, and best briber. In this chapter, Calabresi's four criteria are also applied to the allocation of heath care costs. For example, according to his criteria, are health care costs a legitimate cost of business?

Fourth, Coase used the Kaldor-Hicks which compares total gains to total losses as his efficiency criterion whereas Calabrasi criteria are consistent with Pareto efficiency.

Fifth, the work of both theorists has a similar flaw: Coase argued that in disputes between individuals involving mutually interfering activities, the questions of whether bargaining between individuals will achieve efficiency despite an inefficient legal rule is separable from the distribution of the cooperative surplus that results from such bargains; that is, he argued incorrectly that questions of efficiency and distribution can be separated in the context of bargaining where an inefficient legal rule gives one party monopoly power over the other. In the previous chapter, game theory and bilateral monopoly were used to show that contention is false.

Calabresi also argued the efficiency, that is, how best to deter behavior that subjects others to unreasonable risks of injury such as driving faster than the speed limit, can be separated from distribution, whether to compensate those who are injured in accidents. Calabresi argued that a no-fault system of accident compensation plus penalties for risky behavior such as strict enforcement of speed limits on roads, which he called specific deterrence, is more efficient than a system of accident compensation based upon fault. His argument ignores the way in which no-fault insurance may affect a victim's behavior: insured individuals often act more carelessly than uninsured individuals. This effect, which is known as moral hazard, will be discussed in chapter 16. However, unlike Coase, the truth of whose theorems depend upon the separability of efficiency and distribution, Calabresi's four criteria for the efficient allocation of accident costs common to several activities are independent of his contention that a no-fault system plus specific deterrence is more efficient than a fault system. Moreover, since Calabresi's four criteria are applied on a case-by-case basis, they are consistent with both the methodology of economics and of law.

EFFICIENT PRICING JUSTIFICATION FOR ENTERPRISE LIABILITY

Calabresi said that it was customary to say that enterprise liability is a form of loss spreading whereby the cost of accidents is passed on to consumers in the form of higher prices. Enterprise liability often spreads losses, but often does not. The conditions under which enterprises can and cannot spread losses will be discussed below. He set himself the task of providing a theoretically sound basis for enterprise liability as opposed to justifying it based on a vague notion of fairness.

The theoretical basis for enterprise liability type of loss spreading can be called the efficient pricing or cost internalization justification, which is based upon two principles: 1) consumer sovereignty: a market economy should produce what consumers want as determined by the goods and services they choose to buy; and 2) all costs of producing a product or service, both private and social (externalities), should be reflected in its price so that individuals will make informed choices.[2] Calabresi's view is consistent with Pareto efficiency (chapter 1), which requires that transactions be voluntary, make parties better off in their own estimation, and that indirect negative effects on third parties are internalized either in the cost of products or by a tax equivalent to the damage caused.

Parenthetically, it should be noted that we usually take efficient pricing for granted because most prices in the American economy are reasonably efficient in terms of reflecting actual costs of producing goods and services. However, the United States is experiencing a great upheaval in the way medical care is provided, which is discussed later in this chapter, due to the conflict between the current, inefficient medical pricing system and the huge increase in demand for medical services from an aging population, the uninsured, and the tremendous advances in medical technology during the last thirty years. A striking example of an inefficient pricing system and its consequences occurred in the Soviet Union under communism, which system collapsed in large part due to a disastrous misallocation of resources due to pricing that was unrelated to costs.

Pricing of Automobile Insurance: USA versus Sweden

Calabresi used a simple hypothetical example to clarify the efficient pricing justification for cost spreading. In the United States, the cost of accidents is paid for by private insurance; in Sweden, the cost of accidents is covered by social insurance, which is funded through taxes. Compare the decision of two families, one from the US, the other from Sweden, who are both considering whether to buy a second (used) car or to take taxis when both husband and wife need transportation.

In the United States, the annual cost of the used car is $400 plus $400 for automobile insurance, which is a total of $800, as opposed to an annual cost of $500 for taxis, so, rather than buying a second car, the family takes taxis and saves $300.

In Sweden, accident losses are funded through taxes so that the Swedish family pays insurance costs whether or not they buy a car. The cost of automobile insurance, which is a fixed cost, does not affect their decision whether or not to buy a used car, so the relevant choice is whether to spend $400 for a second car versus $500 for taxis. The Swedish family chooses to buy a car even though, if given a choice between buying a car versus taking taxis and having an additional $300 to spend, they would prefer the latter "bundle" of goods to the former. There is a misallocation of resources in Sweden. The demand for cars in Sweden is inefficiently large. Having automobile insurance funded by government is similar to a subsidy for cars.[3] Throughout his work, Calabresi stresses that cost shifting, subsidization, and cost externalization are the enemies of efficient pricing.

However, when the cost of accidents is common to several activities, the efficient allocation of those costs common among several activities is often unclear. Calabresi developed four criteria with which to decide what activity should bear accident risks that are common to several activities.

EFFICIENT ALLOCATION OF ACCIDENT COSTS COMMON TO SEVERAL ACTIVITIES

The exposition of Calabresi's criteria for the allocation of risk common to several activities will be made both more concrete and easier to understand after three examples are discussed: *O'Brien v. Muskin*, the shift from negligence-based liability for employee injuries to workers' compensation, and *Vincent v. Lake Erie Transp. Co.*

O'Brien v. Muskin

In *O'Brien v. Muskin,*[4] a twenty-three-year-old man, who arrived uninvited (trespasser) at a homeowner's property, jumped from an eight-foot-high garage into a 3 ½ foot deep, backyard, above-ground swimming pool and was seriously injured. The manufacturer's warning, "DO NOT DIVE," on the outer wall of the pool was in letters roughly ½ inch high. The young man sued the manufacturer of the swimming pool to recover damages for his injuries. The Supreme Court of New Jersey held that the trial court erred in taking away from the jury two issues: whether manufacturing a pool with a slippery, non-embossed vinyl liner constituted either a design defect or manufacturing defect; and, whether the risk posed by the pool liner outweighed its utility (risk-utility analysis).

In the context of the Judge Hand test (chapter 3), which attempts to deter careless behavior, the young man is the most negligent party by far and should bear most if not all of the risk. However, from a broader perspective, the risk of someone being injured in a backyard swimming pool is a cost that is common to three activities: swimming in backyard pools, production of pools, and liability of homeowners for general risks on their property, which include those posed by swimming pools. Which of these three activities, production of pools, landowners' responsibility for property, or recreation by individuals, should bear the risk of this common cost is not an easy question to answer.

Vincent v. Lake Erie Transp. Co.

Calabresi agrees with Coase that there are circumstances where accident costs may be placed upon either of two activities because they are associated by bargaining.[5] As an example, he discussed the famous case of *Vincent v. Lake Erie Transp. Co.*[6] In *Vincent*, a ship was moored to a dock while unloading its cargo when a severe storm occurred. The ship's master added more lines in order to prevent the ship from drifting from the dock and possibly being driven on the shore and destroyed. The ship, which was continually thrown against the dock by the gale, did substantial damage to the dock. The court held that the ship owner was liable to the dock owner for the damage. Calabresi says that if the ship owner is liable, dockage fees will be less; if the ship owner is not liable, dockage fees will be more. This analysis, which Calabresi calls the pure theory loss-distribution argument, employs Rottenberg's corollary for liability rules. He says that *Vincent* can be distinguished from the nuisance cases analyzed by Coase because the two parties are already bargaining so that there is no added cost when the loss is placed on either party.[7]

However, whether or not the ship owner is liable to the dock owner for this common cost often does make a difference even in competitive markets due to strategic considerations. Consider *Ploof v. Putnam*,[8] which is a companion case to *Vincent* in many torts casebooks. Under stress of weather, the ship in *Ploof* moored to a private dock without permission. A person for whom dock owner was legally responsible unmoored the ship in order to prevent the ship from damaging the dock by being thrown against it in the storm. The dock owner was held liable for the damage the ship sustained when it drifted upon the shore. If ship owners are *not* liable for damage to docks, dock owners may refuse to allow ships to dock when storms are impending, may charge exorbitant rates in emergency situations, and will have an incentive as in *Ploof* to unmoor vessels or prohibit them from docking so as not to incur uncompensated damage to their docks, whether or not dockage fees have adjusted upward under the no liability rule. Thus, it does make a difference whether ship owners are liable to dock owners for damage whether or not they can bargain in advance. The efficient liability rule is for ship owners to be liable for damages to docks.

CALABRESI'S CRITERIA FOR ALLOCATION OF COMMON ACCIDENT COSTS

Allocation of Common Costs within a Business Is often Arbitrary

Calabresi says that just as there is no formula for allocating overhead cost among activities that share the same facilities, there is no precise formula for allocating accident costs among several activities and parties in imperfect competition. An example of common costs internal to a business will illustrate his point.

Assume there is a retail department store, which has two departments: a jewelry department and a furniture department. The managers of the store wish to set the prices for their products. They decide to use the *fully distributed* cost method of pricing, which requires that every cost that is incurred by the store must be allocated to and included in the price of some product.[9] In short, the price of each product must cover all the costs that are attributable to it.

First, consider *direct costs*, that is, costs wholly attributable to one product or another. The wholesale cost of jewelry is allocated to the price of jewelry, just as the wholesale cost of furniture is allocated to the price of furniture. If the jewelry department has one sales person while the furniture department has three because it is spread over a much larger area in the store, the price of furniture must cover the salaries and commissions of three sales persons, while the price of jewelry must cover the cost of only one sale's person. Assume further that the store has several trucks it uses to deliver furniture (buyers of jewelry take their purchases with them when they purchase them). The cost of the trucks as well as their operating and maintenance expenses must be borne solely by the furniture department.

Now consider the problem of joint or common costs, that is, costs that are common to both departments, such as rent for the store (or amortization and taxes if the building is owned by the store), heat, electricity, accounting, management costs, etc. Discussion of the allocation of one common cost, say electricity, will be sufficient to explain the problem of common costs. The common monthly electric bill could be allocated in several ways: 1) Electricity cost could be split in half; one-half allocated to the jewelry department and one-half allocated to the furniture department. 2) Electricity cost could be allocated on the basis of space. If the furniture department covers an area of the store that is twice as large as the area of the jewelry department, two-thirds of the electric cost would be allocated to the furniture department and one-third to the jewelry department. 3) Electricity cost could be allocated on the basis of sales. Assume that by dollar volume the jewelry department has three times the sales of the furniture department because the jewelry department sells a great deal of expensive jewelry. Under this method of allocation, three-fourths of the electric cost would be allocated to jewelry and one-fourth to furniture. There is no precise formula that dictates which allocation of common costs is efficient. The allocation of accident costs, which are common to two or more activities, as in the *O'Brien* case, has similar difficulties.

Four Criteria for Allocating Common Accident Costs

Since competition is often imperfect and accident costs are often common to several activities, which of several parties bears accident costs (e.g., employer-employee, automobile manufacturer-driver-pedestrian, product manufacturer-consumer) affects efficiency. Calabresi suggests several guidelines for the efficient pricing of common accident costs.[10]

Cheapest Cost Avoider or Cheapest Preventer

First, one party may be the cheapest cost avoider or cheapest preventer. For example, circular saw table manufacturers are better able than consumers to decide how to minimize the sum of all costs, manufacturing plus accidents. If circular saw table manufacturers bear accident costs that result from manufacturing and design defects, then the total costs of manufacturers of cheap, unsafe saws will be pressured to make their saws safer in order to reduce total costs. On the other hand, a person who is injured while using a saw without the saw blade's protective cover is guilty of product misuse as is the young man in *O'Brien* who jumped into a 3 1/2-feet deep swimming pool from the top of a garage; in cases of product misuse, the users are the cheapest cost avoiders. If careless or accident-prone individuals are responsible for accidents they cause, the higher premiums they must pay for private insurance will give them an incentive to be more careful. In general, if we are concerned with giving parties efficient incentives to prevent accidents, that is, if we focus on the cheapest cost preventer or cost avoider, a fault system (negligence) is usually efficient.

Cheapest Insurer

Second, one party may be far better able than others to evaluate the risk of accidents, that is, better able to evaluate expected accident costs. If buyers of circular saw tables must bear the cost of circular saw accidents, the probability of which is usually unknown to them, but is in fact quite high, they will neither purchase insurance nor buy a more expensive but safer saw. On the other hand, circular saw table manufacturers are more likely to be aware of the expected accident costs from using saws. Such informational asymmetry means that the manufacturer is usually the cheapest insurer because it can more accurately determine the need for and cost of insurance. The party with the information advantage is also usually in a better position to bargain with insurance companies or, if a business is large enough, it can self-insure. On the other hand, if consumers can evaluate the accident potential of products from different manufacturers more cheaply than manufacturers can evaluate the accident potential of different consumers, it might be more efficient for consumers to bear accident costs rather than manufacturers. In the context of employment relations, employers usually have more accurate information on the number, probability, and cost of on-the-job injuries than do employees and so are usually the cheapest insurers. In general, if we are more concerned with spreading losses from accident-causing activities, that is, we focus on the cheapest insurer, a no-fault system is more efficient.

If manufacturers bear accident costs, the price of circular saw tables will include both manufacturing costs and accident costs of using saws. In potential bargaining situations like those of manufacturer-consumer, allocating accident costs to manufacturers rather than to consumers means that consumers are compelled to buy insurance. A careful consumer may prefer to pay a lower price for the product because she will be overcharged because insurance companies base premiums on average accident rates. So, too, a manufacturer of safe products who is charged premiums based on average industry accident rates will be overcharged. On the other hand, a careless consumer will be undercharged.

Sometimes the cheapest insurer and the cheapest cost avoider are different parties as in many on-the-job accidents in which the employer is the cheapest insurer and the employee is the cheapest cost avoider. Due to possible conflicts in the two guidelines, cheapest insurer and cheapest cost avoider, an efficient liability system may require a combination of both fault and no-fault systems. For example, employee injuries are covered by workers' compensation unless employees are injured under the influence of alcohol, controlled substances, or injuries are self-inflicted.

Whether particular costs should be classified under the cheapest cost avoider guideline or the cheapest insurer guideline may depend upon how broadly the accident-causing activity is defined. For example, in *O'Brien*, the Supreme Court of New Jersey held that a jury might determine that above-ground swimming pools are so inherently dangerous, even if they are designed as safely as possible, that a manufacturer should bear the risk of all foreseeable accidents, even to a person who jumps into a 3 ½ feet deep, swimming pool from the top of the garage. Moreover, the Court said that it is possible that enterprise liability for all above-ground swimming pool accidents may dissuade manufacturers of this product, which is a luxury as opposed to a product that fills a critical need, from placing the product on the market; that is, the court said that if above-ground swimming pools cannot bear cost of foreseeable accidents, maybe they should not be produced.[11] On the other hand, if manufacturers are only held liable for pool-related accidents not due to negligent misuse, manufacturer's liability will be more narrowly circumscribed. What if automobile manufacturers were held liable for all automobile accidents regardless of whether drivers are at fault?

Do Not Externalize Accident Costs

The third, and maybe the most important guideline is that the cost of accident-causing activities should not be placed on parties or activities which are unrelated to the activities that generated accidents or risk. For example, if the cost of circular saw table accidents is paid for out of general social insurance funded by taxes, such loss allocation would be equivalent to a subsidy to manufacturers. In other words, the third guideline is that there should be no cost externalization or cost shifting.

Best Briber

Fourth, Calabresi also constructed a guideline, called the best briber, which incorporates Coase's emphasis on the possibility of bargaining and mutual accommodation among persons affected by externalities. Under the best briber guideline, accident costs should be allocated to the party who can initiate an exchange of entitlement, if necessary, at least cost.[12] The best briber is the activity or party who is aware of the relevant risks, has information concerning the party with whom it is most advantageous to make transactions, and can enter into such transactions with the least use of coercion. In *Vincent* and *Ploof*, discussed above, neither ship nor dock owners have an advantage in terms of the cheapest cost avoider or cheapest insurer guidelines. However, ship owners are better bribers than dock owners because less coercion is involved if ship owners are liable for damage to docks caused by acts of God. If ship owners are *not* liable for damage to docks, dock owners may refuse to allow ships to dock when storms are impending.

SAINT BARNABAS V. COUNTY OF ESSEX

In *Saint Barnabas Hospital Center v. County of Essex*,[13] an inmate in a county jail, who was committed for a fifteen-day sentence, set himself on fire and was sent to a burn center where he was hospitalized for seven weeks, well beyond the termination date of his sentence. The total hospital bill was nearly $54,000. The county agreed to pay nearly $5,500 for the first four days of hospitalization until the termination of his sentence. The issue in the case was whether the county or the hospital should bear "the cost associated with the medical treatment of an indigent initially requiring such treatment while a ward of the county?" The Supreme Court of New Jersey held that there was no implied contract between county and hospital to pay for the services, since the circumstances did not sufficiently negate the idea that they were gratuitous. It said that Saint Barnabas would be compensated from the legislatively mandated Uncompensated Care Trust Fund, which is a uniform statewide "uncompensated care add-on" that is used in rate-setting at all hospitals. Hospitals whose indigent health care costs exceed the amount collected from patients receive payments from the fund and hospitals that with the add-on collect more than necessary to cover costs remit the net difference to the Fund.

Discussion

1. As between the Uncompensated Care Trust Fund (UCTF) and Essex County, which maintains the jail, which is better able to prevent prisoners from setting themselves on fire? Only the county, not UCTF, can protect against fires set by inmates. The county is the cheapest cost avoider.

2. Consider county versus UCTF from the perspective of risk spreading; that is, who actually bears or should bear the cost of caring for the man who set himself on fire? If the county bears the cost, the ultimate bearers of the cost depend on its tax base. Is the county government financed by property taxes, sales taxes, or income taxes, or some combination thereof? For example, sales taxes are usually regressive. If the county includes a shopping mall within its territory, a sales tax would be borne by many who are not county residents.

   If the UCTF bears the cost, it would be spread among those who are hospital patients. Since most hospital bills are paid by patients' employers rather than by patients out of their own pockets, the cost would be borne by businesses within and beyond New Jersey, some of whom would be regional, national, or international businesses. Should the cost of a fire in a county jail be paid by county tax payers or should it be borne by business in general? If American business has to bear the cost of fires in a local jail and indigent patients in general, how will it fare in international competition against firms who reside in countries with national health insurance?
3. In *Saint Barnabas*, are the cheapest cost avoider and the cheapest insurer the same party or different? The county is clearly the cheapest cost avoider. Depending upon your answer to the previous question, the cheapest insurer could be either the county or UCTF.
4. According to the criterion of do not externalize costs, the county, which is responsible for the injury to the inmate, should bear the cost of his hospitalization.
5. The Court said: "The implausibility of imposing insurer-like duties on a county is evinced by positing the case of an indigent prisoner who during the pendency of a county jail term, discovers he has cancer or AIDS, requiring extensive medical treatment for the rest of his life." Are cancer and AIDS distinguishable from setting oneself on fire? Are not diseases versus injuries, no-fault versus fault distinguishable categories?

SHOULD ACCIDENT COSTS INCLUDE SUBJECTIVE VALUE?

Assuming that an activity should internalize the accident costs it generates, how broadly should costs be defined is open to question. There is little dispute that the accident costs should include property damage calculated at market values (according to a liability rule), injury costs in terms of hospital bills, and lost wages. However, a market test does not include the noneconomic value it destroys. It is possible to argue that accident costs should also include subjective value, that is, noneconomic values that are converted into dollar amounts. For example, a jury is sometimes allowed to add the cost of pain and suffering to the cost of property damage and lost income. Reasonable persons can differ as to whether and how much pain and suffering should be converted into dollar amounts.

The question what market value should be placed on lives lost in accidents is susceptible to widely varying answers. Courts often place a monetary value on a life based upon the present value of the expected earnings that the deceased would have earned until retirement, adjusted upward for inflation and downward for the living costs and taxes that the deceased will not incur.

Economists often determine the value of a life in relation to the risk premiums that individuals demand to bear above average on-the-job risks (this method will be discussed in chapter 15). This second calculation of the value of a life, which is based upon how individuals behave, is usually substantially higher than the expected earnings approach, which is based upon the deceased's value to his estate. Third, some would argue that the value of a life should be equivalent to what it would take to buy a willing victim (property right). This third value is probably infinite for most people. Whether and what kinds of noneconomic costs should be considered as costs of accident-causing activities involve political, judicial, and jury decisions.

WHEN CAN ENTERPRISES PASS ON COSTS?

Calabresi said that it was customary for writers to justify efficient risk allocation or enterprise liability by vague statements that activities *should* bear the costs they engender and that such cost will be passed on to consumers as higher prices. He said that efficient risk allocation sometimes spreads losses, but sometimes it does not. The conditions under which enterprises can and cannot pass losses on to consumers will be discussed below. Enterprise liability may cause losses to be passed forward to consumers, backward to suppliers of labor and materials, or remain on the enterprise and its owners.

The Three Pass-On Cases

There are two extreme and one general pass-on cases. Case 1: If *one* firm in a competitive industry has higher costs than other firms, it cannot pass on its higher costs to consumers because price is determined by the market; all firms are price takers. Buyers will not buy from a firm which charges a higher price than the competitive price because they have alternative firms with which to deal at the lower price. A single firm with higher costs will have lower profits and output.

Case 2: If *all* firms in a competitive industry have higher costs, and the industry sells products that consumers greatly desire, the quantity purchased will be little changed after a price rise (e.g., perfectly inelastic demand), so the higher costs will be totally passed on to consumers. In the real world, the case where higher costs are totally passed forward to consumers is approximated by government levies of sin taxes on alcohol and cigarettes. The objective of sin taxes is more often to raise revenue than to discourage consumption; it is a mere incident of the tax on alcohol that it may also serve to reduce consumption slightly; with respect to cigarettes, the motives of revenue raising and demand reduction seem to be competing. The classic case of unchanged demand for a product after a price increase is the quantity of insulin demanded by a diabetic or a drug by an addict.

Cases 3 is the general case where firms in a competitive industry can pass on some but not all higher costs to consumers: how much of the higher costs are passed on depends upon how responsive quantity demanded and quantity supplied are to a rise in price (relative price elasticities of supply and demand). The more responsive quantity supplied and demanded are to price rises (more elastic), the less likely costs will be passed on to consumers; the less responsive quantity supplied and demanded are to price rises (more inelastic), the more likely costs will be passed on to consumers. After a price increase, the likelihood that quantity demanded by consumers will be little changed, so that higher costs will be passed forward to consumers (inelastic demand), is higher 1) the smaller the proportion of consumers' budgets accounted for by that product (e.g., salt), 2) the fewer the substitutes available for that product (e.g., gasoline for automobiles), and 3) the more difficult it is to increase supply of substitute products (e.g., nuclear energy as substitute for fossil fuels). In the general case, some but not all, higher costs will be passed on to consumers.[14]

To the extent that firms with higher costs cannot pass on the entire increase to consumers, some costs are passed backward to the factors of production, land, labor, materials, and management in the form of lower wages, input prices, or returns to owners. Higher costs will be passed backward to factors of production that have supply curves that are not responsive to price (inelastic), that is, to factors of production that are specialized; for example, if workers in an industry have particular skills that cannot be transferred to other industries or if an industry has specialized equipment that has no other uses, higher costs will result in reductions in wages or profits. In sum, in the general case, Case 3, the percentage of an increase in costs that can be passed forward or backward depends upon the relative responsiveness of supply and demand to a rise in price (price elasticities).

POLITICAL VERSUS MARKET DECISIONS CONCERNING ACTIVITIES THAT CAUSE INJURY OR DEATH

Although economic theory suggests that an activity which is profitable when accident costs are internalized should be allowed, the decision whether or not to legalize an activity is not made solely by cost-benefit analysis. The sale of fireworks is prohibited even if manufacturers can make a profit after paying all resulting injuries.

We justify government restrictions on market decisions by external social costs and benefits generated by activities, concepts which can be defined either restrictively or expansively. The concept of external social benefits is used to justify subsidization of activities like universal public education that cannot pay its own way; external social costs is use to prohibit some activities like prostitution that can pay its own way.

Decisions with respect to significant activities such as whether to allow the sale of fireworks can easily be decided by political choice. But whether to use steel or plastic in constructing an automobile is not easy to answer by a political process because there are too many choices to make. Since almost any product can be made safer or stronger in myriad ways, every product involves thousands of cost-benefit decisions in which the costs of added safety must be weighed against the benefit of providing the public with a cheap product. Products that occasionally cause injury or death are usually allowed if the manufacturer can make a profit after paying compensation.[15]

HEALTH CARE, COST SHIFTING, AND COST EXTERNALIZATION

In this section, a first look at health care in the United States will be given from Calabresi's perspective. (The health care system of the United States cannot be fully analyzed until the concepts of adverse selection and moral hazard are discussed in chapter 16.) The medical system in the United States is grossly inefficient from Calabresi's efficient pricing perspective because it is rife with cost externalization or cost shifting.

Cost Shifting Due to Lack of Universal Coverage

First, notwithstanding the passage of the Affordable Care Act of 2010 (Obamacare), health care coverage in the United States is not universal.[16] When the uninsured get sick, they go to emergency rooms at hospitals. If one has to wait until there is an emergency before receiving health care for a life-threatening condition, the likelihood is high that the condition will not get treated until it is at an advanced stage, thus increasing treatment expense and reducing the likelihood of recovery. Individuals who go to emergency rooms either bear the cost of their health care directly, which drives many to bankruptcy, or the cost of their health care, inadequate as it is, is shifted or externalized to paying patients, their employers (recall *Saint Barnabas*, discussed above), or to the government.

Second, prior to the Affordable Care Act, those individuals who have *preexisting conditions* and are not covered by employer-based health care plans often could not buy private health insurance at affordable rates. They, too, ended up in emergency rooms of hospitals, which are often paid for by those with health insurance, as discussed. (In chapter 16, these problems will be discussed in the context of adverse selections and moral hazard.)

The Affordable Health Care Act addresses these problems by requiring insurance companies to offer health insurance to all individuals at nondiscriminatory rates, regardless of preexisting conditions, and mandates that all individuals either have health care insurance or pay a "tax" or "penalty" if they do not; the Act also includes insurance cost subsidies for low income individuals.

Cost Shifting Due to Third Party Payment

A second reason for externalization or cost shifting of medical expenses is *third party payment*; a majority of those who are covered by health insurance do not pay for it directly. This cost externalization problem is exacerbated because the insurance premiums paid by employers are tax-deductible to employers and not treated as part of employees' income; that is, health insurance contains a large component of government subsidy. In addition, for retired persons and the poor, health care is largely paid for by social insurance financed through taxes (Medicare and Medicaid). Moreover, as discussed in *Saint Barnabas*, some of the cost of providing health care to the uninsured is shifted to patients with health insurance and to their employers. This means that most individual decisions about how much health care to buy are made by patients and physicians, but most of the payments are by a third party, so that those who make the decisions have no direct incentive to save money. This third party payment or cost externalization causes patients to be insensitive to the price of medical care. Because much of the cost of health care is subsided from general government funds financed by taxes, the price of medical care to patients is lower than it "should" be from an efficient pricing viewpoint.

Cost Shifting of Administrative Expenses

Third, a high percentage of medical expenditure in the United States is due to cost shifting of administrative expenses. Estimates of administrative expenses in the private sector of United States health care range from a low of 13 percent to a high of 31 percent.[17] A great deal of the high administrative expense in the United States is due to cost shifting in a decentralized system. Insurance companies have large bureaucracies whose job it is to question, limit, and deny coverage, so that medical costs can be shifted back to patients or the government. In order to counter potential denial of coverage by insurance companies, physicians hire staffs whose job it is to make sure that the services physicians perform on patients will be covered by insurance. If insurance coverage in a decentralized system were standardized with mandatory coverage for all or there were a single insurance payer, there would be much less dispute about the extent of coverage so a great deal of cost-shifting administrative expenses could be eliminated.

Another problem with the current decentralized system is that insurers have little incentive to provide preventive care because insurance is provided to consumers by employers who often change their insurance providers; an insurance company that invests in preventative care bears the cost of such care, while other companies may reap the benefit (a positive externality).

The three cost-shifting considerations discussed above, lack of universal care, third party payment, and high administrative expenses, are major reasons why health care in the United States absorbs a higher percentage of GDP than in other advanced industrial countries. Despite the higher health care costs in the United States, statistics on population longevity and outcomes in different medical specialties in the United States are not superior to, and by some measures, are inferior to those of other advanced countries. United States total spending on health care has grown from nearly 14% GDP in 2001 to 17% of GDP in 2013. France, 11.7%; Germany, 11.3%; Canada, 10.9%; Japan, 10.3%; Sweden, 9.7%; and Great Britain, 9.1%.[18]

SHOULD HEALTH COSTS BE A COST OF BUSINESS?

In the context of health care costs, consider Calabresi's question, "What is the cost of what?" Is health care a cost that business should bear? First, the United States is one of the few advanced industrial nations where health care is primarily paid for by employers (so too is France). Second, the reason why US employers pay for health care is due to a historical accident. Employers agreed to pay for health care as a way around post–World War II wage controls. At that time, health care costs were a much smaller percentage of GDP, so having employers pay for health care was a small fringe benefit. It has also been alleged that the issue of race in America played a major role in the failure to enact universal health care after World War II because those who favored segregation assumed that a government-run health program would mandate racial integration in hospitals.

Let us apply Calabresi's criteria, cheapest cost avoider or preventer, cheapest insurer, don't externalize costs, and best briber with respect to health insurance. With respect to the first criterion, cheapest cost avoider, employers have little control over their employees' health (they can force employees not to smoke on the job and may give incentives to engage in exercise).

Although employers are a cheaper insurer than individuals because employers can buy policies that cover all their employees regardless of individual health risks or preexisting conditions (employers do not allow insurance companies to refuse coverage to "adverse selections"), government has a significant advantage over employers with respect to the cheapest insurer criterion due to significantly lower marketing and information costs and savings from the elimination of battling bureaucracies. Consumer health care cooperatives may also be cheap insurers.

With respect to the do-not-externalize criterion, having employers pay for health costs, externalizes a purely personal expense having nothing to do with the product or service being produced by the firm. With respect to the best briber criterion, here again employers have superior bargaining power to that of individuals who have to deal one-on-one with insurance companies, but government has superior bargaining or bribing power to that of employers. The Affordable Health Care Act sanctions consumer collectives as purchasers of group insurance. In sum, Calabresi's criteria provide support for either consumer cooperatives or single-payer plans of the United States (Medicare and Medicaid) and other industrialized countries, which have government-financed medicine provided by private physicians. Whether socialized medicine, government provision of health care is efficient, is another question, which is not discussed here.

From an efficient pricing perspective, does it make a difference whether individuals pay for their health care expenses directly or whether employers pay for health care? If there were no tax subsidy considerations, Rottenberg's corollary says that if liability rules are clearly specified one way or the other, markets will produce efficient results regardless of the specification of liability rules. However, labor markets are imperfectly competitive and transaction costs are significant. American wages tend to be "sticky," that is, workers resist downward pressure on wages even if medical costs are increasing faster than they have in the past.

Second, unlike the cost for workers' compensation insurance, which is both stable and a small percentage of the cost of labor, health care cost is a large and growing part of wage compensation. Moreover, the stagnation of median wage for over a decade has in significant part been caused by health care costs which have consistently exceeded the rate of inflation.

Third, in this age of global competition, if the market mechanism of compensating wage differentials does not work perfectly so that American manufacturers cannot pass backward to workers the entire cost of health care, they are put at a severe competitive disadvantage with respect to firms in other advanced countries in which health care is supplied by the government or firms in third world countries where there is no health care. Employers may seek to outsource work to other countries.

NOTES

1. Guido Calabresi, *The Cost of Accidents: A Legal and Economic Analysis* (New Haven, CT: Yale University Press, 1970); "Some Thoughts on Risk Distribution and the Law of Torts" (hereafter, Risk Distribution), *Yale Law Journal* 70 (1961): 499–553; "The Decision for Accidents: An Approach to Non Fault Allocation of Costs" (hereafter, Decision for Accidents), *Harvard Law Review* 78 (1965): 713–745; "First Party, Third Party, and Product Liability Systems: Can Economic Analysis of Law Tell Us Anything about Them," *Iowa Law Review* 60 (1984): 833–851.
2. "Risk Distribution," note 1, 517–518.
3. Ibid., 502.
4. 94 N.J. 169 (1983).
5. Calabresi, "Decision for Accidents," note 1, 725–26.
6. 109 Minn. 456 (1910).
7. Calabresi, "Decision for Accidents," note 1, 726, 728.
8. 81 Vt. 471 (1908).
9. The opportunity cost of capital will be omitted for simplicity.
10. Calabresi, "Decisions for Accidents," note 1, 725–732; Calabresi, "*The Cost of Accidents*," note 1, 60, 139, 144–146, 150.
11. 94 N.J. 184–185.
12. Calabresi, *The Cost of Accidents*, note 1, 150.
13. 111 N.J. 916 (1988).
14. The question of whether costs at one level of production or distribution can be passed on or spread forward to consumers has been extensively litigated in antitrust law where the issue is whether a monopoly overcharge at one level of production has been passed on to other levels and to consumers. *Illinois Brick Co. v. Illinois*, 431 U.S. 720 (1977); *California v. ARC America Corp.*, 490 U.S. 93 (1989).
15. Calabresi, "*The Cost of Accidents*," note 2, 19–20.
16. Jessica Smith and Carla Medalla, *Health Insurance Coverage in the United States 2013* (Washington, DC: US Census Bureau, 2014).
17. (25%) David Himmelstein, Miraya Jun, et al., "A Comparison of Hospital Administrative Costs in Eight Nations: US Costs Exceed All Others by Far," *Health Affairs* 33 (2014): 1586–1594.
18. World Bank Data, *Health Expenditure, Total (% of GDP)* (Washington, DC: World Bank Group, 2015).

Chapter 14

Posner's Economic Analysis of the Common Law

ECONOMIC PRINCIPLES UNIFY THE COMMON LAW

In his classic book, *Economic Analysis of Law*, which was first published in 1972 and is now in its 9th edition (2014), Judge Richard Posner was the first to argue that economic analysis of law is not limited merely to antitrust law, taxation, the Coase Theorem, and Calabresi's analysis of the cost of accidents, but that it can be used to analyze the entire common law, which is mostly judge-made law, as opposed to statutory law enacted by legislatures. He says: "[E]conomics is the deep structure of the common law and the doctrines of that law are the surface structure."[1]

In Posner's economic analysis of the common law paradigm, he shows that a tort (civil wrong) problem can be translated into a contract problem by asking what enforceable promises individuals would agree to in advance concerning precautions to prevent injuries that should be taken, or if injuries do occur, what compensation should be paid to injured parties. Conversely, a contract problem can be translated into a tort problem by asking what penalties are necessary to prevent the party who is yet to perform her or his promise from engaging in opportunistic behavior with respect to the party who has performed her or his promise first. Both tort and contract problems can be translated into property law by asking what collection of rights and liabilities create incentives to prevent wasting resources.[2]

COASE THEOREM VERSUS BILATERAL MONOPOLY

His lawyer-like insistence that bilateral monopoly is a more accurate analysis of bargaining power in one-on-one disputes than is the Coase Theorem is consistent with legal methodology (chapter 11) where an economic theory which conflicts with hard cases that the legal community generally considers to be correctly decided may be insightful but is not useful in specific cases unless there is a principled way to distinguish cases in which a theorem is correct from those in which it is not.

Posner begins with a Coasian theme: "when transaction costs are low, common law doctrines create incentives for people to channel their transaction through the market. . . They do this by creating property rights (broadly defined) and enforcing them . . . to prevent coerced transfers." The law of property establishes rights, the law of contracts regulates exchange of such rights, the law of torts protects these rights against unreasonable interference, and the criminal law prohibits coerced transfer of these rights. In situations in which "transaction costs are prohibitively high," judges decide cases so as to "mimic the market."[3]

However, Posner parts company with the Coase Theorem in the sense that he believes that the law also has a second function to perform, which is to reduce transaction costs by setting limits on strategic behavior and bargaining power in circumstances of bilateral monopoly, which he argues are ubiquitous (as discussed below).

He points out that there are many doctrines in the common law that would be unnecessary if the Coase Theorem were true. As an example, he discusses the common law doctrine, mitigation of damages. Assume that a retailer-buyer of a product has a contract to buy 10,000 shirts. Assume further that after the manufacturer-seller has produced and delivered 4,000 shirts, the retailer's market for shirts collapses due a change in taste by consumers, so she informs the seller that she does not need and will not accept the 6,000 shirts, which have not yet been produced. After the buyer has breached the contract, seller may threaten to complete the contract, produce the unwanted 6,000 shirts, and sue for the entire contract price.[4]

The doctrine of mitigation of damages deters contract completion by disallowing damages for costs of production (labor and materials) of shirts that should not be produced after notice of termination; the doctrine limits seller's damage remedy to the contract price for 4,000 shirts that were accepted, plus lost profits on the unwanted and unproduced 6,000 shirts, plus a proportionate share of fixed or overhead costs. A legal remedy that encouraged seller to complete the contract and produce unwanted goods would inefficiently waste resources.

> But doesn't the Coase Theorem imply that the danger is unreal. There are only two parties, and there is a range of prices for the supplier's forbearing to enforce his contract rights that will make both parties better off. But this is just another example of bilateral monopoly; transaction costs will be high *because* there are only two parties.[5]

The myriad topics which Posner lists as involving bilateral monopoly that require potential judicial intervention in order for parties to bargain cooperatively include eminent domain, nuisance, and pollution; divided estates in land including landlord-tenant, joint tenancy, life tenant-remainderman, adjacent property owners, property rights in water law, and lessor-lessee relations (e.g., auto rentals); as a source of rent seeking; duress; relations between salvers and disabled ships; mitigation of damages, penalty clauses, and specific performance issues in contracts; transfer of cable television franchises; monopoly-monopsony labor-management negotiations, closely held corporations and settlement negotiations; and as a cause of increased transaction costs in buyer-seller relations, bilateral negotiations, and externalities; and as an impediment to the Coase Theorem. Although these examples are in the 9th edition, they are not specifically listed in its Index as they are in the Index of the 5th edition.[6]

In all these circumstances, parties cannot be expected to bargain cooperatively as required by the Coase Theorem unless there is an implicit threat that if the parties do not agree, courts will determine a binding price via liability rules, will craft common law rules (e.g., mitigation of damages), or use their equity powers to dictate a reasonable and efficient solution.

With respect to the prevalence of bilateral monopoly in the real world, an economist, Edwin Mansfield, wrote: "Clearly, there are few cases of this sort, except perhaps in labor markets."[7] On the other hand, Judge Posner gives myriad examples of bilateral monopoly, which he employs more often than the Coase Theorem. The difference in Mansfield's and Posner's perspectives on the prevalence of bilateral monopoly is that Mansfield, an economist, probably saw the world from the point of view of economics' general case methodology whereas Posner, a lawyer and federal appellate judge, is sensitive to law's hard case methodology.

PROPERTY

Posner states that a system of absolute or unqualified property rights would maximize value if three conditions are satisfied: 1) every scarce resource is owned by someone, the criterion of *universality,* that is, there are no commons problems; 2) ownership connotes the unqualified power to exclude everybody else from using a resource, the criterion of *exclusivity*, that is, there are no mutually interfering activities (no externalities); and 3) the bundle of distinct legal rights that comprise ownership of a resource are *freely transferable* in whole or in part, that is, no transaction costs, hold-up, or bilateral monopoly problems.[8] However, if one or more of these three criteria is not true, that is, there are commons problems, externalities, and transferability is restricted due to high transaction costs and bilateral monopoly, a system of unqualified property rights is untenable. In a world in which any or all of the three criteria are violated, the benefits from a system of property rights must be weighed against costs of such a system. Posner argues that property rights are less than universal because they are costly to enforce. For example, because many uses of real property necessarily interfere with the use and enjoyment of the property of others, it is not possible to have truly exclusive (absolute and unqualified) property rights.

CONTRACTS

The law of contracts is primarily concerned with agreements in which parties perform their obligations sequentially rather than simultaneously. Posner argues that contract law performs three functions: First, the deterrence of opportunistic behavior, such as when a landowner refuses to pay a contractor for a house after it is built or when due to substandard construction, the walls of the house crack due to shoddy construction a year after the land owner paid the contractor for the house.

Second, the interpolation of missing clauses for contingencies that were either unforeseen or even if foreseen were so unlikely to occur that the costs of including such clauses would exceed the benefits discounted by the probability that the contingency would occur (Judge Hand test). A court will attempt to imagine how the parties would have provided for these contingencies if they had done so; it will ask which party could more easily prevent the contingency from occurring (cheapest cost avoider) or has better knowledge of the magnitude and likelihood of the contingency occurring (the cheapest insurer). With respect to the example of a house accidentally destroyed by fire during construction, the cheapest cost avoider points to having the builder bear this risk, whereas neither the land owner nor the builder is necessarily the cheapest insurer. In this and similar cases, where it is not clear which party is the superior risk bearer, it is important that the law be clear one way or the other, so that the parties will know in advance which one is required to buy insurance.

The third function of contracts is risk shifting or insurance. For example, in the typical contract to build a house, after the agreement to build the house is made but before it is performed, the builder bears the risk of a rise in the price of building materials. This is an example of an ordinary commercial contract that shifts risks. Other examples of this third function are provided by explicit insurance contracts and sales of common stock where the buyer accepts the risk that the stock price will decline and the seller accepts the risk that the stock price will rise.[9]

Posner's analysis of the contract doctrine of impossibility, which doctrine excuses performance by a party technically in breach of contract, provides a good example of the distinction between the second and third functions of contract law. He says a party to a contract may be held liable for an occurrence over which it has no control or which may be physically impossible to perform or prevent. For example, a party may agree to pay a sum of money if it rains tomorrow[10] (third function). Conversely, even if physically possible, a contract may not be enforced if it becomes commercially impracticable. For example, a farmer may be excused from delivering on an output contract for a crop if a major portion of the crop in a region is destroyed by an unexpected blight (second function).

TORTS (NEGLIGENCE)

Judge Posner's analysis of negligence is based on the Judge Hand test, which he applies to unilateral care, alternative care, and joint care. The unilateral care case involves a straightforward application of the Judge Hand test: in circumstances in which the marginal burden of taking care in dollars is less than the probability an accident will occur if a precaution is not taken multiplied by the marginal loss in dollars if the accident occurs, that is, $\Delta B < \Delta PL$, an actor, who does not take care is negligent; if $\Delta B > \Delta PL$, an actor is not required to remove the risk of injury (chapter 3).

In alternative care circumstances, injury and damage can be avoided if either plaintiff or defendant takes care; both need not take care. First, consider circumstances where plaintiff is the least cost accident avoider: Assume plaintiff-pedestrian, who is jaywalking, is injured by defendant's automobile, which does not have time to stop. Plaintiff-pedestrian is least cost avoider; defendant-driver, who is high cost avoider, is not negligent (neither contributory nor comparative) for not taking care (looking for jay walkers) because defendant reasonably assumes plaintiff will take care and not jaywalk. Consider circumstances where defendant is least cost avoider: Assume plaintiff-pedestrian, who is walking on a crosswalk with a green light, is injured by an automobile driven by defendant, who ran a red light. Defendant is least cost avoider and plaintiff is not negligent for not taking care (looking both ways) because by being in the crosswalk he or she reasonably assumes defendant will take care.

In joint care circumstances, both parties are required to take the efficient level of care, which level of care is determined by the assumption that the other party is exercising due care; this means that in joint care circumstances, if one party takes less than the efficient level of care, that is, one party is negligent, the other party may not be required to compensate for the former's lack of due care with a higher level of care than due care. In other words, in this joint care model, parties are not required to compensate for another's negligence. This conclusion is qualified in chapter 18's Defensive Care model.[11]

NOTES

1. Posner, 9th ed., 297 (footnote omitted).
2. *Ibid.*, 298–299.
3. *Ibid.*, 297–298.
4. *Ibid.*, 129–130.
5. *Ibid.*, 130.
6. Compare the number of bilateral monopoly entries in 5th ed., Index on page 767 (p. 62, 68–69, 80–81, 82, 85, 88, 129–130, 131, 142, 145–146, 273–274, 350, 353, 358, 421, 428n2, 465, 608, 609), to the number of entries in the 9th ed., Index on page 1006 (§§ 3.5, 3.7, 3.9–3.11, 4.8. See also Monopoly).
7. Edwin Mansfield, *Microeconomics: Theory and Applications*, 6th ed. (New York: W. W. Norton, 1988), 300.
8. Posner, 9th ed., 41.
9. Ibid., 97, 99–100, 104–114.
10. "It may be said that when a man covenants that it shall rain to-morrow [*sic*], or that shall paint a picture, he only says, in short form, I will pay if it does not rain, of if A does not paint a picture. . . . A promise, then, is simply an accepted assurance that a certain event or state of things shall come to pass." Oliver Wendell Holmes Jr., *The Common Law* (Cambridge, MA: Harvard University Press, 1963), 234–235.
11. Posner, 9th ed., 199–204.

Part IV

RISK, INSURANCE, AND INCOMPLETE INFORMATION

Part IV consists of two chapters on risk and insurance. Game theory has a natural framework which is based upon two key variables which concern the temporal order of action and the state of players' information. As to action, do players act simultaneously or sequentially? As to information, do players have complete or incomplete information?

In Part IV, the division of the material in the two chapters reflects game theory's second variable, whether there is complete information or incomplete information. Chapter 15 is concerned with decisions involving risk where there is complete information. The concepts of expected monetary value, risk preference, and the normal probability distribution are explained; these concepts provide the basis for the theory of insurance, the Judge Hand test, and the method economists use to place a monetary value on a statistical life.

In chapter 16, the effect of incomplete information, where some players have knowledge that other players do not have, on insurance and product markets is considered under three headings: adverse selections (e.g., buyers in a market cannot observe the "type" or quality of products being offered for sale, good or bad, before purchase), moral hazard (e.g., people with insurance tend to be more careless), and principle-agent problem (e.g., if one hires someone to do a job and pay by the hour, he may prolong the job more than necessary whereas if he is paid first, getting him to complete the job may be difficult).

Chapter 15

Risk, Insurance, Judge Hand Test, and Value of a Statistical Life

This chapter begins with concepts related to decisions involving risk: expected monetary value, risk preference, and the normal probability distribution. These concepts provide the basis for the theory of insurance and its practical applications, such as fire, home, medical, and life insurance as well as workers' compensation, disability insurance, and social security. The Judge Hand test, which was discussed in chapters 1 and 3, is an expected monetary value test. The method economists use to place a monetary value on a statistical life is also explained. The material in this chapter and the next one, which explains the theory of incomplete information, is necessary to understand the effect of insurance on negligence and products liability in the law of torts.

EXPECTED MONETARY VALUE: E(V)

When decisions are made under conditions of risk, we generally want to estimate the consequences of prospective decisions (profits, losses, penalties, or rewards) in relation to the probabilities that such consequences will occur. For example, if a person is deciding whether to buy a new car, just knowing that his old car will soon require some repairs is not enough to make a rational decision; he must compare an estimate of the probable frequency and cost of repairs on the old car versus the cost of a new car minus the trade-in value of the old car. Or suppose that a person wants to decide whether to make an investment that she estimates has three outcomes: a profit of $1 million with probability of 10%, a loss of $30,000 with a probability of 40%, or a loss of $100,000 with a probability of 50%. How can she evaluate this investment opportunity? These examples demonstrate the need for a method of combining probabilities and consequences, which method is known as expected monetary value or expected value, E(V).

The concept of expected monetary value arose from attempts to understand games of chance; in its simplest form, expected value is the amount a player stands to win times (discounted by) the probability that she will win (assuming the payoff of not winning is zero). For example, in the event that a person will receive $10 if a coin comes up heads after being tossed, the expected value is equal to $10 multiplied by ½, which is the probability that the coin will come up heads: E(V) = $10(½) = $5. Consider another example: a person will receive $12 if a six-sided die (1, 2, 3, 4, 5, 6) comes up with 3 or 5 on top after being rolled once; the expected value is equal to $12 multiplied by 2/6 = 1/3, which is the probability that 3 or 5 will come up on top: E(V) = $12(1/3) = $4.

The probability of an event that cannot happen, that is, an event that is impossible, is zero, $p_i = 0$, where p stands for probability and subscript i stands for impossible; the probability of an event that is certain to happen is 1, $p_c = 1$, where subscript c stands for certainty. Probability of all other events involving risk is between 0 and 1; for example, the probability of a coin landing heads-up after a toss is, $p_h = ½$, according to the *counting rule*, which is defined as the number of outcomes in which one is interested divided by the total number of equally likely outcomes. In the coin flip example, we are interested in one outcome, a head, divided by 2, the total number of equally likely outcomes (head or tail). The counting rule is used to calculate classical probabilities in card games, dice, and theoretical simulations; it is also used to calculate empirical or statistical probabilities, which are based upon relative frequencies. For example, if a car dealer sold 3 cars on each of 10 days in a 30 day month, for that month, the probability that he or she sold 3 cars on any given day that month was 10/30 = 1/3. When the counting rule is used to calculate empirical probabilities, it can be continuously updated as new information is obtained. If this car dealer sells two cars on the first day of the next month, the car dealer now has 31 days of experience. The probability that three cars will be sold on the next day is now 10/31 = .03225, which is slightly less than 1/3. As more experience occurs, probability of an event occurring is continuously updated. "Fans" of a sports team implicitly use the empirical version of the counting rule over the course of the playing season to update the evaluation of their team's performance and the probability that it will win the next game.[1]

In an experiment, say either a toss of a coin or roll of a die, the sum of the probabilities of all possible outcomes that can occur equals 1; that is, $p_1 + p_2 + p_3 + \ldots + p_n = 1$. For example, if a coin is tossed, only two outcomes can occur, either head or tail: the probability of a head occurring equals ½ and the probability of a tail occurring equals ½; that is, $p_h + p_t = ½ + ½ = 1$. The outcomes (head or tail) are *collectively exhaustive* (no other outcome is possible) and are *mutually exclusive* (head and tail cannot occur together).

More formally, the calculation of expected monetary value requires three steps,

1. List the dollar value of all possible outcomes;
2. Multiply (discount) the dollar value of each outcome by its probability;
3. Sum the results.

In the context of the formal definition of expected monetary value, the coin toss example has two possible payoffs, $10 and $0, which are mutually exclusive (if one occurs, the other outcome cannot occur on that toss) and collectively exhaustive (no other outcomes), (p_h = ½) + (p_t = ½) = 1; therefore, the expected value is, E(V) = $10(½) + $0(½) = $5.

What would the expected value of 100 coin tosses be where the payoffs are $10 for each head and $0 for each tail? The answer is: E(V) = 100[$10(½) + $0(½)] = 100($5) = $500. An assumption underlying this multiplication of $5 by 100 flips is that each toss of a coin is independent of all others. Two events are *independent* if the occurrence or nonoccurrence of one event does not affect the occurrence or nonoccurrence of other event(s). For example, if the first time a coin is flipped it comes up heads, this outcome has no effect on and give no information as to whether the second flip will be head or tail; that is, the probability of head on the second toss is ½, so the E(V) of each of the 100 tosses is always $5.[2]

It is important to note that expected value should not be interpreted to mean that if a coin is tossed 100 times a person would receive exactly $500, which is 100 times $5, the expected value per toss. In fact, it is unlikely that he or she would receive exactly $500; however, it is highly probable that the total amount would be close to $500, a little more or a little less. (The variation of the 100 tosses around the mean conform to a normal distribution, which is explained below.) Expected value should be interpreted as a long-run average from an experiment that is repeated many times: the greater the number of coin tosses that are pooled together, the closer the average (the total money won divided by the total number of tosses), will approach $5 per toss. The *law of large numbers* says that as an experiment is repeated over and over, empirical (actual) probability approaches the theoretical probability.

Of course, expected value is not limited to games of chance or experiments with only two outcomes. The expected monetary value of the investment example described at the beginning of this chapter, which had three possible outcomes, is: E(V) = $1,000,000(.1) + (–$30,000)(.40) + (–$100,000)(.5) = $38,000. It should be noted that the E(V) = $38,000 does not correspond to any of the three actual outcomes; again, expected value should be interpreted as the long-run average return that would be earned from pooling the results of many investment opportunities with outcomes identical to this one.

RISK PREFERENCE

A natural question to ask is whether individuals actually use the concept of expected monetary value when they make decisions involving risk. Assume that an individual is asked to rank the three following choices in order of preference:

1. Sum certain of $60,000 in cash.
2. Lottery in which there is a 10% chance of winning $600,000 and a 90% chance of winning $0; that is, a person is blindfolded and is allowed to choose one piece of paper at random from a hat containing ten pieces of paper, where $600,000 is written on one piece of paper and $0 is written on each of the other nine. As shown below, the expected monetary value of this lottery is $60,000. (Choosing an object at *random* means that each object in the population under consideration has an equal chance of being chosen; each of the ten pieces of paper in the hat has an equal chance of being chosen.)

3. Lottery in which there is a 1% chance of winning $6,000,000 and a 99% chance of winning $0. The hat in this example has 100 pieces of paper in it, one with $6,000,000 written on it, 99 pieces of paper, each of which has $0 written on it. As shown below, the expected value of this lottery is also $60,000. The expected values corresponding to the three choices are:
   a. $60,000 sum certain
   b. $E(V_b) = (.1)(\$600{,}000) + (.9)(\$0) = \$60{,}000$
   c. $E(V_c) = (.01)(\$6{,}000{,}000) + (.99)(\$0) = \$60{,}000$

When asked to rank the three alternatives in order of preference, most individuals rank them (a), (b), and (c). This ranking indicates that most people are risk adverse; that is, they would prefer a sum certain of $60,000 to a small probability of a large gain. The larger the probability that the outcome will be $0, the less a risk adverse person prefers that choice. A few individuals rank the three alternatives in reverse order; these individuals are risk preferrers. Given these three alternatives, they like to gamble; they focus on the possibility of the large gain, $6,000,000, or big score, even though the probability of getting a large gain, 1/100, is small.

Although very few individuals are indifferent among the three alternatives, businesses and corporations often are. The reason is that individuals usually must make a one-time choice among these alternatives, whereas someone who took one hundred such risks, each of which is independent of the others, such as alternative (c), would win 1% of them on average and over the long-run her or his earnings would average $60,000/risk.

Consider another example: while an individual may be reluctant to file *one* lawsuit for fear of losing and having to bear the expense of litigation, a law firm may be willing to finance one hundred lawsuits on a *contingency* basis (it only gets paid if it wins), knowing that on average it will win between 60 to 70 of the suits because it will not litigate suits for which the probability of winning is low. So, the theory of expected monetary value and risk preference gives theoretical support for contingent fee arrangements in law.

We can now define the terms risk neutral, risk preference, and risk adverse. Someone who is *risk neutral* makes decisions involving risk on the basis of expected monetary value only; such persons are not concerned with the distribution or variance of risk. Persons or businesses who are risk neutral usually pool many decisions involving similar, independent risks and so can afford to take a long-run, average view. Someone who is *risk adverse* is concerned with avoiding variable outcomes, especially the worst possible outcome. (Recall the maximin solution in game theory.) A risk adverse person will choose the alternative with the least variance in outcomes. Someone who has is a *risk preferrer* will choose an alternative with a large payoff even if the chance of getting that payoff is small. Those who have a preference for risks are optimists or people who like to gamble.

THE NORMAL PROBABILITY DISTRIBUTION

In order to calculate expected mathematical value, one must know the probability distribution of outcomes or at least the probability of each outcome needed to calculate

E(V). As shown above, the counting rule is used to calculate objective probabilities of outcomes in card games, coin flips, roulette, computer simulations and other deterministic systems in which there is complete information. Probabilities in these systems do not change over time (unless built into the model).

However, in many situations of empirical uncertainty (incomplete information), decision makers rely on the empirical counting rule (automobile sales and sports team examples discussed above), similar past investment or litigation experience, or sophisticated theoretical and statistical analyses. In these circumstances, the probability of events not only changes over time, the past may or may not be a good guide to the future. What is the justification for assuming that the past is a good guide to the future, or at least for assuming that present experience will not deviate greatly from past experience?

The assumption that the past is a good guide to future experience is based upon the assumption that many characteristics of living things (humans, animals, fish, insects, bacteria, etc.), such as height, weight, intelligence, speed, length, and longevity, are normally distributed. For example, scores on the scholastic aptitude test (SAT) and IQ test are normally distributed. In addition, most items produced, filled, or packed by machines as well as many attributes in service industries are normally distributed; quality control theory is based largely upon the normal distribution. Although there are numerous other probability distributions, the normal probability distribution is used for more phenomena than any other probability distribution.

The most important attribute of the normal probability distribution is that *the farther a value is from the mean,*[3] *the less likely it is to occur*; that is, if outcomes are normally distributed, extreme values (outliers), either high or low, have a much lower probability of occurring than do values that are closer to the mean.

In Figure 15.1, height of men in the United States is represented on the horizontal axis from 5'1" to 6'7"; the number of men in millions is represented on the vertical axis. The height of US men is normally distributed and has a mean of 5'10". The area *under* the normal probability distribution (bell-shaped curve) has the following four characteristics.

First, the total *area* under the normal distribution accounts for 100% of the men (or whatever data is under consideration). Second, the normal distribution is symmetric around the mean, which in Figure 15.1 is 5'10", so that one side of the distribution is the mirror image of the other side; the symmetry characteristic implies that 50% of men (outcomes) are taller than the mean and 50% of men are shorter than the mean. Third, normal distribution is unimodal (the mode is the most frequent number); this characteristic plus symmetry implies that normal distribution's mode (most frequent number), mean (average), and median (middle number) are identical, which in this example is 5'10".

Fourth, the left and right "tails" of the normal distribution are "asymptotic" to the horizontal axis; that is, in theory, the tails of the distribution get continuously closer to the axis, but never touch it. Because the bulk of the probability-area is near the center and the probability-area in either tail end is increasingly small, this implies that the farther a value is from the mean, the lower the probability that it will be chosen randomly from the population; that is, *extreme values are much less likely to occur than those that are close to the mean.* In the normal probability distribution, average outcomes are very likely and extremely high or extremely low outcomes are unlikely.

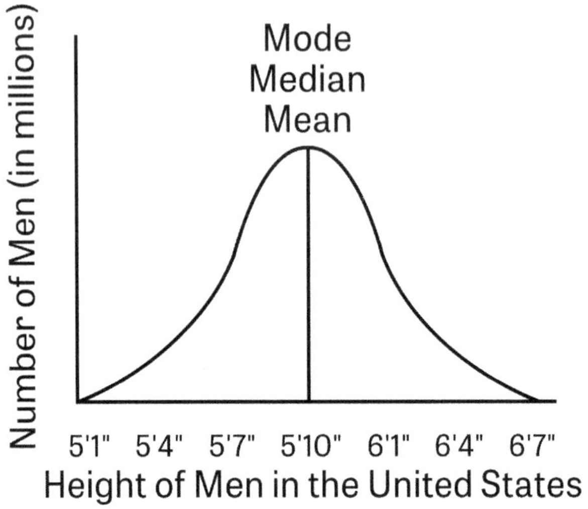

Figure 15.1 Normal Probability Distribution. *Source*: Created by the author..

Consider this conclusion with reference to the height of one man chosen randomly from the US population. Actually, as opposed to in theory, choosing a man at random from the population of the United States is difficult. If we choose a man based on Internal Revenue Service (IRS) records, the choice would not be random because not all men are represented on the IRS rolls, so all men do not have the same probability of being chosen. Choosing a man from US Census records would be closer to a random choice, but still not totally random because the Census does not count everyone.

Assuming that we choose one man at random from the US population, which we know is normally distributed, what is the probability that he will be 5'10" or taller? Answer: 50% (right half of Figure 15.1a). Compare the probability that this randomly chosen man's height will be around 5'10" to the probability that his height will be around 6'4". From Figure 15.1, the probability that his height will be around 5'10" is much greater than his height will be around 6'4" because the area under the normal curve around the height of 5'10" is much larger than the area under the normal curve around 6'4". Once again, in a normal distribution, extreme values are much less likely to occur than values closer to the mean. (This characteristic is important for the theory of insurances, which is discussed below.)

If a population under consideration is not distributed normally (bell shaped), but say, has fatter left and right tails than a normal distribution, then the probability that a data point or event chosen at random will have an extreme value is much higher; that is, extreme events are much more likely. So, knowing whether a given population (data set) is normally distributed or not, is crucial to predicting whether extreme outcomes are likely to occur.

Financial Asset Deviations from the Mean Are Not Normally Distributed

The assumptions that the dispersion or fluctuation of stock prices and other financial asset values around the mean are normally distributed, so that extreme declines in value of traded financial assets are very unlikely to occur, proved to be disastrous in the financial meltdown of 2007–2008. The distribution of financial asset fluctuations is not normally distributed.[4] Extreme values in financial markets occur more frequently than predicted by a normal distribution due to herd-like behavior in financial markets. (Recall the discussion in chapter 7 of the difficulty of determining a rational course of action with respect to the future when the success of one's decision is interdependent with the decisions of others.)

A second failure contributing the financial meltdown of 2007–2008 concerned the assumption that the probability of default on individual mortgages, which were combined and sold as securitized investments, were *independent* of each other, or at least that mortgages having the different probabilities of default were grouped in different "tranches." The independence assumption proved to be false because the severity of the recession after 2008 reduced most housing prices and caused a drop in employment, which affected home owners' ability to make mortgage payments; in addition, many home mortgages were granted fraudulently to those without verified income, credit worthiness, or ability to make mortgage payments, especially on adjustable interest rate (subprime) mortgages that automatically reset to higher interest rates several years after the repayment period began. Before 2008, many in the financial community assumed that housing prices would only rise over time and never decline (housing prices hadn't declined for over sixty years); they thought erroneously that Figure 15.1 or something close to a normal probability distribution described the variation in prices of securitized home mortgages.

In addition to the normal distribution, there are a number of well-known probability distributions, discussion of which is beyond the scope of this book. Knowing or having a sense whether a probability distribution is normal is of utmost importance in estimating the probability of outcomes. In many cases, as with human characteristics, quality control, etc., there is much empirical verification of the assumption that data are normally distributed; in others, like the distribution of values in stock and financial markets, estimates of probabilities based upon the assumption that variations around the mean are normally distributed can lead to seriously flawed predictions. In sum, if the probability distribution of events is normally distributed, the frequency of past events is often a good estimate of the probability of future events.

THEORY OF INSURANCE AND GAMBLING

In considering insurance against risks such as unexpected death, illness, or accidents involving persons or property, two questions naturally arise: one, on what principles is the theory of insurance based; and two, why do people buy insurance? In this section, we also consider why people gamble. Assume that a forty-year-old man is contemplating buying a *term-life* insurance policy of one year's duration. The face value of the policy, that is, the amount that his named beneficiary will receive in the event of his death during the year he is insured is $5 million. If he buys term insurance and lives more than one year, he and his beneficiary get nothing. He can, of course, always buy another insurance policy next year. Assume that a forty-year-old man chosen at random from the population of the United States has a 1/10,000 probability of dying during the year of coverage according to actuarial tables, which are based upon the life experiences of thousands of men.[5] These tables support the prediction that the number of deaths of forty-year-old men is normally distributed around a mean of 1/10,000 and that the death of one person can be assumed to be independent of all others; for example, the probability one man will suffer a heart attack is independent of whether another man will suffer a heart attack, unless most of the population of men are obese. (No insurance company will insure against damages from a nuclear attack because the risk that any individual would be damaged is not independent of the risk that others would be damaged.)

What is the expected monetary value implicit in this life insurance policy? There are two possible outcomes, either he dies during the year in which case his beneficiary gets $5 million or he lives out the year, in which case neither he nor his beneficiary receives any money; the probabilities attached to these outcomes are 1/10,000 and 9,999/10,000; therefore, the expected value of the one-year life insurance policy is,

$$E(V) = (\$5{,}000{,}000)(1/10{,}000) + (\$0)(9{,}999/10{,}000) = \$500.$$

Another way of looking at the concept of insurance is to say that for every 10,000 40-year-old men who buy this policy, on average, one of them, although we don't know which one, will die during the year; therefore, if 10,000 men each contribute $500 to a fund, the fund's total will be $500(10,000) = $5 million, which will be received by the beneficiary of the deceased. (Many insurance companies started out as non-profit or "mutual" insurance companies.) This insurance policy is an example of *actuarially fair insurance*; that is, insurance where 100% of the proceeds is paid out in benefits to the policy holders. The insurance company can be risk neutral because it combines thousands or even millions of normally distributed, independent risks.

What is the maximum insurance premium that a risk neutral person would pay for this one-year insurance policy? A risk neutral person would pay at most $500. Assume that because of administrative costs and the need to make a profit, the insurance company is only willing and able to sell this insurance for an annual premium of $1,000. If a person is willing to buy this insurance for $1,000, what does this tell us about his attitude toward risk? The person must be risk adverse; that is, a risk adverse person is someone who dislikes small risks of large losses so much that he is willing to pay more than the expected monetary value of the risk in order to shift the risk to someone else. If the insurance company offered to sell this insurance policy to a person for an insurance premium of $250 and she refused to buy it, what does this tell us about her risk preference? She is a risk preferrer; she prefers to accept the risk of death rather than shift it to the insurance company for substantially less than the actuarially fair price (expected value).

Now, consider gambling: Assume that there is a $20 million lottery where each ticket costs $1 and that the probability of winning the lottery is 1/100-million. What is the expected monetary value of the lottery? There are two possible outcomes, either the person wins the lottery of $20 million or loses and gets $0; the probabilities attached to these outcomes are 1/100,000,000 and 99,999,999/100,000,000; therefore, the expected monetary value of the lottery is

$$E(V) = (\$20{,}000{,}000)(1/100{,}000{,}000) + (\$0)(99{,}999{,}999/100{,}000{,}000) = 20¢.$$

Would a person who is risk neutral play this lottery for $1 per play? No. A risk neutral person would pay at most 20¢ to play this lottery. If a person buys a lottery ticket for $1, what does it tell us about that person's risk preference? He must be a risk preferrer, at least with respect to this transaction, because he is willing to pay 80¢ more than the lottery is worth in terms of its expected value. Under what circumstances would a risk adverse person play this lottery? The price would have to be less than 20¢ per play; the more risk adverse the individual, the lower the price would have to be before she would play (closer toward 0¢).

Now we can explain why the same forty-year-old man who buys insurance might also play the lottery. *Individuals are usually risk adverse with respect to a small probability of a large loss*; they prefer to make contracts to shift such risks to other parties, usually insurance companies, who are risk neutral because they can pool thousands of these independent, normally distributed risks. These same *individuals are often risk preferrers with respect to a small probability of a large gain*; they prefer to accept such risks by making contracts with those who run lotteries, who are risk neutral because they can pool thousands of these risks.

JUDGE HAND (EXPECTED MONETARY VALUE) TEST

In torts suits where the purpose of the law is to prevent careless behavior, the question of whether one person is culpable for the unintentional interference with another's person or property is often determined by Judge Hand test for negligence, $B < PL$, where P is the probability that an accident will occur if precaution to prevent it is not taken, L is the loss (in dollars) if the accident occurs, and B is the burden (in dollars) of taking a precaution that could have avoided the accident. (As discussed in chapter 3, the proper application of the Judge Hand test requires that it be applied marginally, $\Delta B < \Delta PL$, that is, for small changes in cost burdens and injury reduction benefits. For simplicity, it will be assumed that the values B and PL refer to marginal values.)

Consider the Judge Hand test in the context of expected monetary value, risk preference, and insurance. Each day, as millions of individuals go about their daily activities, each person makes thousands of decisions concerning whether and how much precaution to take against accidents. As examples, should a driver of a car drive defensively or aggressively; should a driver come to a complete stop at a stop sign, merely slow down, or ignore the sign? In every one of the millions of potential accident situations, there are two possible outcomes: either an accident will occur, in which case the loss in dollars is L, or an accident will not occur, in which case the loss is zero, 0. If precaution is not taken, the probability that an accident will occur is P, which implies that the probability an accident will not occur is $(1 - P)$ because the sum of the probabilities from all possible mutually exclusive and collectively exhaustive outcomes must add to 1; that is, $P + (1 - P) = 1$. Thus, in any potential accident situation, the expected monetary value (loss) is $E(V_L) = PL + (1 - P)0 = PL$, which is the right side of the Judge Hand test.

If an accident occurs when the burden of precaution, B, that an individual could have taken, is less than the expected value of the accident, that is, if $B < PL$, the actor is legally negligent. In terms of risk preference, the Judge Hand test means that the reasonably prudent person is legally required to behave in either a risk neutral, B = PL, or risk adverse, B > PL, manner with respect to activities which significantly affect the health, safety, and welfare of others. Alternatively, in terms of the theory of insurance, when engaging in activities that pose risks to others, a person is legally required to "spend," in terms of precaution or care taken, at least the equivalent of what a risk neutral person would pay to buy actuarially fair insurance against such risks. Although every person may have the right to be a risk preferrer with respect to risks such as whether or not to buy life insurance, climb mountains, or hang glide, when a person engages in activities that pose significant externalities to health, safety, and welfare of others (like, driving), the Judge Hand test means that he or she does not have the legal (or moral) right to be a risk preferrer. Moreover, the Judge Hand test also has a very strong moral foundation because it is consistent with the golden rule: "Do unto others as you would have them do unto you." The Judge Hand test version of the golden rule is: do not subject others to risks of injury and death to which you would not want others to subject you.

MONETARY VALUE OF A STATISTICAL LIFE

If an automobile accident occurs, the term L in the Judge Hand test represents the loss in dollars, which includes 1) the value of property damage, for example, damage to an automobile, 2) personal injury damage as measured by hospital bills, 3) loss of income during convalescence, 4) pain and suffering, sometimes called noneconomic loss, and 5) the monetary value placed on lives lost. The losses 1), 2), and 3) are measured in terms of objective market values; determining a monetary equivalent for 4) and 5) is a much more difficult and speculative exercise. With respect to 4), some states allow juries to speculate freely on the dollar value that is to be placed on pain and suffering; other states have placed a legislative cap on damages that can be awarded for pain and suffering.

With respect to the value that is placed on life in torts cases involving wrongful death, the jury is usually asked to make the following determinations: one, the sum of estimated income that the deceased would have earned until retirement; two, estimated income is adjusted downward by the expenditure that would have been necessary to keep the deceased clothed, housed, and fed and by the amount of income that would have been paid as taxes; three, estimated income is adjusted upward by the expected rate of inflation; and four, the present discounted value of the estimated income stream (discussed in chapter 10) is calculated to account for the fact that the representatives of the deceased are receiving a lump sum of money now, which sum they would have received only in the future if the deceased had remained alive.

Economists believe that the method used to determine the value of life in wrongful death cases is not the appropriate value to use in the Judge Hand test or other cost-benefit analyses, which are implicit in public policy debates, because wrongful death calculation asks how much a person's life is worth to others rather than how much a person values her or his own life. The value that people place on their own lives is implicit in all their decisions involving a risk of death. As examples, does a person wear a seat belt while driving, does he or she drive aggressively, smoke, engage in extreme sports like mountain climbing or hang gliding, choose a risky occupation, and so on? After determining the value of life that is implicit in people's decisions concerning the risk of death, economists use this monetary value as L in the Judge Hand test in cases in which lives have potential to be lost.

We begin a discussion of the economic value of a statistical life, or at least the value placed on the risk of death, by considering the question of whether a life, at least a statistical life, has a price or is priceless. A statistical life is different from the life of a particular named person in the following sense: if a ten-mile tunnel is going to be built, we do not know the specific individuals who will die, but based on past experience with tunnel construction, we can predict with a high degree of probability that a certain number of workers will be killed during construction. If we decide to construct the tunnel notwithstanding the near certainty that one or more persons will be killed during its construction, it must mean that we are willing to sacrifice life for, as Judge Calabresi would say, the money price of convenience.

Consider the following thought experiment. Assume that you were allowed to set the speed limit on highways. When confronted with this choice, most people are against having no speed limit. The reason that most people usually give for not being in favor of unlimited speeds on highways is that it is easy to lose control of a car at very high speeds; moreover, should the driver be confronted with an emergency situation while driving at very high speed, the stopping distance is often too long for a driver to be able to take evasive action. Most people neither intend to drive at very high speed nor want to be subjected to the risk created by those who drive at very high speeds (externalities). If given a choice, most people are comfortable with a speed limit between 55–75 mph.

These same people usually reject a speed limit on the highways as low as 40 mph even when they are told that thousands of lives could be saved by reducing the speed limit on highways to 40 mph. (On average, 35,000 deaths occur annually in the United States as a result of automobile accidents.) They are not willing to reduce the speed limit that low because of the much higher costs and longer time for travel, the higher cost for moving goods, and inconvenience (time is money). Most people are willing to accept a slightly higher risk of death on the highways at speeds of between 55–75 mph, instead of 40 mph, in order to save time and money.

In other words, if given the opportunity to set the speed limit, most people would decide the question based on an implicit cost-benefit analysis in which they are willing to make marginal trade-offs between the risk of death and money; that is, each of us is willing to accept small risks of death for a price.

The value of a statistical life was estimated from a study of the way in which automobile drivers responded to an increased highway speed limit. Due to the Middle East oil embargos of the 1970s, the US government mandated a maximum speed limit on all interstate highways of 55 mph in order to save gasoline. In 1987, states were allowed to raise speed limits on rural interstate highways from 55 mph back to 65 mph, but not on urban interstate highways. In 2002, a study on the effect of the speed limit increase in 21 states found that although speeds increased only 3.5% (2 mph), death rates rose by 35%.[6] The authors estimated that due to the higher speed allowed, 125,000 hours were saved in driving time per lost life. Since they valued a saved hour of driving time at the prevailing wage rate of $12.33/hour, the benefit from driving faster was ($12.33)(125,000) ≈ $1,540,000 per death. As determined by this study, the value of a statistical life, $1.5 million in 1987 dollars, was implicit in drivers' response to the higher speed limit.

Economists also estimate the value that people place on life based on the market decisions people make with respect to risks to health and safety that are acceptable to them in the manner described below. *If* 1) individuals have good or complete information of the health and safety risks implicit in their market decisions, 2) all health and safety risks are voluntarily assumed (no coercion or externalities), 3) markets are competitive, 4) individuals are risk neutral, 5) the risk premiums that must be paid in different occupations can be determined accurately, and 6) the distribution of income meets minimum fairness requirements, so that the poor don't have to take a disproportionate share of life-threatening risks, *then* markets will deal efficiently with health and safety risks. The greater that divergence between these assumptions and reality, the stronger is the case for government intervention in on-the-job health and safety conditions.

Consider the following example: during the construction of a modern sky scraper, how many individuals would choose to be employed as steel workers on the fiftieth floor? With respect to information, it is assumed that potential workers are aware of the number of deaths that have occurred on similar jobs and know whether there are nets and other safety precautions to catch workers who slip and fall. With respect to whether risks are voluntary, it is assumed that potential workers have alternative safe construction jobs (alternative opportunities) where they can find employment. Notwithstanding complete information, alternative employment possibilities, and competitive markets, most people would probably not want to become steel workers on the fiftieth floor. However, those who accept work on the construction of such buildings would in all likelihood demand a risk premium over and above the salary paid for comparable but less risky construction work, which risk premium will be considered below.

As an example of a situation in which the complete information assumption was not fulfilled, consider those workers who worked in asbestos factories several decades ago and were unaware of the risk of lung damage and death to which they were exposed; these workers did not demand appropriate risk premiums because they were unaware of the risk to their health. As an example in which the assumption that risks are voluntary is violated, consider the case of illegal immigrants who work at dangerous jobs but are afraid to complain about working conditions or demand the appropriate risk premiums for fear of being fired or deported. An historical example of an involuntary risk would be the exposure to coal dust in mines incurred by miners who lived in the rural Appalachian towns where there was little or no alternative source of employment.

Assume the assumptions of complete information and voluntarily risk taking are satisfied. Since more workers want to work at safe jobs than want to work at risky jobs, in order for supply to equal demand in risky jobs, the wages in risky jobs must be higher than the wage in safe jobs due to a risk premium. In a competitive labor market, an employer must either pay the risk premium differential or be able to make the job safer so that the risk premium is unnecessary. If a profit maximizing employer has a choice between paying a risk premium for dangerous work and making the factory safer, he or she will choose that which is cheaper. If investing in safety precautions is cheaper than paying workers a risk premium, employers will choose to make the work place safer. Thus, competitive labor markets force employers to choose the efficient level of safety.[7]

Economists estimate risk premiums in various occupations and use these estimates to calculate the expected monetary value of life as follows: If workers demand a $2,000 annual risk premium in order to take a job that has an 1/1,000 risk of death greater than the risk in similar, relatively safe jobs, this implies that for every thousand workers, one of whom will on average die each year (assuming a normal probability distribution), the value of a statistical life is worth 1,000($2,000) = $2,000,000. In other words, in the Judge Hand test, B = PL, if the risk premium of $2,000 is B and 1/1,000 is P, which is the increase in probability of death from engaging in this risky occupation, then the value of a statistical life, L, in the equation, $2,000 = (1/1000)L, is L = $2,000,000. Conversely, estimates of the value of a statistical life are implicit in L when judges use the Judge Hand test to determine the socially appropriate level of care, B, when lives are at risk.

When the Judge Hand test is used to determine whether there is legal liability in accident cases which involve death, whether or not someone is found to be negligent is extremely sensitive to the value of L that is used. If L is very high, then an actor is negligent unless a high level of care, high B, is used; if L is low, then the actor is only required to use a low level of care in order to avoid culpability. Estimates of the value of life in 2015 are between $7 and $9 million per life.[8]

Two caveats are in order: First, the relation between risk and the value of life as determined by the expected monetary value or Judge Hand test is not linear; that is, just because a person would accept a small risk, say a 1/1000 probability of death if he were paid an annual risk premium of $2,000, which implies that he values his life at $2,000,000, does not mean he would accept certain death for $2,000,000 [B = PL; ($2,000,000) = (1)$2,000,000]. The amount of money that would a person would demand in order to voluntarily accept the risk of death increases exponentially in relation to the rise in probability of death, quickly becoming infinite.

Second, the variation in estimates of the value of a statistical life (between $7 and $9 million) are due to the difficulties in determining risk levels in different occupations, what part of the total wage represents a risk premium, and whether people who choose risky occupations are risk preferrers, so that they accept lower risk premiums than individuals who are risk neutral or risk adverse. As a result of these problems, the value of life may be underestimated in some occupations and overestimated in others.

GRIMSHAW V. FORD MOTOR CO.

In *Grimshaw v. Ford Motor Co.*,[9] the plaintiffs were traveling on an interstate freeway in a Pinto hatchback when the carburetor failed, causing the car to stall in the middle lane where it was hit. The impact of the collision drove the Pinto's gas tank forward, causing it to rupture and spill fuel into the passenger compartment. Both the driver and passenger were engulfed in flames and sustained severe burns, from which the driver died.

Defendant-manufacturer was aware that the preferred placement of the gas tank in Europe and Japan was above rather than behind the rear axle as in the Pinto. The appellate court said that the Pinto had less "crush space" than any other American automobile of the period, the rear "bumper was little more than a chrome strip, less substantial than the bumper of any other American car produced then or later," and its rear structure lacked reinforcing members. The plaintiffs were awarded slightly more than $3.5 million in compensatory damages and $125 million in punitive damages; all but $3.5 million of the punitive damages were remitted (eliminated).

Grimshaw attracted a great deal of comment because defendant's in-house cost-benefit studies concluded that it was not cost efficient to redesign or reposition the gas tank in its Pinto automobile, or to reinforce the Pinto's rear structure to make it more crashworthy. The defendant did not use the Judge Hand test in its cost-benefit studies, possibly because it may have believed that the probability that an accident would occur if redesign was not undertaken, the P variable in the Hand formula, was unquantifiable. One study is summarized as follows:

Benefits

Savings—180 burn deaths, 180 serious burn injuries, and 2,100 burned vehicles.
Unit Cost—$200,000 per death, $67,000 per injury, and $700 per vehicle.
Total Benefit—180 × ($200,000) + 180 × (($67,000) + 2,100 × ($700)) = $49,500,000.

Costs

Sales—11 million cars, 1.5 million light trucks.
Unit Cost—$11 per car, $11 per truck.
Total Cost—11,000,000 × ($11) + 1,500,000 × ($11) = $137,000,000.

The study concludes that total benefit of $49.5 million from preventing the occurrence of an estimated 180 serious burn injuries, 180 burn deaths, and 2,100 destroyed vehicles would be significantly less than the total dollar costs of $137 million required to strengthen or reposition the Pinto's fuel tank on the estimated 11 million cars and 1.5 million trucks that it expected to produce.

Three estimates used in the study invite discussion: 1) estimate of the number of serious burn injuries, deaths, and burned vehicles; 2) estimate of the value of a statistical life, and 3) estimate of cost to redesign or to reinforce gas tanks. Defendant's estimate of the number of serious burns, deaths, and burned vehicles appears to be reasonable when one considers that 500 to 1000 persons die from traffic accident fires in the United States each year, and that the Pinto fire fatality rate was apparently within the norm for subcompacts. The defendant claimed that the $200,000 figure assigned to the value of a life was supplied by the National Highway Traffic Safety Administration. However, the value of a statistical life in the study of the effect of a 10 mph increase in the speed limit, discussed above, was determined to be $1.54 million. In 1983, a leading expert on safety recommended that for policy purposes a life should be valued at $2 million.[10] Had the defendants valued life at $700,000 or more per person in its study, total benefits from redesign or repositioning the gas tank would have outweighed total costs. In other words, if a higher value of life had been used, redesign would have been cost justified, which shows how important the value of a statistical life is in determining the level of care that should be required.

Another issue is whether the cost of $11 per vehicle for redesign or reinforcement of the gas tank was the proper amount to use in defendant's cost-benefit analysis (or as the burden, B, in a Judge Hand test). The small redesign cost per car reflects the fact that defendant expected to be able to spread the redesign cost over 12.5 million vehicles. One can argue that many hundreds of design choices are made by engineers in the production of a single complex product—one more cross beam can always be added here, strengthened there, a stronger material can be used, and so forth—so that each such decision involves a range of trade-offs between cost, weight, appearance, performance, and safety. Defendant could argue that engineering decisions should not be judged entirely by hindsight, which is perfect.

On the other hand, one can argue that if the defendant had designed the Pinto properly in the first place, relying on common knowledge and industry custom, it could have shielded the fuel tank by placing the tank above the rear axle and farther forward in the chassis. Moreover, should not the public be entitled to believe that the Pinto's fuel tank was as safe as that of other automobiles of its size and price? Given the latter argument, the benefits of redesign would clearly outweigh the costs, which in a social sense would be zero. Moreover, if one were to employ the standard economic model of negligence with its assumption of complete information, the cost of redesign would be zero, since the defendant would be held to know all relevant costs and legal standards.

NOTES

1. The formal theory concerning the revision of prior probabilities based upon new information is known as Bayes' theorem.

2. Events are dependent if knowing the first event has occurred effects the probability that the second event will occur. For example, if a deck of ordinary playing cards was repeatedly "cut" to reveal a card, after which the chosen card is discarded from the deck before the next card is chosen, the card chosen on the second cut and all cuts thereafter would be *dependent* on all previous cuts. For example, before the first cut, the probability of a red card being chosen is 26/52 = ½. If a red card is chosen and discarded after the first cut, the probability of choosing a red card on the second cut is 25/51, which is less than ½.

3. The mean is the total sum of all observations divided by the total number of observations.

4. Benoit Mandelbrot and Richard L Hudson, *The Misbehavior of Markets: A Fractal View of Risk, Ruin, and Reward* (New York: Basic Books, 2006); Taleb Nassim, *The Black Swan* (New York: Random House, 2007).

5. The probability of death of a 50-year-old man over the course of a year would be greater, say, 10/10,000.

6. Orley Ashenfelter and Michael Greenstone, *National Bureau of Economic Research, Working Paper No. 9094* (Cambridge, MA: NBER, 2002).

7. Douglas Greer, *Business, Government, and Society*, 3rd ed. (New York: Macmillan, 1993), 414.

8. US Department of Transportation, *Guidance on Treatment of the Economic Value of a Statistical Life in U.S. Department of Transportation Analyses* (Washington, DC: DOT, 2015).

9. 174 Cal. Rptr. 348 (Cal. Ct. App. 1981).

10. W. Kip Viscusi, *Risk by Choice: Regulating Health and Safety in the Workplace* (Cambridge, MA: Harvard University Press, 1983).

Chapter 16

Incomplete Information

Adverse Selection, Moral Hazard, and Principal-Agent Problem

The previous chapter assumed that the probability distribution of all risks is known, that is, there is complete information. In this chapter, the effect of incomplete information in insurance and product markets is considered under three headings: adverse selections, moral hazard, and principal-agent problem. Adverse selections refer to circumstances where buyers in a market cannot observe the type or quality of products being offered for sale, good or bad, before purchase or where sellers of insurance do not know whether a person seeking insurance is a good or bad health risk. Adverse selections are the bad or defective products among mostly good products and sick persons among a mostly healthy population.

Moral hazard refers to the fact that people with insurance tend to be more careless; it also refers to unaccountable conduct in general, as where someone uses other people's money to invest in high-risk, high-reward investments; that is, an investor who has no "skin in the game."

Adverse selection and moral hazard are also considered in the context of signaling, screening, and pooling or separating Nash equilibria (discussed below), which are used to reveal good from bad products and services.

The principal-agent problem, which is a generalization of moral hazard, involves incentive incompatibility among contracting parties; for example, if you hire someone to do a job and pay by the hour, he may prolong the job more than necessary, whereas if he is paid in advance, getting him to complete the job may be difficult.

In 2001, the Nobel Prize in Economics was shared by three men, Akerlof, Spence, and Stiglitz, for their work on the effect of incomplete information on market efficiency. Their contributions will be discussed in this chapter.

ADVERSE SELECTIONS: "BAD APPLES IN THE BARREL"

Market for "Lemons"

The seminal work on incomplete information is George Akerlof's article, "The Market for Lemons."[1] If product quality cannot be observed by buyers prior to purchase, products are bought infrequently so that the seller's reputation is unknown, and warranties are too costly in terms of transaction costs and uncertainty as to whether warranties will be honored, Akerlof showed that buyer's inability to determine whether a product is good or defective before use may result in total market failure. His "paper relates quality and uncertainty . . . in particular, a structure is given for determining the economic costs of dishonesty."[2]

Assume that 50% of used cars are "lemons," that is, low-quality used cars, which are worth $6,000 to buyers, and 50% of used cars are "plums," that is, good-quality used cars, which are worth $10,000 to buyers. A risk neutral buyer, who knows the proportions of plums to lemon cars but cannot tell the difference between plum and lemon used cars prior to purchase and use and who makes decisions based upon expected monetary value, would offer $8,000 for a used car of unknown quality: E(V) = .5($6,000) + .5($10,000) = $8,000. The market price is the *average* value of used cars in the market. However, sellers of plums will not accept offers of $8,000 because their cars are worth $10,000, so only lemons would be offered for sale; buyers, as they learn from experience that only lemons will be offered for sale, will either not buy a used car or offer a reduced price which will over time approach $6,000. In this simple model, there is either total market failure or bad (lemons) used cars will drive out good used cars from the market due to the incomplete information. Buyers will not get an *unbiased* selection of used cars; rather they will get *adverse selections*, lemons, or, to mix the fruit metaphor, the bad apples (bad used cars) in a barrel consisting of good and bad apples (all used cars).

Akerlof states that the market for lemons is analogous but not identical to Greshman's Law: bad money drives out good money (buyers prefer to pay with paper currency of dubious value and hold onto gold dollars or highly valued currency). The difference between lemons market and Gresham's law is that in the lemons market, buyers cannot distinguish between good and bad used cars, whereas in Greshman's law, both buyers and sellers know which currencies are good and which are bad.[3]

Market failure, which is by definition inefficient, could be avoided if sellers have reputations or it is possible to offer a warranty so that sellers of defective cars agree to buy back used cars or pay for repairs. If sellers offer a warranty, sellers can charge $10,000 for high-quality used cars, plums. However, for the warranty to be effective, the buyer must be reasonably certain that the seller will be available after the sale, is financially sound, and will make good on the warranty. Alternatively, some states have passed *lemon laws*, which mandate that used cars must pass inspection before sale or that sellers of defective used cars must pay compensatory damages to buyers. But, since enforcement of warranties or legal remedies is not costless, a buyer is less than fully protected; so, due to high transaction costs, these remedies cannot fully eliminate the inefficiency resulting from incomplete information.

Assume that warranties or lemon laws were costless to enforce; that is, transaction costs are zero. Then, buyer's best offer for cars would be $10,000, even though they don't know whether they are getting a plum or a lemon; sellers of plums would get the true worth of their cars, $10,000; and sellers of lemons, who would have to pay $4,000 in repairs, would get what their cars are worth, $10,000 – $4,000 = $6,000.

Recall that a strategy combination is a Nash equilibrium if, given the strategy choices of the other player, neither player has an incentive to switch strategies.

A *pooling Nash equilibrium* is one where quality or type of a good or service, whether high or low quality, cannot be determined before use, so both good and bad types are offered for sale. A *separating Nash equilibrium* is one where qualities or types, good or bad, can be determined before contracting, so that the efficient price is paid for each quality. Often buyers are able to separate good from bad goods by testing, such as where a used car is inspected by a mechanic before sale. Second, sellers of high-quality goods are often able to separate themselves from sellers of low-quality goods by *signaling*, say, by offering to sell a used car with a warranty. Akerlof's model appears to be a partial separating equilibrium because only defective quality used cars, lemons, are sold, not a complete separating equilibrium where both good and bad types are sold and each is paid its value. If warranties and lemon laws are costless to enforce, the result, $10,000 for good used cars and $6,000 for bad used cars, is an efficient separating equilibrium where the price of each car is equal to its worth.

Pooling and Separating Nash Equilibria

Separating equilibria, where prices of high or low-quality goods and services each reflect equal marginal cost production, are usually more efficient than pooling equilibria, where prices reflect the average cost of high and low-quality goods or services; however, when society wishes to socialize risk, say by guaranteeing that health insurance is affordable for individuals with preexisting conditions or those who are genetically prone to life-threatening diseases, pooling equilibria are socially desirable to protect adverse selections. Society does not want adverse selections to be identified or have their hidden information revealed.

Consider the Beer-Quiche Game, in which "Tough guys don't eat quiche."[4] Patrons of a restaurant are of two types: tough guys with a probability of 90 percent, and wimps with a probability of 10 percent. Bullies come to the restaurant to start fights with wimps but do not like to fight tough guys. Neither tough guys nor wimps like to fight; but, if tough guys have to fight, they usually win. Bullies think of wimps as adverse selections. Patrons must decide whether to order beer or quiche for breakfast. After seeing what a patron orders, a bully will decide whether to pick a fight. Tough guys prefer beer for breakfast; wimps prefer quiche but will disguise themselves as tough guys by ordering beer instead of quiche. Thus, wimps will not reveal their hidden information or *signal* their type. Since a bully cannot distinguish tough guys from wimps by their breakfast orders, a pooling solution results in which bullies cannot tell tough guys from wimps and so will not pick fights because they would be fighting tough guys 90 percent of the time.

This pooling solution is a Nash equilibrium because no one can do better by choosing a different strategy: tough guys are eating their preferred breakfast and do not have to fight, so they have no reason to choose a different strategy; wimps get more satisfaction from not being beaten up than from eating a breakfast of quiche, so they have no reason to choose a different strategy; and, bullies do not start fights because they cannot differentiate tough guys from wimps. A separating equilibrium in which a bully was able to identify and pick fights with wimps would not be socially desirable. This game shows that there are some circumstances involving incomplete information in which pooling equilibria, where adverse selections are not revealed, are more socially desirable than separating equilibria where adverse selections are revealed.[5]

Adverse selections are an important topic in insurance theory, where a pooling equilibrium sometimes increases and sometimes decreases social welfare. Insurance companies do not want to insure high risk individuals whom they regard as adverse selections. Insurance companies prefer separating equilibria. High-risk individuals want to get into insurance pools with low risk individuals and so prefer pooling equilibria; low risk individuals want to get out of insurance pools with high-risk individuals and so prefer separating equilibria.

For example, a person who has acquired immune deficiency syndrome (AIDS) would like to be in an insurance pool with individuals who are healthy because the insurance premiums would be lower than if he or she were in an insurance pool in which everyone either has AIDS or has a high probability of contracting AIDS; on the other hand, insurance companies do not want to insure anyone who has AIDS, or is a member of a group that is at higher than average risk to get AIDS. Before the Affordable Health Care Act, persons who have a medical condition such as heart disease or cancer and who leave employment where they were covered by their employers' group health coverage often found that no insurance company would sell them individual health insurance policies at affordable rates because they have a preexisting condition that makes them adverse selections.

Most people do not object to insurance premiums that discriminate based on life style choices, such as choices to smoke cigarettes, not to wear seat belts, or engage in extreme sports like hang gliding; that is, most individuals believe that these adverse selections (bad apples) should not be a burden to others, so separating equilibria are preferred. Status categories based on age, even if discriminatory, as, for example, unmarried drivers under twenty-five years old, who pay higher than average premiums for automobile insurance even if they have good driving records, are also usually acceptable separating equilibria because they are experienced by all drivers over the course of a lifetime.

However, separating equilibria based on classifications over which a person does not have control are usually rejected as being inimical to social welfare. For example, the Genetic Information Nondiscrimination Act (GINA) of 2008 prohibits discrimination in health insurance and employment based upon genetic information. Thus, GINA officially mandates a pooling equilibrium and uses health insurance to socialize the risk that someone may, through the luck of the draw, have a higher genetic probability of contracting a disease than the average person. Medicare and Medicaid protect the aged and the poor from being classified as adverse selections. The Affordable Health Care Act prohibits insurance companies from rejecting applicants based upon preexisting conditions. There seems to be a consensus forming that homosexuality has a genetic component and may not be a life style choice, in which case pooling equilibria are justified. There is disagreement over whether being overweight is a life style choice, a genetic predisposition, or both; if it is a life style choice, higher health insurance premiums are justified; if not, higher health premiums are no more justified than for any other genetic tendency.

Question: Should insurance companies be required to sell insurance to individuals who build homes in flood, hurricane, and tornado prone areas? If so, should this risk be completely socialized or pooled; that is, should people, who build homes in areas where the risk of such damage is great, pay the same premium as home owners in less risky areas? Or should these high risks be lumped together in one government-mandated high-risk category? Or, should these home owners have to negotiate as individuals in the private insurance market (separating equilibria)? One of the major benefits of getting health insurance through one's employer is that employers demand a group rate for all employees; that is, employers do not allow insurance companies to reject persons who are adverse selections.

In sum, the problem of adverse selections in insurance markets requires government intervention. Only government, as the people's representative, can determine which risks should be socialized, that is, pooled, and which should be borne by individuals, that is, separated.

There are two ways in which government can require insurance companies to insure adverse selections: compulsory community rating and social insurance. Under compulsory community rating, an insurance company which holds itself out as selling a given type of insurance, say health, home, or automobile insurance, must sell an insurance policy at the same premium to all members of a non-discriminatorily defined community. Two examples of possible nondiscriminatory community rating categories are geographic-residence or age-family status. With respect to geographically based community rating, the appropriate community may be city, regional, or state-wide. Of course, the size of geographically defined communities is crucial and must be regulated; otherwise, gerrymandering or "redlining," whereby a high-risk area is singled out and isolated as a geographic adverse selection. For example, insurance companies have been known not to sell fire insurance in urban ghetto areas.

With respect to community rating categories based on age or family status, issues concerning discrimination against the young, old, or untraditional families must be settled. Social insurance is like compulsory community rating except that the community is defined to be national and government fiat replaces the market as the setter of premiums and benefits.

SIGNALING AND SCREENING

Signaling and screening models are concerned with ameliorating or eliminating market inefficiencies that result from incomplete information. If buyers cannot determine the quality of products or services that are being sold in a market, whether high or low quality, a risk neutral buyer will pay no more than an average price between high and low-quality types. The result in the lemons markets is partial market failure, only low-quality used cars are offered for sale; in employment situations, where employers cannot distinguish between high- and low-quality workers, workers get an average salary that underpays high-quality workers and overpays low-quality workers. However, if high-quality sellers of products and services can give a signal that demonstrates their superior quality, a separating equilibrium is possible in which each quality receives its efficient price or wage.

Job Market Signaling

A. Michael Spence shared the 2001 Nobel Prize in Economics for his work on the effect of signaling on market efficiency.[6] In his model, he assumes that there are two types of workers, high and low quality, who, if their quality was easily observable, would be paid accordingly. If it is not possible to distinguish the two types or qualities, both types are paid an average wage (pooling solution) due to incomplete information. However, Spence assumes employers use education as a signal that workers are high quality and also assumes that workers of different quality have different costs of acquiring education in terms of the sum of (explicit) dollar costs, (implicit) opportunity costs, and effort required. Moreover, the extreme assumption is also made that education is only a signal; it has no effect on worker productivity. The rationale for this assumption is that it allows the model to isolate and consider only the effect of the signal.

If the costs of getting an education are substantially higher for low-quality workers than for high-quality workers, high-quality workers will get an education but low-quality workers will not get an education; that is, even if both types know that employers are determining a workers' quality or type based upon the signal of whether or not they get an education, low-quality workers do not acquire education because the cost of education to them exceeds the benefit of an increased wage; and, high-quality workers acquire education because benefit exceeds cost even though education does not change their productivity. Since neither type of worker has an incentive to change her or his behavior, signaling results in a separating Nash equilibrium, in which high-quality workers are paid more than low-quality workers.

This signaling, separating Nash equilibrium is an extreme case: the amount of output produced is the same as when there was no signaling, so the investment in education produces private benefit but no social benefit; the only difference is that instead of all workers being paid an average wage, the wage of high-quality workers is not depressed because employers have the ability to distinguish between high- and low-quality workers.

Empirical studies have determined that even though the effect of the last year of a high school education on productivity is small, the effect of graduation from high school on income is large in comparison to the effect on income of just three years of high school. The small increase in productivity from the last year of high school is unlikely to account for the huge increase in income that occurs as a result of graduation. This disparity has been called the "sheepskin effect." Empirical studies have also shown that high school graduates tend to stay with firms longer and have fewer absences. Thus, graduation from high school may be a signal for reliability, which results in increased productivity. The same discontinuous jump has been observed for college graduation over merely attending college.[7]

"Talk is Cheap"

In order to increase their chances of being accepted to a college of their choice, many high school students can and do file many college applications almost simultaneously on the internet, which medium has reduced the cost of applying to college. However, colleges, who are often evaluated at least in part based upon the percentage of students who enroll after being admitted to their college, know that students who apply to multiple colleges may get multiple acceptances, so that the probability they will enroll after being accepted may be low. In order to determine whether to accept a student, they will want to know whether a student will enroll after being accepted. They use signaling and screening. Early admission programs, whereby a student agrees in advance to enroll if accepted, screen out many applicants who are shopping among many colleges. Whether a student took the time and effort to visit the college's campus is a signal that a student is seriously considering this college. Requiring an essay with a specific or unusual topic, which cannot be used in other applications, provides another signal. In other words, since filing an application is inexpensive, that is, talk is cheap, colleges use signals and screens in order to separate applicants who are likely to enroll from those who are unlikely to enroll.

Similarly, job applicants can send their résumés to myriad employers at the push of the "send" button. Employers know that the overwhelming majority of such résumés, especially if unaccompanied by a cover letter that addresses the specifics of the job and why the applicant would be an asset to this specific employer, are not worth the time of a reply. Again, talk is cheap. Due to being inundated by unsolicited résumés, many employers have begun to rely on networking and employee referrals.

Does Signaling and Screening Increase Efficiency?

Signaling can increase or decrease market efficiency. For example, in the lemons market, if owners of used cars can offer warranties as a signal that their cars are of good quality, such signaling results in an efficient separating equilibrium that eliminates market failure. If a used-car warranty is considered in the context of Spence's signaling model, offering a warranty is so much more expensive for owners of defective used cars than for owners of good used cars that only the latter will offer warranties; the result will be an efficient signaling, separating equilibrium in which good used cars are sold with warranties at a higher price than defective used cars sold without warranties at a lower price. Trademarks, branding, and extensive advertising are also signals of quality that are used to create separating equilibria. A firm that spends a great deal of money advertising its product is likely to protect that investment by producing a high-quality product.

In the Beer-Quiche model, wimps drink beer for breakfast rather than eat quiche in order not to signal their type to bullies. In this context, lack of a signal, which results in a pooling equilibrium, increases social welfare because wimps are not beaten up and there is no breach of the peace.

The Family and Medical Act[8] is a less fanciful example in which the prohibition of signaling or screening may increase efficiency.[9] Assume there are two types of workers: those with a high probability of becoming parents and those with a low probability of becoming parents. If there is complete information and no discrimination against those who wish to become parents, those workers with a high probability of becoming parents will accept lower wages in return for a parental leave clause in their contracts and those with a low probability of becoming parents will bargain for higher wages without a parental leave clause. This separating Nash equilibrium is the first best solution.

But, suppose that employers do not want to hire workers who will become parents because they believe, whether true or not, that workers who become parents will be less devoted to the job, will not work as hard, or will be less willing to work overtime.

Given such discrimination, the first best solution is not possible. Workers with a high probability of becoming parents will mimic the strategy of workers with a low probability of becoming parents so as not to signal their type, as in the Beer-Quiche model, thereby increasing their chance of being hired. No one will negotiate for parental leave benefits.

On the other hand, if there is a law both prohibiting employees from signaling that they do not want to become parents and preventing employers from screening workers based upon the probability that they will become parents, the choice is between two pooling equilibria: one in which there is no parental leave because the issue is not discussed due to fear of discrimination and one in which parental leave is mandated by law. If the benefit to prospective parents of having parental leave is large and the cost in terms of lower wages to workers who are unlikely to become parents is small, the pooling equilibrium in which parental leave is mandatory may be an improvement over a pooling equilibrium in which parental leave is left entirely to private bargains, which will not occur due to the fear of discrimination. The pooling model supports laws that prevent some forms of discrimination by mandating incomplete information; however, laws preventing discrimination on the basis of race, sex, or age, that is, observable characteristics, require other justifications.

Employers screen applicants for many characteristics. For example, employers often give prospective employees personality, skill, or drug tests as a precondition of employment. To the extent that such tests can detect whether a prospective employee has the requisite skills, mental stability, and ethics, they may enhance social welfare; however, to the extent that they test characteristics that are irrelevant to the job, screening may be burdensome or discriminatory.

MORAL HAZARD: UNACCOUNTABLE ACTION

If risk adverse individuals were prohibited from buying insurance against liability for careless behavior that results in economic damage, they would be too (inefficiently) careful. For example, if risk adverse individuals were held financially responsible for automobile accidents for which they are the cause but were not allowed to buy automobile collision insurance, they would use their automobiles less than is economically efficient due to the fear of crushing financial liability in the event of an accident. On the other hand, if individuals have total insurance coverage for all losses caused by their carelessness, so that risk of loss is shifted completely to an insurance company, the effective cost of accidents for those individuals is zero. Individuals who are insured against risk tend to become more careless; they have reduced incentives to take care. This tendency of insured persons to reduce the level of accident preventing care for risks whose expected cost has been shifted to an insurance company is known as *moral hazard*. Since accidents are in a sense free to a person who is insured, moral hazard causes accident costs to be higher than expected.

As examples, people with theft insurance are more likely to leave their houses and cars unlocked and their possessions unguarded; people with health insurance use physicians' services more often than those without such insurance; people who receive unemployment benefits or are on welfare are less likely to be actively looking for work; and, bankers whose deposits are federally insured are more likely to choose high-risk, high-return investments. More generally, the principal-agent problem, in which the agent's actions are unobserved or uncontrolled by the principal, which is a generalization of moral hazard, is discussed below. As examples, chief executive officers (CEO) of corporations often serve their own interests at the expense of the shareholders; and, salaried workers expend less effort than those who work on commission.

The existence of moral hazard means that the goal of complete shifting of the accident risk from risk adverse individuals to risk neutral insurance companies usually cannot be accomplished at affordable rates. In order not to incur losses due to moral hazard, insurance companies must either shift some of the risk back to insured individuals in order to control moral hazard or they must closely monitor the insured's actions and set premiums based upon the actual level of care.

Deductibles, Co-Insurance, Experience Rating, and Social Insurance

Unlike the problem of adverse selections, moral hazard on the part of buyers of insurance is manageable in unregulated markets, except for individuals who refuse to buy health insurance because they think of themselves as having a low probability of getting sick and yet expect to be taken care of at hospitals in the event of an emergency. The moral hazard by those who refuse to buy health insurance voluntarily is currently addressed but not solved by the Affordable Care Act, which mandates that all persons with incomes over a certain threshold must buy health insurance or pay a penalty (tax). Those persons with incomes below the threshold are subsidized or covered by Medicaid.

Insurers employ several methods to control moral hazard by shifting some of the risk back to insured persons. First, insurance policies often include *deductible provisions*, which allow the insurance company to withhold or deduct a stipulated amount of money from the total dollar loss of any occurrence covered by insurance. For example, if an insurance policy has a $1,000 deductible, the insured pays the first $1,000 of any occurrence of loss and the insurance company pays the balance up to the policy's limit. Second, insurance contracts often contain a *co-insurance provision*, by which the loss from any occurrence covered by insurance is shared by the insured and the insurance company in stipulated proportions. For example, if home insurance policy has a 20% co-insurance provision, the insured will bear 20% of the loss and the insurance company will pay 80% for any occurrence covered by insurance. Third, the insured bears the risk of any monetary loss which exceeds the *insurance policy's limit*. By stipulating policy limits below market or replacement value for the insured item, an insurance company can shift some of the risk back to the insured.

Insurance companies often require insured persons to choose from a menu of deductible and co-insurance provisions; insurance policies with low deductibles and low co-insurance provisions cost more than insurance policies with high deductibles and high co-insurance provisions. By choosing from a menu of deductible and co-insurance provisions, insured persons "reveal" some information or signal their "type," that is, their attitude toward risk and whether they consider themselves to be careful or careless persons. And by stating the premiums for such choices, the insurance company reveals the cost of covering such risks. Such a menu of choices more nearly aligns the incentives of both parties to the insurance contract, called *incentive compatibility*, but usually does not completely eliminate the conflict between complete insurance and moral hazard.[10] The difficulty of completely aligning incentives, that is, the difficulty of eliminating moral hazard, will also be discussed below in the context of the principal-agent problem.

Moral hazard is often called a problem of hidden or unaccountable action. If the amount of care is observable, then the moral hazard problem is reduced because insurance premiums can be adjusted to reflect some risks. Consider examples of observable levels of care or precaution against accidents that reduce moral hazard: businesses that install sprinkler systems are given lower fire insurance rates; and, drivers who agree to abide by specified parameters with respect to speed, acceleration, and braking intensity, which parameters can be monitored by on-board computers that transmit performance data via geographic positioning systems (GPS) to insurance companies for evaluation of compliance, pay lower automobile insurance premiums. Are you in favor of ubiquitous video cameras at street intersections and in public spaces? Does the right to privacy extend to individuals who use public roads and subject others to unreasonable risks of bodily injury according to the Judge Hand test (B < PL)?

Insurance companies often attempt to affect the level of care taken by an insured by basing insurance premiums on *experience rating*, which means that insured individuals who make more insurance claims pay higher premiums than those who make fewer claims. As examples, automobile insurance premiums often rise after a person is involved in an automobile collision; or, the premium which employers pay for workers' compensation insurance is related to the number of claims filed by their employees. Of course, the effect of experience rating upon behavior depends crucially upon the percentage of the loss that is recaptured in premiums and the number of years over which recapture is stretched. To take an extreme example, if an insurance company which pays $50,000 as a result of an automobile accident is allowed to recapture the entire loss in the form of higher premiums over a two year period, the risk will be completely shifted back to the insured in a short time. Such an extreme recapture provision might cause risk adverse drivers to be too careful or drive too little.

At the other extreme, a major automobile insurer has advertised that its policy includes "accident forgiveness" protection (for the first accident); that is, being involved in an automobile accident will have no effect on insurance premiums. What effect would complete accident forgiveness protection have on driving behavior and on the deterrent effect of negligence liability? Do careful drivers want to be in an insurance pool with drivers who want to buy automobile insurance with total accident forgiveness protection? Would a healthy person want to buy life insurance from a company that advertises that it will sell life insurance policies without screening applicants via a medical exam? Are individuals who opt for insurance policies that do not attempt to control moral hazard more likely to be adverse selections?

If automobile insurance was complete and moral hazard was totally uncontrolled through deductibles, co-insurance provisions, policy limits, or experience rating, then insurance would significantly undermine or vitiate the deterrence effect of negligence liability in civil suits. At the other extreme, if the level of care is completely monitored and premiums are set to recapture losses in a short time, then the existence of insurance would have little effect on negligence liability because moral hazard would be almost completely controlled.

The Affordable Health Care Act also attempts to control moral hazard by requiring all individuals to have health insurance rather than allowing uninsured individuals to wait until they become seriously ill before buying coverage or depending on emergency care which is financed by surcharges on those who have health insurance.

INSURANCE COMPANIES AND MORAL HAZARD

There are three major problems with insurance that require government regulation of insurance companies: One, self-interested insurance companies do not want to insure high-risk individuals who are adverse selections (bad apples in the barrel). As discussed above, if unregulated, insurance companies would not sell health insurance to women whose families had a higher than average risk of having breast cancer, those with a history of heart problems, or those with other preexisting medical conditions.

Two, in unregulated insurance markets, consumers will have incomplete information about what the policy covers or does not cover, which is often buried in the fine print until a catastrophic occurrence, after which they may learn coverage is lacking. Thus, what illnesses or events that are covered by insurance policies must be regulated by government.

Three, since most people pay insurance premiums for many years or decades before events occur that trigger coverage (health, home, and retirement plans), insurance companies in unregulated markets may invest in high-reward, high-risk investments or not retain enough monetary reserves so that when economic downturns or financial panics occur, they may go bankrupt and not be there when needed in the long-run. When someone gambles with someone else's money, it raises the problem of moral hazard, or unaccountability.

The possibility of moral hazard by insurance companies with respect to risky investments requires government regulation of the types (riskiness) of their investments, the amount of reserves that they must maintain, and the amount of borrowing by which they can *leverage* their investments (since higher reserve requirements limit the amount of leverage, they are substantially interchangeable).

Contrast the riskiness of an unleveraged investment with that of a leveraged investment. Unleveraged investment: if a person puts \$10 in an investment that will earn 20% in profits annually, the investment of \$10 will result in \$2 in profits/year. Leveraged investment: assume a person with the \$10 to invest can also borrow an additional \$300 to invest, at say 10% interest on the loan, which investment the lender holds as collateral in case of default, so that the person has a total investment of \$310 (\$10 + \$300), which is 30 to 1 leverage. If the total investment earns profit of 20%(310) = \$62; after paying 10%(\$300) = \$30 of the profit as interest, the investor has earned \$32 on her or his investment of \$10, for a gain of 320%. This is the magic of leverage. However, if instead of making profit, the investment loses money so that it goes down in value, for example, by a mere 4%, then total investment is now worth only 96%(310) = \$298. The lender will now want the borrower to put up an additional \$2 as collateral or security for the \$300 loan, the collateral for which is now only worth

$298. If the borrower does not have reserves with which to put up the additional collateral, the borrower is bankrupt because the lender will sell the collateral and sue the borrower for the difference between loan value, $300, and collateral value, $298. High leverage produces tremendous profits on the up side but can produce nearly instant bankruptcy if investment value declines even a small amount. In sum, the moral hazard with respect to the possibility of high-risk investments by insurance companies must be regulated through reserve requirements and restrictions on leverage and riskiness of investments.

MORAL HAZARD-ADVERSE SELECTION DEATH SPIRAL

Consider an example of how adverse selection and moral hazard may interact in what is called a "death spiral." Assume that owners of bicycles have to decide whether or not to buy locks as protection from theft. If there is no theft insurance, most owners will buy strong locks. On the other hand, if there is complete insurance for bicycle theft, some owners will, due to moral hazard, buy weak locks, not use their locks, or not buy locks at all. These owners will experience higher theft rates, which will cause the cost of insurance to rise. As the price of this insurance rises due to unexpected losses from moral hazard, bicycle owners, who buy strong locks and are more careful about locking their bicycles, will not want to buy theft insurance for their bicycles because they believe the premiums for such insurance are substantially higher than their expected loss due to the carelessness of adverse selections in the insurance pool. As careful bicycle owners leave the insurance pool, the cost of bicycle theft insurance premiums will rise further because only individuals at high risk of having their bicycles stolen remain in the insurance pool; that is, the insurance pool contains a large number of adverse selections.

This additional rise in premiums for bicycle theft insurance may cause other insured persons who are low risk relative to the most careless members in the insurance pool to drop out. Due to the higher risk profile of remaining members of the insurance pool, there will be a further round of insurance rate increases and more insurance drop outs. The moral hazard-adverse selection spiral may continue until the price of insurance rises so high that the insurance pool collapses because too few people are willing to buy bicycle theft insurance at very high premiums.

A moral hazard-adverse selection spiral seems to have occurred in the "insurance crisis" of the 1980s, during which many diving boards were removed from pools and public playgrounds were closed due to dramatic rise in the premiums for accident insurance. Insurance companies raised premiums and reduced coverage partially due to fear of large personal injury awards by juries, who were giving away "other peoples' money"—a form of moral hazard or unaccountable behavior. In circumstances in which there is a moral hazard-adverse selection spiral, there is no Nash equilibrium, either pooling or separating.

If participation in the federal Social Security Insurance program was not mandatory, a moral hazard-adverse selection spiral would almost assuredly cause it to collapse. Many low income people might not contribute voluntarily due to moral hazard; many upper income people would not contribute voluntarily because their access to private investments with higher returns causes them to view middle class and low income earners as adverse selections.

The Affordable Health Care Act attempts to protect adverse selections and control moral hazard: insurance companies are required to offer health insurance to individuals with preexisting conditions (adverse selections) on a nondiscriminatory basis. Moral hazard is controlled by requiring all individuals to buy health insurance; healthy individuals are not allowed to wait until they become sick to buy health insurance or get care at emergency rooms of hospitals for which they may not be able to pay.

PRINCIPAL-AGENT (MORAL HAZARD) PROBLEM

Moral hazard, or unaccountable action, is a particular instance of the more general *principal-agent problem*, which arises whenever anyone asks, how do I compensate someone who is doing something for me so that he or she will be given incentives to complete the job in an efficient manner? As examples, if one hires a lawyer by the hour, how does the client know whether the lawyer will bill for more than the efficient number of hours to do the job? If the lawyer is paid a lump sum, especially if a substantial retainer is paid "up front," how does the client know whether the lawyer will put in more than minimal effort into the job? If someone is hired as CEO of a corporation, how do the stock holders know whether he or she is serving their interests or is more interested in maximizing salary, stock options, prerequisites, and business expense accounts? If a restaurateur hires a bar tender, how does the owner know whether the bar tender is putting too much alcohol into drinks, giving free drinks to friends, or skimming money from the cash register? For employees working in a large corporation, how does management know which employees are making an effort commensurate with their salary and which employees are "goofing off?"

Resolving the principal-agent problem requires a determination of how sensitive the agent's compensation should be with respect to the amount of effort expended and output produced. Is a lump sum or annual salary for a worker, which is chosen independent of the amount of output or effort expended, efficient? No, because the worker has little incentive to put in more than a minimal amount of effort. Is a salary that was based entirely on the amount of output sold efficient? Not in circumstances where employees are risk adverse and there is "noise" in the system, that is, random elements such as weather, quality of other inputs, strength of competition, technological change, and "herd" behavior of consumers (unforeseen trends), all of which can affect performance and output. For example, a sales person, who is given a low-quality product to sell, will probably not earn a high income even with a great deal of effort.

Risk Sharing and Moral Hazard

Joseph Stiglitz shared the 2001 Nobel Prize for his model in which he investigated the efficiency of different risk sharing arrangements on incentives and moral hazard.[11] His sharecropping model, where output is shared in some proportion between landlord and tenant, also applies to sharing and incentive arrangements in industrial and service sectors involving mixed compensation packages of wages, commissions, profit sharing, contingent fee contracts, franchise arrangements, and other incentive payments. In general, Stiglitz's model sheds light on risk sharing and incentive payments in any principal-agent setting.

Stiglitz asked two questions: 1) Why is it that some economists have argued that risk sharing (sharecropping) in agriculture is inefficient because of negative incentives on the amount of labor supplied, but that risk sharing arrangements and "commissions" in manufacturing and services are said to be efficient due to positive incentive effects? 2) Why is it that less-developed economies tend to utilize sharecropping in agriculture but not in the manufacturing whereas modern economies use risk sharing agreements in manufacturing or services (e.g., commissions) but not in agriculture? Stiglitz said that the answers to the two questions are related.

Classical economists, who argued that sharecropping in agriculture is inefficient, reasoned that if tenant's share of the crop is less than 100%, tenant will supply less than the optimal amount of labor.[12] Assume tenant receives 50% of the crop and is deciding whether to work an additional hour of work per week. If an additional hour of work costs the tenant $75/week in effort for 52 weeks or $75(52) for the year and would increase total revenue $100/week or $100(52) for the year, then tenant would not supply the additional hour because he would receive only half the increased output, $50(52), which is less than the cost to him, $75(52). This decision is inefficient because the tenant's marginal cost, $75(52), is less than marginal revenue $100(52); in this situation, efficiency requires that the tenant supply the additional hour of work/week, but, so goes the argument, sharecropping does not supply the efficient incentive.

Stiglitz argues that the flaw in this argument is that it fails to consider the context in which risk sharing arrangements are employed: a) production processes subject to great uncertainty as to whether low output is due to lack of effort by workers or circumstances beyond their control (e.g., bad weather, crop disease, recessions, selling products of poor quality, etc.), b) supervision of workers is difficult, time consuming, and prohibitively expensive (many small farms), and c) specifying a contract covering all contingencies is impossibly complex (prohibitive transaction costs).

Stiglitz argues that in the context of substantial uncertainty, sharing arrangements apportion risks and provide efficient incentives. Under sharing arrangements, workers bear a greater share of risk, have greater incentive, and less moral hazard than if they received wages only.

Modern Agriculture Economies Do Not Employ Risk Sharing

Stiglitz gives two reasons why modern economies do not employ risk sharing in agriculture. When in addition to land and labor, a third factor of production, capital, plus advances in technology are added to the production process, *economies of scale* often result; that is, larger farms have lower costs per unit of output. When there are scale economies, the cost of supervision declines, so that a wage system becomes more efficient. Moreover, when workers are closely supervised, individual effort can be more accurately assessed.

Corporation as a Form of Risk Sharing

In the context of Stiglitz's model, a corporation's equity may be characterized by several types of risk sharing arrangements: bondholders, who want to limit their risk of loss, usually received fixed interest on bonds and stock holders, who take most of the risk, have no guaranteed return; however, in mixed compensation forms, bond holders, in addition to a fixed interest, get a share based upon risk (e.g., convertible bonds or participating bonds), and stock holders may accept less risk and a fixed dividend (preferred stock with fixed dividends).

With respect to worker remuneration, depending upon the factors a) uncertainty of the relation between worker effort and output, b) whether supervision of workers is difficult, and c) how complicated a complete contingency contract is to stipulate, worker remuneration can be salaries, commissions, or mixture of the two.

NOTES

1. George Akerlof, "The Market for Lemons: Uncertainty and the Market Mechanism," *Quarterly Journal of Economics* 84 (1970): 488–500.
2. *Ibid.*, 488.
3. *Ibid.*, 489–490.
4. In-Koo Cho and David Kreps, "Signaling Games and Stable Equilibria," *The Quarterly Journal of Economics* 102 (1987): 179–222.
5. Question: What is socially more efficient, requiring committees in the US Congress to report the vote of each member on each proposed law or merely reporting the total for and against a resolution? Separating or pooling equilibrium? What arguments support each side?
6. A. Michael Spence, "Job Market Signaling," *Quarterly Journal of Economics* 87 (1973): 355–374.
7. Thomas Hungerford and Gary Solon, "Sheepskin Effects in the Returns to Education," *Review of Economics and Statistic* 69 (1987): 175–177.
8. Family and Medical Leave Act of 1993, P.L. 103–03 (February 5, 1993).
9. Douglas Baird, Robert Gertner, and Randall Picker, *Game Theory and the Law* (Cambridge, MA: Harvard University Press, 1994), 142–147.
10. Jack Hirshleifer and John Riley, *The Analytics of Uncertainty and Information* (Cambridge: Cambridge University Press, 2003).
11. Joseph Stiglitz, "Incentives and Risk Sharing in Sharecropping," *Review of Economic Studies* 41 (1974): 219–255.
12. Alfred Marshall, *Principles of Economics*, 8th ed. (New York: Palgrave Macmillan, 1920), Book VI, Chapter X.

Part V

LAW AND ECONOMICS OF CIVIL OBLIGATION

Part V consists of five chapters. Chapter 17 sets out the game theoretic framework for the law and economics of civil obligation (torts, contracts, and unjust enrichment), which is based upon two key variables, whether players act simultaneously or sequentially and whether players have complete or incomplete information. Since each of the two variables has two states, $2 \times 2 = 4$ models of torts (negligence) and four models of contracts and unjust enrichment result.

Chapter 18 on negligent torts is divided into three parts: 1) where only one person must take care (unilateral care); 2) where two people must take care (bilateral care) in the context of four game theoretic models (classic bilateral care, last clear chance, defensive care, and foreseeable use and misuse); and 3) two types of products liability, manufacturing defects and design defects. Analytically, bilateral care's foreseeable use and misuse negligence model includes design defect, products liability; and, negligence involving unilateral care includes manufacturing defect, products liability.

Chapters 19 and 20 explain two contract models. Chapter 19 begins by distinguishing decision-making by strict rules versus discretionary standards; then, Gilmore's strict rules contract model is distinguished from his discretionary standards contract model; third, Gilmore's strict rules contract model is jointed with competitive markets to form the strict rules, competitive contract model. In chapter 20, Gilmore's discretionary standards contract model is joined with imperfect competition to form the discretionary standards, imperfect competition contract model.

Chapter 21 concerns decision theory, suit, and settlement. Decision theory is used to evaluate plaintiff's willingness to sue under the American rule versus the British rule for allocating litigation costs. Then, the optimism suit and settlement model is discussed; it attributes settlement of a lawsuit to a convergence in plaintiffs and defendants' estimates as to how likely plaintiff is to win the suit. Finally, the effect of a contingent fee arrangement on a lawyer's willingness to sue and the lawyer-client conflict of interest are analyzed.

Chapter 17

Game Theoretic Framework for Law and Economics of Civil Obligation

CONTRACTS AND UNJUST ENRICHMENT

This chapter provides the outline for the next three chapters. In game theory, the two key variables, which concern whether players act simultaneously or sequentially and whether players have complete or incomplete information, result in $2 \times 2 = 4$ models, which are game theory's natural taxonomy. The four-model game theoretic framework is: 1) simultaneous games of complete information, 2) sequential games of complete information, 3) simultaneous games of incomplete information, and 4) sequential games of incomplete information.[1]

A four-model game theoretic framework will be used in chapter 18 for negligence requiring bilateral care and products liability involving design defects;[2] another four-model game theoretic framework will be used in chapters 19 and 20 on contracts and unjust enrichment.

Four Negligence Models Involving Bilateral Care (Plus Defectively Designed Products)

Chapter 18 on negligent torts is divided into three parts: 1) where only one person must take care (unilateral care); 2) where two parties must take care (bilateral care) in the context of four game theoretic models (classic bilateral care, last clear chance, defensible care, and foreseeable use and misuse); and 3) two types of products liability, manufacturing defects and design defects. Analytically, bilateral care's foreseeable use and misuse negligence model includes design defect, products liability; and, negligence involving unilateral care includes manufacturing defects type products liability.

In Figure 17.1, the four negligence models involving bilateral care (plus design defects) are:

1. Classic bilateral care negligence model assumes complete information and simultaneous action by all persons; it incorporates the Judge Hand test. In the classical negligence model, all persons are justified in assuming that all other persons will

Four Bilateral Care Negligence Models

	Complete Information	Incomplete Information
Simultaneous Action	I. Classical Bilateral Care Negligence Model	III. Defensive Care Model
Sequential Action	II. Last Clear Chance Model	IV. Foreseeable Use and Misuse Model (plus Design Defects)

Figure 17.1 Four Bilateral Care Negligence Models. *Source*: Created by the author.

use due care and all-or-nothing negligence rules and comparative negligence is equally efficient.

2. Last clear chance model assumes complete information and sequential action. In this model, the second mover knows that first mover has not used due care; first mover has a significant strategic advantage in this model. Efficiency requires that two goals be achieved: first mover must be given an incentive not to engage in strategically negligent behavior; and, second mover, who is aware that first mover has not used due care, must be given an incentive to drive defensively, that is, adopt a higher level of care than due care. In the last clear chance model, comparative negligence is more efficient than are all-or-nothing negligence rules.
3. "Defensive driving" or defensive care model assumes incomplete information and simultaneous action. In this model, a level of care higher than due care is efficient when significant doubt exists as to whether the party, who is the low cost avoider, will take care and cost of accident prevention by either party is relatively small. Redundant or defensive care is efficient in this model. In the defensive care model, comparative negligence is more efficient than are all-or-nothing negligence rules.
4. Foreseeable use or misuse model assumes incomplete information and sequential action. In the foreseeable use or misuse model, the first mover (manufacturer) often reasonably foresees that second mover (product user) may not use due care. Foreseeable use or misuse model includes defectively designed products which involve bilateral care because manufacturers have a duty to prevent injury from foreseeable uses or misuse of products and consumers have a duty not to be contributorily negligent. First mover has a significant disadvantage in this model. In the foreseeable use or misuse model, comparative negligence is more efficient than are all-or-nothing negligence rules.

FOUR CONTRACT AND UNJUST ENRICHMENT MODELS

Just as one economic model cannot describe and explain accurately all economic market structures, that is, economists have developed the model of competition and several models of imperfect competition—monopoly, monopsony, bilateral monopoly, and oligopoly, which different models imply different public policy prescriptions; so too, one contract model cannot be expected to explain all contracts made under different market structures; nor can one contract model rationalize the wide range of judicial responses in contract cases.

Chapter 19 begins by explaining decision-making by strict rules versus discretionary standards; then, Grant Gilmore's strict rules contract model is distinguished from his discretionary standards contract model;[3] third, Gilmore's strict rules contract model is associated with competitive markets to form the strict rules, competitive contract model. In chapter 20, Gilmore's discretionary standards contract model is associated with imperfect competition to form the discretionary standards, imperfect competition contract model. (The unilateral contract model and the unjust enrichment models of promissory estoppel and quasi contracts are stated in this chapter as part of the four-model framework but are not discussed in chapters 19 and 20.)

In Figure 17.2, the four contract and unjust enrichment models are:

1. Strict rules, competitive market contract model assumes complete information (all firms know each other's payoffs, costs, and profits) and simultaneous promises (bilateral contracts). In this model, strict classical contract rules are more efficient than are discretionary standards. Rottenberg's corollary (chapter 11), where a change in liability rules or property rights in free markets is irrelevant to efficiency, is often true; for example, if products are *fungible*, the four direct damages rules for breach of contract—expectation damages, reliance damages as measured by opportunity cost, specific performance, and liquidated damages—are all equally efficient. Specific performance is the preferred breach of contract remedy in this model on grounds of risk allocation and equity. The lost-volume seller rule is irrelevant to efficiency. Consequential damages are foreseeable.
2. The unilateral contract model assumes complete information and sequential action. In this model, a promise is followed by an act or an act is followed by a promise. The strategic problems inherent in the unilateral contract model are similar to those of its tort counterpart, the last clear chance model. In both of these models, the first mover has a significant strategic advantage, which the law attempts to control or negate. For example, after an offer for a unilateral contract is made, the first mover is not allowed to revoke the offer once second mover has made preparation for or has begun an act of acceptance. This model is not discussed in chapter 19.

Four Contract and Unjust Enrichment Models

	Complete Information	Incomplete Information
Simultaneous Action	I. Strict Rules, Competitive Market Contract Model	III. Discretionary Standards, Imperfect Competition Contract Model
Sequential Action	II. Unilateral Contract Model	IV. Unjust Enrichment (Promissory Estoppel or Quasi Contract Model)

Figure 17.2 Four Contract Models. *Source*: Created by the author.

3. Discretionary standards, imperfect competition contract model assumes incomplete information and simultaneous (promises). In this model, a regime of discretionary standards is more efficient than a regime of strict contract rules. Courts are concerned with issues of honesty and good faith, mistake, asymmetric information, monopoly, bilateral monopoly, and unconscionability. In general, money damages is the efficient remedy for breach of contract. Over reliance on promises can occur. The narrow interpretation of consequential damages, known as the rule of *Hadley v. Baxendale,*[4] is efficient in this model because it has an information forcing function to perform and prevents over reliance on contracts. The doctrine of mitigation of damages also prevents over reliance on contracts.
4. Unjust enrichment model (promissory estoppel and quasi contract) assumes incomplete information and sequential action. In this unjust enrichment model, a promise is followed by an act which does not meet the formal requirements of contract (promissory estoppel) or an unsolicited act is followed by a promise which does not meet the formal requirements of a contract (quasi contract). This model is concerned with how much second mover strategic advantage it is efficient for courts to allow. For example, promissory estoppel is concerned with questions such as, in what circumstances should second mover be allowed to rely on first mover's promise which was not the supported by a bargain? In quasi contract, should second mover be allowed to accept first mover's unsolicited benefit without compensation? This model is not discussed in chapter 20.

NOTES

1. Eric Rasmussen, *Games and Information: An Introduction to Game Theory*, 4th ed. (Cambridge, MA: Blackwell, 2006).
2. Negligence involving unilateral care and defectively manufactured products, which are analytically similar, is not covered by the four-model game theoretic framework.
3. Grant Gilmore, *The Death of Contract* (Columbus, OH: The Ohio State University Press, 1974).
4. 9 Exch. 341, 156 Eng.Rep. 145 (1854).

Chapter 18

Torts

Negligence and Products Liability

This chapter on negligent torts is divided into three parts: 1) where only one person must take care (unilateral care); 2) where two people must take care (bilateral care) in the context of four game theoretic models (classic bilateral care, last clear chance, defensible care, and foreseeable use and misuse); and 3) two types of products liability, manufacturing defects and design defects. Analytically, bilateral care's foreseeable use and misuse negligence model includes design defect, products liability; and, negligence involving unilateral care includes manufacturing defect, products liability.[1]

UNILATERAL CARE

The unilateral care case is illustrated by *Blyth v. Proprietors of Birmingham Waterworks Co.*,[2] in which a connection between a fire plug and water main opposite plaintiff's house burst, causing water damage to his premises. The apparatus had been installed and maintained for twenty-five years without incident by the defendant waterworks. The main pipe opposite plaintiff's house was buried more than the eighteen inches below the surface, as required by statute. A severe frost caused a connection between a water main and a pipe to spring a leak when frozen water expanded. The issue was whether the defendant was negligent.

The Court of Exchequer held that even if defendant's precautions proved insufficient "against the effects of the extreme severity of the frost of 1855, which penetrated to a greater depth than any which ordinarily occurs south of polar regions," defendant was not negligent because, by providing for the "average circumstances of temperature in ordinary years," it had "provided against such frosts as experience would have led men, acting prudently, to provide against."[3]

Unilateral Care and Judge Hand Test

Under conditions of unilateral care, complete information, and a risk neutral actor, the Judge Hand, $B < PL$, test explains *Blyth*. Two relevant social costs must be considered: the cost of damage that results from accidents and the cost of preventing accidents. In Figure 18.1, any point on the curve labeled marginal expected damage represents the marginal change in expected monetary damage, ΔPL, as measured on the vertical axis, when one additional unit of care is taken to prevent accidents, as measured on the horizontal axis. For example, marginal expected damage is the change, Δ, in one unit of care, say from burying the pipe one foot deeper, multiplied by the reduction in probability, P, of pipes bursting at that increased depth, multiplied by loss, L, from water damage should it occur. The marginal expected damage curve is downward sloping because additional precaution yields only diminishing returns; that is, increasing the level of precaution initially prevents major accidents, then less serious accidents, and so forth.

In the Judge Hand test, B < PL, any point on the curve labeled marginal prevention cost, ΔB, represents the marginal change in cost of accident prevention, as measured on the vertical axis in dollars, when one additional unit of precaution is taken to avoid accidents, as measured on the horizontal axis. For example, marginal prevention cost equals the change, Δ, in the burden, B, in dollars, say from burying the pipes one foot deeper. The marginal prevention cost curve is upward sloping because of the assumption that persons increase precaution first by using the least expensive method of care, the second least expensive method next, and so forth. For example, if the cost of burying pipes deeper is less than the cost of making pipes, water mains, and connections thicker, the former would be undertaken first.

In terms of the Hand test, the efficient level of precaution is that level at which the marginal expected damage curve intersects the marginal prevention cost curve, level E (where ΔB = ΔPL). A level of precaution greater than E is inefficiently high because (B > PL); a level of precaution less than E is inefficiently low because (B < PL).

Two important generalizations are implicit in the Judge Hand unitary care model of negligence. First, it is not socially desirable to eliminate all accidents because doing so would be too costly. Put starkly, not only are some injuries or deaths from accidents acceptable, it is efficient that some accidental injuries or deaths occur.

A second generalization is implicit in the Judge Hand test. In a unilateral care case, even if a court were to hold a person to a standard of care higher than the efficient level, E, a rational person would not subsequently adopt a level of precaution greater than the efficient level. Since the marginal prevention cost curve is higher than the

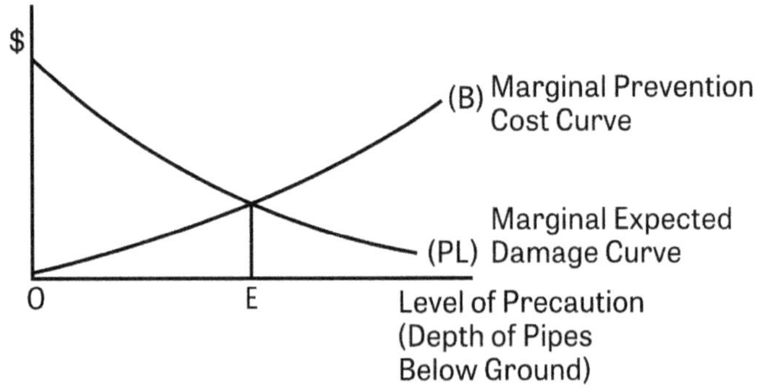

Figure 18.1 Unilateral Care Negligence Model. *Source*: Created by the author.

marginal expected damage at all levels of precaution greater than the efficient level, paying for any additional damage that occurs to the right of the efficient level of precaution would be less expensive than paying for the cost of taking precaution to prevent such damage. If held to a standard of care greater than the efficient level, say a standard of strict liability without regard to fault, a rational person, as indicated, would not take precaution beyond the efficient level.

Let us now use the economic model to explain *Blyth*. The Court of Exchequer held that the defendant acted as a reasonably prudent person by providing against such frosts as experience would have led such a person to do. In the context of the Judge Hand test, the court found that defendant waterworks adopted the efficient level of care, level E in Figure 18.1. Providing for the effects of an unusual and extreme frost would require an inefficient level of precaution greater than the efficient level, E.

Moreover, as implied by the economic model of negligence, even if the court had found the defendant waterworks strictly liable, a rational defendant would neither incur the cost of providing for thicker pipes nor burying pipes deeper in order to prevent damage from an extremely severe frost which occurs very infrequently; rather, he would realize that its least expensive course of action would be to pay for the infrequent damage when it occurs.

Under negligence, homeowners pay slightly lower water bills but have a choice whether or not to buy insurance for the cost-justified damage. Under strict liability, water bills are slightly higher because homeowners are buying water plus compulsory insurance. A shift from negligence to strict liability would have little effect on the quantity of water sold because water is a necessity; that is, water has a very inelastic demand within the relevant price range. This conclusion may be viewed as another theorem in which a change from one legal regime to another, in this case from negligence to strict liability for unilateral care, is irrelevant for efficiency.[4] (This conclusion also anticipates similar conclusions in Section C with respect to products liability involving defectively manufactured products, but not defectively designed products.)

BILATERAL CARE

In this section, bilateral care, where two people interact and one or both must take care if injury is to be avoided, is discussed in the context of game theory that consists of four related models. The theory is motivated by two questions: One, when we go about our daily activities, can we assume that other people with whom we interact will use reasonable or due care all the time, some of the time, or never? Two, under what circumstances is all-or-nothing negligence where any contributory negligence on the plaintiff's part is an absolute bar to recovery by plaintiff as efficient as comparative negligence where damages are apportioned based upon relative fault of plaintiff and defendant?[5] These two negligence rules are equally efficient only if the reasonably prudent person can assume that all other persons act in a reasonably prudent manner, that is, all others act with due care. This is another law and economics irrelevance theorem. When we know or should know that persons with whom we interact are not acting with due care, which is frequently, comparative negligence is more efficient than all-or-nothing negligence.

In the context of game theory, whether a reasonably prudent person can assume all other persons always use due care is shown to depend upon two variables having to do with whether all persons act simultaneously or sequentially and whether they have complete information or incomplete information. Since each of these two variables has two possible states (simultaneous action vs. sequential action and complete information vs. incomplete information), there are 2 × 2 or 4 possible models of negligence: the classic negligence model, last clear chance model, defensive care model, and foreseeable use and misuse model. With the exception of the classic negligence model, in which all-or-nothing negligence and comparative negligence are equally efficient, in the latter three models, comparative negligence is more efficient than all-or-nothing negligence.

Complete Information and Simultaneous Action: Classic Negligence Model

The classic negligence model has two key assumptions that are important in its exposition and conclusions. First, all persons know the legal standard of precaution and know the costs of satisfying the legal standard, both their own costs and the costs incurred by all other persons; moreover, the courts are always able to determine accurately whether any person has satisfied the legal standard. Neither persons nor courts ever make mistakes with respect to such calculations. This is the complete information assumption. Second, all persons choose the level of precaution they take at one point in time; that is, if more than one person takes action, all persons act simultaneously. This is the simultaneous action assumption. Given these two key assumptions of the classic negligence model, all-or-nothing negligence is just as efficient as comparative negligence.

The classic negligence model is often explicated in terms of defendant-injurer and plaintiff-victim. Three situations will be considered: 1) alternative care in which plaintiff is the least cost avoider; 2) alternative care in which defendant is the least cost avoider, and 3) joint care in which the efficient solution requires that both parties take care. In the context of the complete information and simultaneous action assumptions, the reasonably prudent person is justified in assuming that all other persons will act with due care and all-or-nothing negligence is as efficient as comparative negligence. However, when either of these two assumptions of simultaneous action and complete information is altered in the three subsequent models, all-or-nothing negligence is less efficient than comparative negligence.

Alternative Care Where Plaintiff is Least Cost Avoider

An alternative care accident is one in which damage can be avoided if either the injurer or victim takes care; both need not take care for the accident to be avoided; moreover, redundant care, where both take care, is inefficient. *Tennessee Trailways, Inc. v. Ervin*[6] illustrates an alternative care case in which plaintiff-victim is the cheapest cost avoider. In *Ervin*, a motorcyclist "spurted" out from a private road which intersected a highway, was struck by defendant's bus, and died. The Supreme Court of Tennessee upheld the trial court's directed verdict in favor of the defendant even though there was evidence that the defendant's bus was being driven at 73.5 miles

		Injurer—Defendant	
		Care	No Care
Plaintiff—Victim	Care	-50, -100	-50, 0
	No Care	0, -100	-1,000, 0

Figure 18.2 Alternative Care Negligence Game. *Source*: Created by the author.

per hour which violated a state statute setting the speed limit of 65 miles per hour on grounds that defendant's negligence was not a cause of the collision.

Consider an alternative care game in which the plaintiff is least cost avoider in the context of complete information and simultaneous action assumptions; then, the conclusions of the model are applied to *Ervin*. Assume that according to the Judge Hand test, expected marginal accident damage is \$1,000 (PL), that plaintiff-victim, who is least cost avoider, can avoid the damage by spending \$50 (B), and defendant-injurer, who is high cost avoider, must spend \$100 (B) to avoid damage. These assumptions are depicted in Figure 18.2.

In terms of the classic negligence model, *Ervin* is an alternative care case because the accident could have been avoided if either motorcyclist or bus driver had taken care to avoid the accident. Motorcyclist is least cost avoider; it is clearly less expensive for a motorcyclist to take care before crossing a highway on which traffic is speeding than it is for a bus (and all other vehicles on the highway) to travel slowly enough to stop in time to avoid colliding with a vehicle crossing the highway recklessly.

Which of the following legal rules, all-or-nothing negligence, comparative negligence, no liability, or strict liability, are efficient in this alternative care case in which the plaintiff-victim is the least cost avoider? The efficient, least cost result is in the upper right cell (–50, 0), where the plaintiff-victim takes care and the defendant-injurer does not take care. Both all-or-nothing negligence and comparative negligence achieve the efficient result (–50, 0) because both plaintiff and defendant know that plaintiff is the least cost avoider and should take care. If an accident results because both plaintiff and defendant adopt no-care strategies, the plaintiff-victim is negligent and defendant-injurer is not negligent under both all-or-nothing negligence and comparative negligence. Defendant bus driver has *no* duty of care with respect to motorcyclist because bus driver is entitled to assume that motorcyclist, as least cost avoider, will use due care. This strong conclusion, that in an alternative care case, high cost avoider has no duty of care, depends upon the complete information and simultaneous action assumptions of the classic negligence model, under which all persons and courts know who the least cost avoider is and reasonably expect that the he will act with due care.

In addition, no liability, upper-right cell (–50, 0), achieves the efficient result; strict liability, lower-left cell (0, –100), does not achieve the efficient result.

Alternative Care Where Defendant is Least Cost Avoider

An alternative care case in which the defendant-injurer is the least cost avoider is illustrated by a hypothetical example in which the defendant-injurer is a driver who runs a red light and collides with the plaintiff-victim who is on the crosswalk. All negligence rules are equally efficient in this alternative care model in which defendant-injurer is least cost avoider; plaintiff and defendant both assume that the defendant who is the least cost avoider will take care. In addition, no liability is inefficient but strict liability is efficient, which is the reverse of alternative care in which the plaintiff-victim is the least cost avoider.

In the context of the classic negligence model, a reasonably prudent person is always justified in expecting that other persons will act in a reasonably prudent manner. As will be shown below, once the classic negligence model's key assumptions of complete information or simultaneous action are altered, the model's implicit assumption that the reasonably prudent person never has a duty to foresee the negligent conduct of others is false.

Joint (Bilateral) Care as Least Cost Solution

LeRoy Fibre Co. v. Chicago, Milwaukee & St. Paul Ry.,[7] illustrates a joint care case in which both parties must take care if the least cost solution is to be achieved. In *LeRoy Fibre*, plaintiff-manufacturer stored flax on a tract of ground located between its factory and the railroad's tracks. The flax was arranged in three-ton stacks in two rows parallel with the rails. When large quantities of sparks and live cinders were emitted from defendant's locomotive, one of the flax stacks in the second row ignited, as a result of which all of the flax stacks were consumed.

Despite substantial evidence that the locomotive was operated negligently, defendant contended that plaintiff was contributorily negligent because of the flammable character of the flax and its proximity to the railroad. Speaking for the United States Supreme Court, Justice McKenna held that "[t]he doctrine of contributory negligence is entirely out of place. Depart from the simple requirement of the law, that everyone must use his property so as not to injure others, and you pass to refinements and confusing considerations." Justice Holmes, concurring in part, was not prepared to decide the issue of contributory negligence because the jury should decide "whether the plaintiff's flax was so near to the track as to be in danger from even a prudently managed engine." The Court's decision in *LeRoy Fibre* can be analyzed with Figure 18.3, in which efficiency requires joint care.[8]

In Figure 18.3, under either all-or-nothing negligence or comparative negligence, the middle-center cell (–20, –15), bilateral care, has least *total* cost (–35) of taking care. Given the assumptions of complete information and simultaneous action, both negligence rules are equally efficient because neither railroad nor LeRoy Fibre will be negligent: LeRoy Fibre reasonably assumes that railroad will install a standard spark arrester, and railroad reasonably assumes that LeRoy Fibre will keep its flax stacks 50 feet from the tracks.

Under a regime of no liability, Nash equilibrium is the top-right cell (–40, 0), where only LeRoy Fibre, plaintiff-victim, takes care; under a regime of strict liability, the bottom-left cell is the solution (0, –80), where only the railroad, defendant-injurer, takes care; both no liability and strict liability are inefficient because they are not the least total-cost solution. When both parties must take care, both no liability and strict liability are inefficient.

		Railroad: Superior Spark Arrester	Railroad: Standard Spark Arrester	Railroad: No Care
LeRoy Fibre	Flax Stacks 100 ft. from RR	-40, -80	-40, -15	-40, 0
LeRoy Fibre	Flax Stacks 50 ft. from RR	-20, -80	-20, -15	-1020, 0
LeRoy Fibre	Flax Stacks 0 ft from RR	0, -80	-1,000, -15	-1000, 0

Figure 18.3 Joint Care Negligence Game. *Source*: Created by the author.

Justice McKenna's opinion in LeRoy Fibre, that the railroad must use its property so as not to injure others under any circumstances, that is, strict liability, is not consistent with the efficient solution. Justice Holmes' concurring opinion, that the jury should decide whether the plaintiff's flax was so near to the track as to be in danger from even a prudently managed engine, is consistent with the cost-efficient joint care solution.

In conclusion, the complete information and simultaneous action assumptions of the classic negligence model dictate that the reasonably prudent person always assumes others will use due care. Moreover, all-or-nothing negligence is as efficient as comparative negligence, which apportions liability according to relative fault.

Complete Information and Sequential Action: Last Clear Chance Model

In this section, we retain the classic model's complete information assumption but alter the simultaneous action assumption so that the model now assumes complete information and sequential action.[9] In this model, the person who acts first has a significant advantage over the person who acts second. In negligence law, the second mover's disadvantage is enshrined in the *last clear chance* doctrine, an anomaly of the all-or-nothing negligence era, which has disappeared in the comparative negligence era.

Pure Versus Threshold Comparative Negligence Standards

Under comparative negligence, a plaintiff's contributory negligence does not operate as a complete bar to his recovery, but serves to reduce his damages in proportion to his fault. A number of comparative negligence standards have been adopted by courts or legislatures. Thirteen states have codified the doctrine of pure comparative negligence, in which compensation depends upon the relative negligence of the parties.[10] Under pure comparative negligence, plaintiff receives damages even if he or she is more than 50% negligent. If plaintiff's damages are $10,000 and plaintiff is 60% at fault, defendant is 40% at fault, plaintiff receives $4,000 and suffers a $6,000 loss; of course, if the defendant suffered injuries, she can counterclaim and receive 60% from the plaintiff.

However, the most common legislative approach for apportioning fault is the *threshold* system, under which a plaintiff's contributory negligence does not bar recovery so long as it remains below a specified proportion of total fault. The two varieties of the threshold comparative negligence approach are the 50% and the 51% standards. Under the *equal fault bar*, or 50% approach, the plaintiff cannot recover anything if his fault is equal to or greater than that of the defendant; in other words, the plaintiff is allowed to recover only if he is less negligent than the defendant. Under the *greater fault bar* system, or 51% approach, the plaintiff is prevented from recovering damages only if the plaintiff's fault exceeds that of the defendant; the plaintiff is therefore allowed to recover if his negligence is equal to or less than defendant.[11] Under both 50% and 51% approaches, plaintiff's contributory negligence is a complete bar if it exceeds the threshold. Since juries tend to apportion fault equally in close cases, plaintiff will recover half the loss under the 51% approach and nothing under the 50% approach. The possibility of this result has led to criticism of the 50% approach.

Alternative Care and Sequential Decisions

We begin by considering two alternative care cases involving sequential action: *Butterfield v. Forrester* (1809),[12] which established the contributory negligence defense; and *Davies v. Mann* (1842),[13] which established the last clear chance exception to the contributory negligence defense. Next, the economic model that assumes complete information and sequential action is discussed and then is applied to these two cases and a joint care case.

In *Butterfield*, the defendant, while repairing his house, which was close by the roadside, put up a pole across part of the road, leaving some free passage in the same direction. The plaintiff left a public house at eight in the evening while it was still light enough to discern the obstruction. He rode "as fast as his horse could go," did not observe the obstruction, rode into it, fell with his horse and was injured. The jury verdict for the defendant was affirmed on appeal. Lord Ellenborough's opinion contains the classic definition of the contributory negligence defense which bars recovery by a plaintiff who was also negligent: "Two things must concur to support this action, an obstruction in the road by the fault of the defendant, and no want of ordinary care to avoid it on the part of the plaintiff."

In *Davies*, the plaintiff fettered the forefeet of an ass belonging to him and turned it onto a public highway to graze off the side of the road. Defendant's wagon, drawn by a team of three horses, came down a slight descent, at a "smartish" pace, struck and ran over the ass, killing it. The jury verdict for the plaintiff was affirmed on appeal. Notwithstanding Lord Ellenborough's holding in *Butterfield* that the plaintiff must be free from negligence, Baron Parke's opinion established the last clear chance exception to the contributory negligence defense:

> [A]lthough the ass may have been wrongfully there, still the defendant was bound to go along the road at such a pace as would be likely to prevent mischief. Were this not so, a man might justify the driving over goods left on a public highway, or even over a man lying asleep there, or the *purposely* running against a carriage going on the wrong side of the road.[14]

Thus, as Baron Parke's opinion makes clear, the second mover is at a disadvantage in a last clear chance situation because a reasonably prudent person, who knows or should know of the first mover's negligence, has a duty to avoid the harm.

Last Clear Chance Model

Thomas Schelling gave the first clear statement of first mover strategic advantage in situations involving sequential action where ambiguity exists as to the right of way as where two vehicles meet on opposite sides of a one-lane bridge, approach an intersection with stop signs at right angle to each other, or where one vehicle on the entrance ramp of a highway is trying to break into a stream of traffic.[15] Schelling showed that there is a rational strategy by which one party can take advantage of the other party. If one driver speeds up so that he enters the intersection first and the other driver realizes it, the latter has to yield or an accident will occur. First mover's *irrevocable commitment* puts second mover in a position where he knows he has the last clear chance to avoid the collision. As is clear from the examples used to illustrate Schelling's analysis, automobile cases involving sequential decisions at intersections occur frequently and are a fertile source of litigation.

Consider the sequential game in which the cheapest preventer can gain a strategic advantage by moving first in situations involving complete information and sequential action. The first mover who is cheapest preventer can shift the burden of avoiding harm to second mover, least cost avoider. Assume that the expected damage from an accident if it occurs is $1,000 (PL) and that it can be avoided by either the least cost avoider at a cost of $50 (B) or the high cost avoider at a cost of $100 (B). If this were a model in which each person had complete information of each other's costs and both persons acted simultaneously, the high cost avoider is justified in assuming that the least cost avoider will act with due care. Now assume that the least cost avoider is the first mover. This model is depicted in Figure 18.4. (This model is a sequential version of the Chicken Game discussed in chapter 7.)

The first mover has a choice of either preventing the accident by spending $50 or not preventing the accident and spending $0. If the first mover chooses a Do Not Prevent strategy (Intimidate strategy in the Chicken Game), the least cost method of preventing the accident cannot occur because branch (2) of the decision tree is no longer possible. Given that the first mover chooses a Do Not Prevent strategy, the second mover confronts a choice between a Prevent strategy, in which second mover spends $100 to prevent the expected damage, branch (3) of the decision tree, or a Do Not Prevent strategy in which second mover spends $0 and allows the expected damage of $1,000 to occur, branch (4) of the decision tree. Given the first mover's failure to prevent the damage, the second mover has been put into a position where he has the last clear chance to avoid the harm. The last clear chance represents the *second best* solution. Thus, in the context of a game that assumes complete information and sequential action, the first mover has a decided advantage.

It is difficult to conceive of second mover advantage in this model because the nearly insurmountable problem for second mover is to find a way to make a commitment in advance not to take care that will be creditable to first mover. In other words, second mover can claim that he will not take care when he moves, but if first

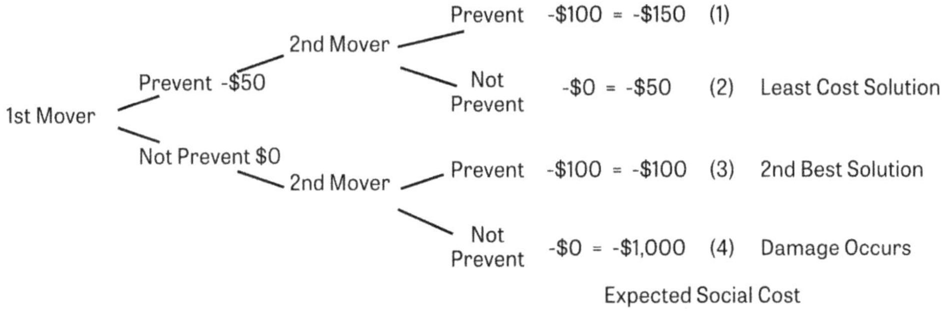

Figure 18.4 Last Clear Chance Game. *Source*: Created by the author.

mover refuses to believe this threat and does not take care, first mover advantage is reinstated.

Assume that both parties, because of negligence or strategic behavior, choose a Do Not Prevent strategy and the expected monetary damage occurs (Intimidate-Intimidate strategies in the Chicken Game). What legal standards will give first mover an incentive to take care in the future and will also give second mover an incentive to prevent damage from occurring if a last clear chance occurs in the future? All-or-nothing negligence, which gives total victory to one party or the other, cannot accomplish these *dual* objectives. Under all-or-nothing negligence, there are only two possibilities, both of which are inefficient. Either first mover is held to be negligent and gets no compensation for his injury, which, as Baron Parke said in *Davies*, would allow second mover to justify driving over a man lying asleep on the highway or purposely running against a carriage on the wrong side of the road. Or, first mover receives a full recovery under all-or-nothing negligence with the last clear chance exception, so that first mover will always have an incentive to act strategically, that is, aggressively and choose to be negligent, knowing that second mover and high cost avoider must either take care or fully compensate first mover.

However, under comparative negligence, first mover, who, is both least cost avoider and is often engaging in strategic (aggressive) behavior, is comparatively more negligent, so she receives either no compensation (threshold comparative negligence) or less than half of her damages (pure comparative negligence). In the future, first mover will have an incentive not to adopt a negligent strategy because she knows that if he is negligent and is injured she will bear the major burden of his injuries. Under comparative negligence, second mover, as in *Butterfield*, who has the last clear chance, has an incentive to avoid a collision because he knows that if an injury occurs he will forfeit at least some compensation.

Although comparative negligence is not fully efficient because partial compensation for either or both parties will not fully internalize the costs of their negligent actions, it is more efficient than the all-or-nothing negligence. Consider an analogy with insurance: Insurance with a co-insurance requirement, which apportions losses between the insured and the insurance company, reduces but does not eliminate moral hazard; nevertheless, co-insurance is more efficient than the extremes of either full coverage or no insurance, that is, loss sharing provides more efficient incentives than all-or-nothing rules. Finally, strategic behavior can also be reduced by criminal sanctions. For example, New York City has instituted a program of surveillance cameras at busy intersections in order to detect and fine motorists who run red lights. Laws against jaywalking can also be enforced.

Joint Care and Sequential Action

In *LeRoy Fibre*, the railroad was actually constructed some years before the flax factory was built, that is, the flax factory "came to" the railroad. Assume that when LeRoy Fibre makes its decision as to how far from railroad tracks it should place its flax stacks, it knows with certainty that the railroad has not installed a spark arrester and refuses to do so. The change in assumption changes the model of negligence from the standard one in which action occurs simultaneously to a model in which action occurs sequentially and in which the railroad is first mover.

The effect of the sequential assumption on the model is illustrated by referring to Figure 18.2. Given that LeRoy Fibre knows that the railroad has not installed spark arresters, would a decision by LeRoy Fibre to place its flax stacks 50 feet from the railroad tracks, which decision was efficient when levels of precaution were chosen simultaneously, be efficient in the case where levels of precaution are chosen sequentially? Answer: no. The combination of no spark arresters and flax stacks 50 feet from the railroad tracks is not one that would avoid $1,000 loss. Given that LeRoy Fibre knows the railroad has not installed spark arresters, LeRoy Fibre, as second mover, should *temporally* place its flax stacks at a distance of 100 feet rather than 50 feet from the tracks so as to avoid destruction of the flax stacks. It should spend $40 rather than $20 to avoid a $1,000 loss. Then, LeRoy Fibre can file a nuisance action against the railroad, requesting that it be enjoined from operating engines without spark arresters. Once the railroad is so enjoined, LeRoy Fibre can then place its stacks 50 feet from the tracks. What legal standard will give LeRoy Fibre, as second mover, an efficient incentive to avoid the damage and effect the efficient result through litigation?

In the classic negligence model, with its complete information and simultaneous action assumptions, both all-or-nothing negligence and comparative negligence provide equally efficient incentives for plaintiffs. However, in a model with sequential action and complete information, only comparative negligence will give the plaintiff, LeRoy Fibre, who is second mover, an efficient incentive to avoid a wasteful loss. If plaintiff's conduct is evaluated with all-or-nothing negligence, a dilemma results: either plaintiff is held not contributorily negligent, even though it had the last clear chance to avoid a $1,000 loss by placing its flax stacks 100 feet rather than 50 feet from the tracks at an additional cost of $20, or plaintiff is held contributorily negligent and receives no compensation even though the level of precaution, placing its flax stacks 50 feet from the tracks, would have been optimal had the railroad taken due care. The first possibility allows plaintiff full compensation even though it knowingly "caused" a $1,000 harm that it could have avoided for an additional $20; the second possibility places the entire loss on plaintiff even though defendant's conduct is comparatively more negligent and plaintiff's conduct would have been optimal if defendant had used due care.

Incomplete Information and Simultaneous Action: Defensive Care Model

We now alter the complete information assumption of the classic negligence model and assume there is incomplete information and simultaneous action. The result is the defensive care model in which redundant care is efficient; additionally, a comparative negligence standard is more efficient than the all-or-nothing negligence rule. *Jackson v. Leach*,[16] will illustrate the defensive care model.

Jackson involved an automobile accident at an intersection of two streets; the plaintiff's vehicle, which was traveling north, collided with the defendant's vehicle, which was travailing west. The plaintiff testified that when he was within a distance of sixty feet from the intersection, traveling at a speed between 15 and 20 miles per hour, he had a clear view to the east for a distance of 180 feet, and there was no vehicle in sight. The plaintiff argued that for the collision to have occurred, the defendants' vehicle must have greatly exceeded the legal speed limit. The defendant argued that the plaintiff was contributorily negligent because he failed to look to his right when he reached the intersection; had he done so, he would have seen the defendant's vehicle in time to have avoided the collision.

In this case, which was decided under all-or-nothing negligence, the Maryland Court of Appeals upheld the jury verdict for plaintiff on the ground that plaintiff was not required to look a second time when he reached the intersection. The Court said:

> While it was incumbent upon plaintiff to respect the rule giving the right of way at street intersections to vehicles approaching from the right, yet, if at the time he looked on approaching and within a short distance of the intersection, the way for a safe distance was clear of traffic coming from that direction, he should not be judicially declared negligent for not taking precautions against the possibility of a collision with a car that could not come into dangerous proximity of his own unless it were unlawfully operated.[17]

Query: Is the court correct in concluding that drivers of automobiles and pedestrians should always assume other drivers use due care? Does one never have a duty to act or drive defensively? In New York City, pedestrians, who upon seeing a traffic light turn green immediately enter a crosswalk without looking both ways on assumption that automobile drivers always obey traffic lights and use due care, are often called the deceased.[18]

The late Nobel laureate William Vickrey wrote that "it is just as important to provide an adequate incentive for driving defensively rather than merely nonnegligently as it is to provide an incentive for driving nonnegligently rather than recklessly."[19] The defensive care model discussed in this section is an alternative care model in which redundant care is often efficient; that is, the model shows that there are situations of incomplete information in which high cost avoider should not rely on cheapest preventer taking care but should act or drive defensively.

Consider the following alternative care model: assume that if an accident occurs the damage will be PL = \$1,000, and that the accident can be avoided either by cheapest preventer at a cost of B = \$50 or by high cost preventer at a cost of B = \$100. If this model employed the classic economic assumptions of complete information and simultaneous action, it is clear that cheapest preventer should always take care and high cost preventer should never take care. However, in this model which assumes a significant probability that one or both parties will be uncertain as to which of them is cheapest preventer, both should take care as a defensive measure whether or not they believe they are the least cost avoider; such defensive care will minimize social cost even though it is redundant most of the time.

Defensive Care Where Both Parties Are Mistaken

Assume that neither party takes care all the time because of incomplete information. Specifically, assume some unspecified probability exists that each party will mistakenly believe that it is the other: there is some likelihood that least cost avoider will mistakenly think that it is high cost avoider and will not take care; there is an equal likelihood that high cost avoider will mistakenly think that it is least cost avoider and will take care. This assumption, that the parties may be mistaken as to who is least cost avoider, is realistic given the myriad automobile accidents involving disputes over right of way. Unless the probability that each party will make a mistake is either very small or very large (in which case parties virtually switch identities), defensive care, which is often redundant, where both parties take care regardless of whether they believe they are least cost avoider, is more efficient than alternative care where they take care only when they believe themselves to be the least cost avoider.

If both parties take care all the time, the cost of care will be \$50 + \$100 = \$150 and no damage will result. Alternatively, if the parties take care only when they think themselves to be the least cost avoider, expected social cost will be higher than \$150 unless the probability that each thinks she is the other (mistaken identities) is either less than 0.06 or greater than 0.925. This conclusion is derived as follows: No accident will occur if the least cost avoider takes care, only the high cost avoider takes care, or both take care; an accident will take place if neither takes care.

These four possibilities are shown in Figure 18.5 where p equals probability that least cost avoider will take care to prevent an accident, and (1 – p) equals probability least cost avoider will mistakenly believe that it is high cost avoider and will not take care to prevent an accident; p also equals the probability that high cost avoider will not take care to prevent an accident; (1 – p) is also the probability that high cost avoider will mistakenly believe it is cheapest preventer and will take care to prevent an accident.

Expected social cost, E(S), will be less than \$150 if,

$$E(S) = (p)(\$50) + (1 - p)(\$100) + (p)(1 - p)(\$1{,}150) < \$150$$

This quadratic equation has two roots, $p < 0.06$, and $p > 0.925$.[20]

The intuitive reason for the conclusion that both parties should take care is that unless the probability of an accident due to mistaken beliefs by the parties is extremely small (or so large that they have effectively switched identities), expected harm from alternative care will exceed \$150. This model may be summarized by saying that where there is a significant probability that parties will be mistaken (or aggressive) as to which one should take care, and where cost of care by both parties is not large, both parties must take care to minimize expected social cost. Thus, defensive care is

		High Cost-Avoider	
		Takes Care	No Care
Low Cost-Avoider	Takes Care	(p)(1 − p)($150)	(p)($50)
	No Care	(1 − p)($100)	(p)(1 − p)($1,000)

Figure 18.5 Defensive Care Game. *Source*: Created by the author.

efficient in this model that assumes incomplete information and simultaneous action. Since these conditions seem to characterize many automobile collisions, this model is unlikely to be trivial.

All-or-nothing negligence is inefficient in *Jackson*. If one party must be given a total victory, either the plaintiff gets total compensation even though he could have avoided the accident had he driven defensively and looked a second time, or plaintiff gets nothing and defendant is not penalized for speeding. All-or-nothing negligence is a single instrument; it cannot achieve two goals. Only comparative negligence can accomplish the two goals: give the defendant an incentive not to speed and give the plaintiff an incentive to look a second time; that is, to drive defensively. Even though comparative negligence will not fully internalize all costs to both parties, it is more efficient than all-or-nothing negligence.[21]

If negligence is defined as the failure to take action that minimizes expected social costs given the reasonably foreseeable level of precaution or lack thereof by other persons, then within the context of the defensive care model, as well as in *Jackson*, a plaintiff who is high cost avoider is negligent for not taking defensive care. Only a comparative negligence standard, which apportioned damages according to relative fault, would give least cost avoider an incentive to avoid speeding in the future by apportioning more than 50 percent fault to it and at the same time give high cost avoider, who is a plaintiff-victim, an incentive to take defensive care by apportioning less than 50 percent to it.

Incomplete Information and Sequential Action: Foreseeable Use or Misuse Model

In this section we begin by considering *Levi v. Southwest Louisiana Electric Membership Coop.*,[22] a negligence case; next, the foreseeable use or misuse model which assumes incomplete information and sequential action is constructed and applied to *Levi*. (This model is also applicable to the products liability design defects, discussed below.)

In *Levi*, plaintiff, an oil company employee, sustained near fatal, permanently disabling injuries when the erectable mast of a truck which he raised while attempting to repair it came in contact with an uninsulated electrical line that serviced a field of oil wells. With the exception of the oil well near the accident site, the power company routed the power line to avoid crossing an oil well driveway or coming close to a well. The power company's employees had warned oil company crews that driving under uninsulated power lines with their rigs partially or fully upright was dangerous.

The jury found for the defendant on the ground that its conduct did not fall below the reasonable standard of care. However, in this comparative negligence state, the Louisiana Supreme Court, reversed and held that the power company's negligence was a legal cause of plaintiff's injuries.

> We do not think reasonable minds can disagree with the conclusion that the power company, particularly with its superior knowledge, skill and experience in electrical safety, should have recognized that its conduct under these circumstances involved a risk of harm to oil field workers. . . . The power company complains that it should not be charged with recognition of any risk that takes effect through a victim's negligence. But the ordinary reasonable person, and even more so the power company, is required to realize that there will be a certain amount of negligence in the world. When the risk becomes serious, either because the threatened harm is great, or because there is an especial likelihood that it will occur, reasonable care may demand precautions against that occasional negligence which is one of the ordinary incidents of human life and therefore to be anticipated.[23]

The power company was not entitled to assume that individuals always use due care. In *Levi*, the power company is first mover and high cost avoider, and plaintiff is second mover and least cost avoider. First mover's alternative care solution is where the power company insulates all overhead wires or reroutes the wires. Second mover's alternative care solution is where plaintiff avoids the accident by carefully scrutinizing the site and perhaps chooses another location for the rig's repair. If probability that plaintiff will be careless is significant and foreseeable, alternative care by plaintiff is not the least cost solution.

Joint care solution is where the parties undertake cooperative measures, such as a warning sign on the power pole and an orange warning ball on the power line combined with greater attentiveness on plaintiff's part. The joint care (cooperative) solution significantly reduces first mover disadvantage inherent in the model by allowing first mover, by warning second mover, to shift a significant portion of burden of care to second mover. Whether joint care is the least cost solution depends upon whether the warning is effective or not: if the warning is not effective, the least cost solution is either where the power company either reroutes or insulates the electric wires.

If the power company had no duty to foresee carelessness by plaintiff, that is, in a world where everyone is presumed to act with due care, plaintiff would be least cost avoider, and the power company would not be liable for his injuries. However, in the context of the foreseeable use or misuse model, all-or-nothing negligence is not as efficient as comparative negligence. Under all-or-nothing negligence, plaintiff, who is cheapest preventer, is negligent and bears the entire loss. But, if negligence is defined as the failure to take action that minimizes expected social cost given the reasonably foreseeable level of precaution or lack thereof taken by other persons, defendant is also negligent for not having at least warned plaintiff. Only a comparative negligence standard, which shares losses according to relative fault, provides defendant with an efficient incentive at least to warn plaintiff in the future. Since Louisiana is a pure comparative negligence state, plaintiff is entitled to some of his damages rather than none. Query: Under comparative negligence, how would you apportion fault between plaintiff and defendant? Who is more at fault?

The foreseeable use or misuse model is one that assumes incomplete information and sequential action. The model contains both alternative care and joint care solutions. With respect to alternative care solutions, if first mover is least cost avoider, the model is trivial. For example, if it is foreseeable that a cheap metal guard on a machine will prevent many injuries to many workers over many years, failure to install such a guard clearly supports a finding that such machine was designed defectively; it is well settled that a manufacturer, as first mover and cheapest preventer, is liable for such injuries.[24] The interesting cases are those where first mover, who is high cost avoider, foresees that second mover, who is least cost avoider, will be negligent. The first mover who is the high cost avoider is often at a disadvantage in this model because it has a duty to minimize expected social cost resulting from second mover's foreseeable use or misuse. How great is that duty? For example, how much costly crash protection must be built into automobiles in order to protect drivers from their own careless and even reckless behavior? (Recall the crashworthy cases in chapter 3.)

Last clear chance model and foreseeable use or misuse model, both of which are sequential models, are strategic opposites. In the last clear chance model, first mover has an *advantage* because it can put second mover in a position where it must take the last clear chance to minimize expected social cost. In the foreseeable use or misuse model, first mover is at a *disadvantage* in cases in which it can reasonably foresee misuse or carelessness by second mover; in such cases, first mover is required to protect second mover against reasonably foreseeable harm in order to minimize expected social cost.

PRODUCTS LIABILITY

In this section on products liability, we begin by distinguishing defective products from defective designs in three ways: by definition, economic theory, and appropriate liability rules. Then, the shift from a negligence standard to strict liability for manufacturing defects is analyzed in terms of asymmetric information, which concerns burden of proof and insurance effects; the effect of strict liability on the number of lawsuits and on demand for products is also discussed. As stated at the beginning of this chapter, manufacturing defects are analytically similar to negligence involving unilateral care and design defects are similar to bilateral care, specifically the foreseeable use and misuse model, because defective designs involve the interaction between manufacturers' designs, foreseeable misuse, and consumers' reasonable expectations or contributory negligence. Major conclusions are that manufacturers should be strictly liable for manufacturing defects, but that in design defects cases, courts should, in general, be wary of extending liability beyond a negligence standard in the context of the foreseeable use and misuse model. (Recall that products liability cases were discussed in chapter 5.)

Defective Products Distinguished from Defective Designs

A defectively (manufactured) product is a unit of output that differs from all other units a manufacturer produces by the same process as, for example, where an automobile's defective steering mechanism causes it to crash, rendering it different from other steering mechanisms produced by the same manufacturing process which did not cause crashes. However, in a design defect case, the issue is whether all products manufactured by the same process subject consumer to unreasonable risks. It may sometimes be difficult to distinguish whether an injury is due to a manufacturing defect, a design defect, or ordinary usage. For example, assume that while striking an object with a hammer, which has been used for several years, the impact causes a sliver of metal from the hammer head to fly into user's eye damaging it. Is the failure of the hammer due to a manufacturing defect, a design defect, or metal fatigue from ordinary wear and tear?

In terms of economic theory, manufacturing defect cases are distinguished from design defects cases in terms of their marginal expected damage curves, which are depicted in Figures 18.6 and 18.7. The essential distinction is that in manufacturing defect cases, total damage under the marginal expected damage curve is bounded, whereas in design defect cases, total damage under the marginal expected damage curve is potentially unbounded.

In Figure 18.6, as the level of care employed in a manufacturing process is increased through a quality control program, marginal expected damage from manufacturing defects can not only be reduced, it can be reduced to zero; that is, near perfect *quality control* is attainable at a level of precaution equal to C. For example, instead of sampling one in a hundred products for accuracy of manufacture and safety, a business could sample one in fifty, one in ten, or at the limit, every product.

The unboundedness of Figure 18.7 illustrates that the possibility of increased safety in design is almost limitless because a product can always be made safer by using stronger or thicker materials, by being strengthened through the addition of cross beams or through the addition of other safety features; the area under the marginal expected damage curve is potentially limitless, that is, unbounded. The same reasoning applies to product warnings which can also be virtually increased to a limitless extent.[25]

Negligence to Strict Liability in Manufacturing Defect Cases

Asymmetric Information, Burden of Proof, Number of Lawsuits, and Level of Safety

If products liability cases involving manufacturing defects are governed by a negligence standard, injured consumers would have the burden of proving what the efficient level of care is, level E in Figures 18.6 and 18.7, and that defendant has taken less than efficient care by a preponderance of evidence (more probable than not). Manufacturers would have an unfair advantage in litigation because they have exclusive access to and control over information necessary for such proof. In economic theory, manufacturers are said to have an advantage over consumers because of this asymmetric information. Thus, one rationale for the shift from a negligence standard to strict liability in manufacturing defect cases is to eliminate the difficult burden of proof problem that victims would encounter in negligence suits.

Under strict liability, plaintiffs have to prove only that seller is a merchant, that their injuries were caused by an unreasonably dangerous, defectively manufactured

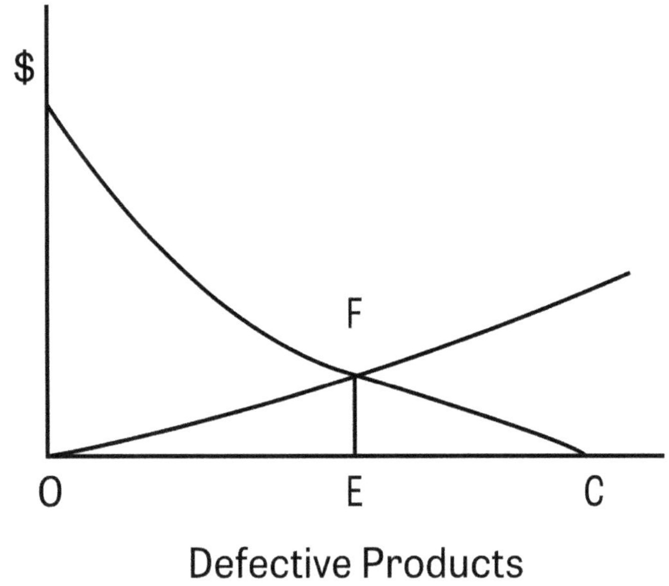

Figure 18.6 *Source*: Created by the author.

product, and that the product is in the same condition as when sold. Since the efficient level of care is not at issue in strict liability (one less element of liability to prove), there should be more lawsuits than under negligence.

Assuming manufacturers are risk neutral, the shift from negligence to strict liability for manufacturing defects should have no effect on the level of safety. This conclusion is easily demonstrated by looking at Figure 18.6. Under both legal standards, manufacturers would not increase the level of precaution to eliminate such accident damage because the cost of eliminating such damage greatly exceeds the damage for which they are liable.

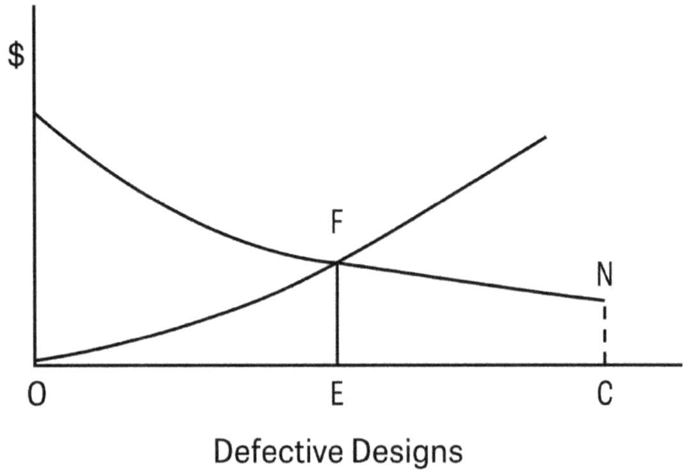

Figure 18.7 *Source*: Created by the author.

Asymmetric Information and Insurance

Manufacturers expressly warrant their products against product failures, which are frequent enough to enter into the buying decisions of consumers. However, because consumers generally have little information concerning the probability that they will sustain serious personal injuries due to defective products, they cannot evaluate accurately the need for insurance against injuries from product failures. On the other hand, manufacturers are usually well apprised of the safety record of their products through reports, complaints, demands for replacements, and products liability suits. As a result of this second information asymmetry that disfavors consumers, manufacturers are, in general, least cost insurers of defective products.

Consider the effect on insurance markets of a shift from negligence to strict liability for manufacturing defects. In terms of Figure 18.6, under a negligence standard, consumers would either have to buy first-party insurance to cover damage equal to triangle-like area EFC for which they are not compensated or accept the risk of such damage. Under a negligence standard, accident costs that are unavoidable by due care are borne by consumers (hence, external to the industry). Consumers pay slightly lower prices for products and can decide whether to buy personal insurance for injuries that occur without manufacturer fault or to accept a slightly higher risk of injury without compensation.

Under strict liability, however, manufacturers have to buy insurance or self-insure for damage equal to area EFC; manufacturers consider that insurance premiums are a cost of doing business which, like any other cost, must be passed on to consumers if the business is to continue operating without a loss. Under strict liability, product price includes a premium for insurance against personal injuries that result from defective products; consumers pay a slightly higher price for products. Thus, under strict liability, consumers must buy both products and insurance even if they would rather pay a slightly lower price for products and accept a slightly higher risk of injury without compensation. Assuming manufacturers are least cost insurers against accidents from defective products because of their superior information concerning safety risks, they can either self-insure or buy insurance at a lower cost than consumers can buy such insurance. Thus, total insurance costs should be lower under strict liability for manufacturing defects.

Demand Effect

Under strict liability in which firms bear the costs of accidents caused by their products, if one firm in a competitive industry has higher accident liability costs than other firms in the industry because its products are less safe than the products of other firms with which it competes (it takes less than efficient level of care, E), this firm cannot pass on its higher accident costs to consumers. The reason is that in a competitive industry, buyers will not buy from the producer who charges a higher price because they have alternative producers from which to buy. As a result, a competitive firm that has higher accident liability costs than other firms in its industry cannot pass on such costs to consumers by an increase in price; its higher per unit costs will result in reduced output and reduced profit.

Under strict liability, if all firms in a competitive industry adopt the efficient level of care, manufacturers will be responsible for area EFC in Figure 18.6, which represents the cost of accident damage from nonnegligently manufactured products. There will be a rise in product prices and a reduction in quantity of output below that which would occur under a negligence standard. Recall the law of downward sloping demand asserts that at higher prices less quantity is bought. However, since the area EFC is bounded and probably relatively small in relation to the total cost of production (assuming efficient quality control), the reduction in quantity demand from a shift from negligence to strict liability for defective products may be small.

If the hazard is relatively small, it will not pay consumers to be attentive to disclaimers; and, for the same reason, manufacturers will not reap significant ill will if they enforce disclaimers in those cases in which someone is injured. The law's refusal to enforce disclaimers of liability for personal injuries caused by manufacturing defects increases efficiency because manufactures are not allowed to externalize this cost through disclaimers.

Strict Liability Versus Negligence in Design Defect Cases

Unlike manufacturing defects which are analytically similar to negligence involving unilateral care, design defects are similar to bilateral care, specifically the foreseeable use and misuse model, because design defects involve the interaction between manufacturers' designs, consumers' reasonable expectations, and foreseeable consumer misuse.

If there is a shift from a negligence standard to strict liability in design defect cases, manufacturers would be responsible for the open-ended area EFNC in Figure 18.7, which represents the cost of accident damage from nonnegligently designed products.

Since the area EFNC is unbounded and probably relatively large in relation to the total cost of production, the effect on demand of a shift from negligence to strict liability in defective design cases would be very substantial (demand may be very price elastic). In some cases, a shift to strict liability may cause the price of nonnegligently designed products to rise so much that manufacturers may cease producing them because there may be no price at which they can make a profit. In defective design cases, if liability is extended beyond negligence, there is no logical stopping place; in other words, strict liability in defective design cases can easily become absolute liability.

Under strict liability, the cost represented by open-ended area EFNC is large and would be passed on to consumers as higher prices. Strict liability for design defects would, in effect, require consumers to buy more "insurance" than they would want if given a choice between buying a product plus insurance to cover the cost of open-ended area EFNC and buying a product under a negligence standard at a lower price without paying for EFNC.

Unlike manufacturing defects, where a shift from negligence to strict liability involves the shifting and spreading of small, known risks whose costs are bounded, a shift from negligence to strict liability in design defects cases involves potentially unbounded liability and therefore has potential to become limitless absolute liability. Thus, in design defects and defective warning cases, courts should be wary of extending liability beyond a negligence standard. The *Restatement (Third) of Torts* has in effect adopted this distinction between strict liability for defective products and negligence for defective designs.[26]

NOTES

1. A version of this chapter appeared in John Cirace, "A Theory of Negligence and Products Liability," *St. John's Law Review* 66 (1992): 1–71.
2. 156 Eng. Rep. 1047 (1856).
3. The Court reversed a jury verdict in favor of the plaintiff homeowner, holding that there was no evidence to leave to the jury. 156. Eng. Rep. 1049.
4. Assuming actors are risk neutral.
5. See, John P. Brown, "Toward an Economic Theory of Liability," *Journal of Legal Studies* 2 (1973): 323, 343.
6. 438 S.W. 733 (1969).
7. 232 U.S. 340 (1914).
8. This table is adapted from Richard Posner, *Economic Analysis of Law*, 9th ed. (New York: Wolters Kluwer, 2014), 199.
9. Steven Shavell, "Torts in Which Victim and Injurer Act Sequentially," *Journal of Law and Economics* 26 (1986): 589–612.
10. Alaska, Arizona, Florida, Kentucky, California, Louisiana, Mississippi, New Mexico, New York, Rhode Island, Washington, and South Dakota.
11. Twelve states had adopted 50% comparative fault; twenty-one states have adopted 51% comparative fault; as of 2012, only these states, Alabama, Maryland, North Carolina, Virginia, and Washington, DC, still employ all-or-nothing negligence.
12. 103 Eng. Rep. 926 (K.B. 1809).
13. 152 Eng. Rep. 588 (Ex. 1842).
14. 152 Eng. Rep. 589 (emphasis added).
15. Thomas Schelling, *The Strategy of Conflict* (Cambridge, MA: Harvard University Press, 1980), 24, 32, 37.
16. 152 A. 813 (Md. 1931).
17. 152 A. 815.
18. "Nearly half (47%) of pedestrian fatalities and severe injuries occurred at signalized intersections; surprisingly, most (57%) of these crashers occurred while the pedestrian was crossing with the signal." *The New York City Pedestrian Safety Study & Action Plan* (August 2010), 23.
19. William Vickrey, "Automobile Accidents, Tort Law, Externalities, and Insurance: An Economist's Critique," *Law & Contemporary Problems* 33 (1968): 464, 467.
20. $50p + 100 - 100p + (p - p2)(1{,}150) < 150$; $-50p + 1{,}150p - 1{,}150\ p2 + 100 < 150$; $-1{,}150\ p2 + 1{,}000x + 100 < 150$; $p < 0.06$, $p > 0.925$.
21. Although insurance cannot totally shift the risk from the insured to the insurance company due to moral hazard which must be controlled by deductible and co-insurance clauses, having less than full insurance is more efficient than having no insurance. So too, comparative negligence, although not fully efficient, is more efficient than all-or-nothing negligence.
22. 542 So.2d 1081 (La. 1989).
23. 542 So.2d 1085–86 (internal quotations omitted).
24. In *Greenman v. Yuba Power Products, Inc.*, 59 Cal.2d 57 (1963); *Phillips v. Kimwood Mach. Co.*, 269 Or. 48 (1974). (Both of these cases were discussed in chapter 5.)

25. Courts have held that manufacturers need not warn against obvious hazards. See, for example, *Posey v. Clark Equip. Co.*, 409 F.2d 560 (7th Cir. 1969) (seller of a forklift platform that had no guard was not required to warn that large cartons might fall off of it), *cert. denied*, 396 U.S. 940)1969); *Fisher v. Johnson Milk Co.*, 174 N.W.2d 752 (Mich. 1970) (manufacturer of wire mesh milk container does not have to warn that bottles will break if container is dropped on sidewalk).

26. *Restatement (Third) of Torts: Products Liability* (St. Paul, MN: American Law Institute, 1998), Section 402A.

Chapter 19

Strict Rules, Competitive Market Contract Model

Many exchanges of goods and services occur simultaneously as where a seller exchanges a loaf of bread with a buyer for cash; such *sales* usually do not involve the law of contract. Contracts involve promises for deferred exchanges, that is, exchanges which will occur in the future either simultaneously, for example, agreement to deliver 500 bushels of wheat for a price of $1000 six months from current date, or sequentially, for example, agreement to build a house for which payment is due upon completion.

A contract is defined as a promise or a set of promises for the breach of which the law gives a remedy, the performance of which the law recognizes as a duty. Contractual promises may be *explicit* and bilateral, for example, I promise to mow your lawn if you promise to pay me $50; explicit and unilateral, a promise for an act, for example, I promise to pay you $50 if you do the act of mowing my lawn; *implicit*, for example, I order a meal in a restaurant; or an obligation or "promise" imposed by law, for example, if Harry delivers a plain unaddressed envelope to Sam containing $100 intended for Ann, law imposes a duty on Sam to give the envelope to Ann.

The law of contract is more likely to enforce promises which represent bargains, for example, Harry promises to sell Mary his 2000 Toyota for $300, as opposed to promises that are more social in nature, for example, Sally promises to come to dinner at Sam's house on Saturday night. However, as shown below, the doctrine that enforceable promises should be limited to those based upon bargains, which are referred to as promises where there is *consideration,* has been attenuated in the twentieth century.

First, the strict rules and discretionary standards approaches to decision-making are compared. Second, Grant Gilmore's strict rules and discretionary standards contract models from his classic little book, *The Death of Contract,* are explained.[1] Two other motivations for this chapter's strict rules, competitive market contract model are economic market structure theory which contrasts competitive markets and imperfectly competitive markets in terms of efficiency and policy prescriptions (chapter 9) and the game theoretic framework for contracts outlined in chapter 17. The third section explains this model.

STRICT RULES VERSUS DISCRETIONARY STANDARDS

The following hypothetical example captures the distinction between strict rules and discretionary standards.[2] Assume that you are a parent who has set an 11 p.m. curfew for your child. Assume further that your child returns home at 11:30 p.m. What are your options? You could say, "Since you've come home after 11 o'clock, you're grounded for the next week," or "cannot watch television," or "get no allowance." This is a strict rule.

Alternatively, the parent could employ the discretionary standards approach under which the child's behavior is compared to a discretionary standard, which in this situation is whether the child made a good faith attempt to arrive home on time. In order to determine whether the child satisfied that standard, the parent would ask the child, "Why are you late?" The child might reply, "The bus broke down," or some other excuse such as "While driving me home, my friend's father stopped unexpectedly to do an errand." Do you believe your child or do you contact the bus company or the friend's father? In the case of the delayed bus excuse, contacting the appropriate dispatcher or person who has such information for verification may be time consuming and difficult. Contacting the friend's father tends to stigmatize the child as untrustworthy. Further questions or information gathering may be necessary in order to determine whether the child's excuse is valid: is the bus delayed often and is your child aware of such delays? Even if the bus was delayed, should your child have taken an earlier bus in order to protect against the possibility of delay? Or, did your child inform the friend's father of the 11 p.m. curfew? Does the friend's father listen to what children say? Is the friend's father reliable? Is your child reliable?

The first response where no questions are asked and no excuses are accepted is the strict rules approach. It has the virtues of being simple, cheap, certain, and consistent: it is simple because there is a only one question that needs to be answered in order to determine whether there has been a violation; was the child home before 11 p.m.? It is cheap because no extensive and expensive investigation and evaluation of facts is necessary. It is certain because no excuses are accepted. It is consistent because all children who violate the curfew will be punished alike.

On the other hand, what if strict rule is adopted and your child is late but without fault? What if the bus did break down unexpectedly, or if your child did tell his friend's father of the necessity of being home before 11 p.m. but was ignored? In these cases, your child will feel that the strict rule as applied to her or him was unfair.

Thus, strict rules, even if they work well in general, may occasionally result in unfairness. Strict rules may be applied in these situations if the benefits of simplicity, cheapness, certainty, and consistency outweigh the costs of an occasional inequity. In most cases in which your child comes home after 11 p.m., he or she has no valid excuse, so using the strict rule saves time and reduces the possibility that your child will make up excuses, that is, lie. Of course, if a rule results in many inequities, such as racial profiling, jettisoning the rule in favor of a standards approach, which justifies stopping and detaining persons only on probable cause, is necessary.

Whereas the strict rules approach is general and outcome-oriented (did the child return by 11 p.m.?), the discretionary standards approach which is individualized and fact sensitive considers all the relevant circumstances; it is a case-by-case approach.

The question to be decided is: did this child make a good faith reasonable effort to get home by 11 p.m.? Whereas the rule approach is precise and strict, the standards approach tends to be imprecise and discretionary. For example, the Fourth Amendment standard for allowable search and seizures of persons and property prohibits only searches that are "unreasonable." The discretionary standards approach has benefits and costs that are the opposite of the strict rules approach. The standards approach has potential to achieve justice in each case; its costs are that a complex and expensive investigation may be required. Moreover, discretionary standards may produce uncertain or inconsistent results as is shown by a continuation of the curfew example.

Assume that two children violate a parent's curfew and that one gives the delayed bus excuse and the other gives the friend's forgetful father excuse. Assume further that after checking, it is determined that the bus was delayed for forty-five minutes due to an unusual break down but that the friend's father does not remember being told about the 11 p.m. curfew even though your child insists otherwise. Under the strict rules approach, both children would receive the same sanction. Under the discretionary standards approach, the delayed bus excuse would be accepted, but what about the "forgetful" father excuse? Did the friend's father actually forget or is your child lying? There may be no way of determining the truth even after taking into account the reputations for veracity and reliability of both your child and the friend's father; that is, there may be no way of resolving uncertainty in this case. If either your child is sanctioned when he was telling the truth and the friend's father actually forgot (Type I error[3]) or is not sanctioned when the child was lying (Type II error[4]), the two children will have been treated inconsistently under the standards approach. Since the second child may be given the benefit of the doubt under the discretionary standards approach, it may be applied more leniently than the strict rules approach.

Consider a well-known application of strict rules versus discretionary standards from federal antitrust law. Under antitrust law, a strict rule is used with respect to agreements among business competitors to fix prices. Under Sherman Antitrust Act §1,[5] price fixing or cartel activity is strictly forbidden and excuses are not allowed. Competitors are not allowed to testify that they only fixed prices at a reasonable level or that price fixing was necessary in order to prevent cutthroat competition. In price-fixing cases, the government need only prove that price fixing took place; no other issues are relevant; no defense or testimony by way of excuse or justification of this activity is allowed. In antitrust law, price fixing is said to be illegal per se (illegal by itself).

On the other hand, a discretionary standards approach is used to determine whether a firm has attempted to monopolize an industry under Sherman Antitrust Act §2.[6] Antitrust law does not make the fact of possessing a monopoly illegal because the relevant market may be so limited or economies of scale so large that the market cannot support more than one firm; changes in taste or costs may drive out all but one firm; or a single producer may be "the survivor out of a group of active competitors, merely by virtue of his superior skill, foresight and industry. . . . The successful competitor, having been urged to compete, must not be turned upon when he wins."[7] This is the so-called "thrust-upon" defense to a charge of monopoly.

What is prohibited is the specific intent to exclude competitors or retain a monopoly by unfair, abusive, or less than honestly industrial means, such as predatory pricing, intimidation of suppliers and buyers through threats to withhold business in order to prevent them from dealing with competitors, product disparagement and misinformation, and so on. Specific intent to monopolize can only be determined with reference to the standard of normal (reasonable) competitive behavior. In order to apply this standard, a court must make an inquiry into any and possibly all of the firm's business dealings. In antitrust law, the discretionary standards approach to the question of whether a firm has attempted to monopolize is called the *rule of reason* approach.[8] Major rule of reason monopoly cases are extensive, expensive, and may last more than a decade.[9]

GILMORE'S STRICT RULES AND DISCRETIONARY STANDARDS CONTRACT MODELS

Nineteenth Century Strict Rules, Classical Contract Model

Gilmore argued that in the last half of the nineteenth century, two trends were apparent in the mutually reinforcing relationship between contract law and economic theory:

First, the domains of contract law and competitive market theory expanded together: the theory of competitive markets provided a rationale for classical contract law; and, conversely, classical contract law provided a rationale for unregulated economic behavior.

Second, in the late nineteenth century, both classical contract law and competitive market theory became highly abstract conceptions: the theory of competition became characterized by a formal model and the common law of contracts became characterized by strict technical rules. Gilmore said that the principal concept that allowed promisors to break their promises without legal liability was the *bargain theory of consideration.* As examples, a promise to pay someone $500 if he or she did not smoke or drink alcoholic beverages for a year was unenforceable because it was not based upon a bargain. A promise said to be held open or irrevocable for a certain length of time was nevertheless revocable for want of consideration, that is, no bargain. And, attempted modifications of an existing contract due to changed circumstances or discharge of existing debts at less than full value were unenforceable for want of new consideration.[10]

The Statute of Frauds, parole evidence rule, and shift from subjective to objective theory of contract were also used to restrict the sphere of enforceable promises: The Statute of Frauds, which dates from seventeenth century England, states that certain contracts—contracts involving land, contracts that cannot be performed within a year, contracts involving goods worth $500 or more, contracts guaranteeing the debt of another, and contracts involving marriage (e.g., prenuptial agreements)—even if the existence of an agreement is admitted by both parties, are not enforceable unless there is a writing signed by the party who refuses to honor the agreement. For example, an employer who orally agrees to hire an employee for two years but subsequently changes her mind is not bound by the agreement because it is within the one-year provision of the Statute of Frauds and is not in writing.

The parole evidence rule states that if a contract is in writing (even if the contract is not within the Statute of Frauds and so does not have to be in writing to be enforceable), evidence that contradicts the provisions of the writing is usually not admissible in court. For example, if a party to a written contract wants to testify that there was an oral guarantee that a product which is the subject of the contract would perform satisfactorily for two years, such evidence would be inadmissible.

The subjective theory of contract focused on whether there was a "meeting of the minds"; it was concerned with the parties' intentions. But, because a person's subjective intentions are difficult to determine, which requires "getting into someone's head," the law of contracts shifted to the objective theory, which is concerned not with what the parties subjectively thought, but with what a hypothetical third person who was observing the transaction would think. For example, under the objective theory of contract, a person who signs a contract without reading it is nevertheless bound even if it contains deceptive provisions hidden in fine print because he or she is "presumed" to have read it.

Gilmore says that the bargain theory of consideration, Statute of Frauds, parole evidence rule, and objective theory of contract narrowed or eliminated the defenses of unfairness, lack of good faith, duress, fraud, mistake, and changed circumstances. "By this process questions, originally perceived as questions of fact, will resolve themselves into questions of law."[11]

Enforceable promises were not only difficult to make under classical contract law, the extent of liability for breach of contract was severely limited; damages were supposed to be compensatory, not punitive, and were often less than actual damages suffered by the victim.[12] Gilmore said that the restricted view of compensatory or consequential damages was typified by the famous case of *Hadley v. Baxendale*,[13] in which a miller was not allowed to recover lost profits for the time his mill was unable to operate due to the delay caused by a negligent carrier, whose job it was to transport the mill's broken crank shaft to and from a manufacturer where it was to be used as a pattern for the new shaft, because the carrier had not been informed of the consequences of delay.

Moreover, the equitable remedy of "specific performance," under which a court requires a party who is in breach of contract to perform the contract as written (rather than pay monetary damages), was to be avoided as far as possible.[14] Exceptions to the rule are for contracts involving land and contracts for unique objects. For example, a court would specifically enforce a contract to sell a Rembrandt painting because money damages would not be an adequate substitute for this unique object; however, specific enforcement would not be allowed for breach of a contract to paint a house because money damages are an adequate remedy.

A leading legal commentator on contracts said: "All in all, our system of legal remedies for breach of contract, heavily influenced by the economic philosophy of free enterprise, showed a marked solicitude for men who do not keep their promises."[15] (In the nineteenth century tort law, the extent of legal liability was similarly severely limited.[16])

Twentieth-Century Discretionary Standards Contract Model

Gilmore argued that the twentieth century witnessed two contrary tendencies: First, the domains of both contract law and unregulated markets diminished through the development of antitrust law, labor law, products liability law, business and environmental regulation, and social welfare legislation. Second, strict contract rules dissolved and became discretionary standards. The bargain theory of consideration, Statute of Frauds, parole evidence rule, and objective theory of contract, all of which restricted the sphere of enforceable promises, were loosened up as courts began to enforce promises that did not meet formal contract requirements on grounds of implied promises, promissory estoppel, quasi contract, and expanded defenses of lack of fairness, fraud, and unconscionability.[17]

Consider the following examples of strict rules loosening. In *Wood v. Lucy, Lady Duff-Gordon*,[18] defendant gave an agent exclusive right to license her clothing designs. When she placed endorsements on clothes without the agent's knowledge or consent, he sued for breach of contract. Defendant argued that since the agent did not affirmatively promise to use his best efforts to license her designs, there was no bargained-for consideration. Judge Cardozo held that there was an implied (unwritten) promise to use reasonable efforts to promote designer's clothes and that this was valuable consideration.

Other twentieth century changes in contract law further diminished the requirement of consideration. Irrevocable offers (e.g., "I promise not to revoke my offer to sell you 50 widgets for $100/widget for two months"), modifications of contracts due to changed circumstances (e.g., agreement to allow an increase in contract price due to a large increase in the cost of materials), or discharge of debts for less than the full amount (e.g., pay $200 to settle a $500 debt), if in writing, are now, by legislative enactments, enforceable without bargained-for consideration.[19] Gilmore wrote: "The effect of such statutory abolition of a consideration requirement will be, I suggest, to drive out into the open what has always been the underlying principle of decision—the distinction between good faith and bad agreements."[20]

Consider also twentieth-century changes in contract law regarding promissory estoppel, quasi contract (agreement implied-in-law), and unconscionability. For example, assume a person, who is getting along in years, says to a relative that if she comes and takes care of him until he dies, he will bequeath his property to her. She faithfully carries out her part of the bargain for ten years until he dies; however, he never changed his will to reflect the agreement that is acknowledged to be true. Under the Statute of Frauds, this contract, which involves land, is unenforceable because it is not in writing. However, this agreement may be enforceable under the doctrine of promissory estoppel, the elements of which are: 1) someone makes a promise, 2) on which another reasonably relies, 3) to her substantial detriment, so that 4) a remedy of either enforcing the promise or restitution equal to the detriment is required by justice. Promissory estoppel can be used to enforce the oral bargain between the deceased property owner and his caretaker because 1) he made her a promise that he would bequeath the house to her, 2) on which she reasonably relied, 3) to her substantial detriment in the form of ten years unpaid for labor, so that 4) enforcement of the promise is required by justice.

In *Weichert Co. Realtors v. Ryan*,[21] a real estate agent informed a potential buyer that some property was for sale and gave him information concerning the property's price, current leases, taxes, zoning, and development plans, along with a letter calling for a 10 percent finder's fee. The agent also scheduled a meeting between the potential buyer and seller which the agent attended. After sale of the property, the buyer maintained that although he intended to compensate the agent, he consistently refused to agree to the requested 10% amount. The court held that there was no mutual assent and so no contract but that the agent was entitled to recover on an implied-in-law contract (quasi contract) for the reasonable value of his services. The case was remanded back to the trial court for a determination of customary brokers' fees for similar transactions.

In *Williams v. Walker-Thomas Furniture Co.*,[22] a consumer purchased sixteen household items on credit over a five-year period. The contracts purported to lease the items to her for monthly payments, seller to retain title until the balance on an item was completely paid. The contracts also contained a pro rata or add-on clause, which kept a balance due on every item purchased until all items, even those subsequently purchased, were paid. If she defaulted on any item, seller could repossess all items previously purchased. She defaulted and seller moved to repossess all sixteen items. The DC Circuit Court held that the pro rata clause was unconscionable. Some economists argue this contract should be enforced because it makes credit more available to people who would otherwise not qualify for credit. (See an amusing caricature of this argument by Leff.[23])

However, when Ms. Williams bought the sixteenth and last item, a TV set, she had no knowledge that if she defaulted on payments, she would lose not only the TV set but also the first fifteen items, so there was no informed consent. Pareto efficiency (consumer sovereignty) requires that she be advised of the add-on clause and its consequences, so that she could make her own informed cost-benefit decision as to whether buying the TV set was worth the increased risk of loss of all previous items.

Gilmore's argument that strict contract rules were gradually replaced by discretionary standards has found qualified acceptance by lawyers. The authors of a contemporary business law text said:

> In the process of evolving to accommodate changing social circumstances, the basic nature of contract rules is change. The precise, technical rules that characterized traditional common law contract are giving way to broader, imprecise standards such as good faith, injustice, reasonableness, and unconscionability. The reason for such standards is clear. If courts are increasingly called on to evaluate private contracts on the basis of fairness, it is necessary to fashion rules that afford the degree of judicial discretion required to reach just decisions in the increasingly complex and varied situations where intervention is needed.
>
> The change in the nature of contact law is far from complete, however. The idea that a contract is an agreement freely entered into by the parties still lies at the heart of contract law today, and contract cases may be found that differ very little in their spirit or ultimate resolution from their 19th-century predecessors. It is probably fair to say, however, that these are most likely to be cases where 19th-century assumptions about the nature of

> contracts are still largely valid. Thus, these cases involve contacts between parties with relatively equal bargaining power and relatively equal knowledge about the subject of the contract.[24]

Thus, an economic theory of contract must incorporate these two, contrary tendencies: in cases in which the two crucial assumptions of the competitive model, *equal bargaining power* and *relatively equal knowledge*, are satisfied, strict contract rules are efficient; and, in cases in which either of these two assumptions are not satisfied, discretionary standards are efficient.

MARKET STRUCTURE AND GAME THEORY MOTIVATIONS FOR TWO CONTRACT MODELS

In addition to Gilmore's distinction between strict rules and discretionary standards, there are two other motivations for the contract models presented in this and the next chapter. Second motivation: economists have constructed several market structure models (competition and imperfect competition—monopoly, monopsony bilateral monopoly, and oligopoly) which different models imply different public policy prescriptions. Two economic market structures, competitive markets and imperfect competition, are combined with Gilmore's strict rules and discretionary standards models to construct the strict rules, competitive market contract model and the discretionary standards, imperfect competition contract model.

Two variables from game theory provide the third motivation for these models: One variable is concerned with whether an actor's information is complete or incomplete; the second variable is concerned with whether the temporal order of action is simultaneous or sequential. These two variables were used in chapter 17's game theoretic framework as the basis for four contract and unjust enrichment models. This chapter is concerned with the contract model that has complete information and simultaneous action (promises) which assumptions are necessary conditions for the strict rules, competitive market contract model; incomplete information and simultaneous action (promises) are conditions for the discretionary standards, imperfect competition contract model, which is the subject of the next chapter.[25] The two contract models allow courts to determine when strict contract rules are or are not efficient in specific cases.

Classical Contract Law Is Inconsistent with the Two Contract Models

As stated above, Gilmore argued that the two essential features of classical contract theory, strict rules for contract formation and limited liability for compensatory and consequential damages, are consistent because both features limited liability for those who did not fulfill their promises.[26] However, in the context of the strict rules competitive market contract model and discretionary standards-imperfect competition contract model, these two features are not internally consistent. The first characteristic, strict rules for contract formation, is consistent with the strict rules, competitive market contract model; but the second characteristic, limited liability for compensatory and consequential damages, is not consistent with the strict rules, competitive market contract model because consequential damages are foreseeable due to complete information and therefore need not be restricted. Limited liability for consequential damages is consistent with the discretionary standards, imperfect competition contract model in which consequential damages are presumed unforeseeable due to incomplete information, unless made known by the buyer at time of contract. In the latter model, the limitation on consequential damages has an information forcing function.

STRICT RULES, COMPETITIVE MARKET CONTRACT MODEL

Complete information and simultaneous promises (action) are assumed by the strict rules, competitive market contract model. Competitive markets are defined by five characteristics: myriad buyers and sellers, products that have good substitutes, complete information, no barriers to entry or exit, and no negative externalities (see chapter 9). Thus, in order for courts to apply this model, that is, for this model to be dispositive of a case, courts must ask whether the assumptions of the competitive model are roughly approximated by the facts of the case: is the contract between parties who have relatively equal bargaining power, and have relatively equal information about the subject of the contract? Many business-to-business transactions and a good many consumer-to-business transactions fit these assumptions.

The salient features of the perfect competition contract model are: The use of strict rules as opposed to imprecise, discretionary standards. Rottenberg's corollary, which states that changes in liability rules in competitive markets have no effect on either efficiency or income distribution is valid. When goods are fungible, four direct damage rules for breach of contract—expectation damages, specific performance, reliance damages as measured by opportunity cost, and liquidated damages—are equally efficient. Due to the assumption of complete information, consequential damages for breach of contract are foreseeable. Lastly, the legal rule regarding the lost-volume seller is irrelevant; this conclusion is a straightforward application of Rottenberg's corollary.

The model is relevant for contracts such as those involving property, services, labor, and securities. Contracts involving goods, which are movable, tangible property, are governed by Article 2, Sales (of Goods) of the UCC. Since many business-to-business contracts involving goods are transacted "on the fly," such as a frantic telephone order from a restaurant to its produce supplier at lunch time for immediate delivery of two cases of lettuce, under the UCC, as long as quantity is stated, open terms as to price, place of delivery, and terms of payment can be filled in later though the use of market prices or UCC default terms (chapter 11). On the other hand, business persons who draft commercial contracts involving goods are strictly held to know and use proper commercial terms due to the inefficiency of having courts spend a great deal of effort construing ambiguous provisions (*Windows v. Jordan Panel Systems Corp.*, discussed below).

Strict Contract Rules

Courts Strictly Enforce but Do Not Make Contracts

In the context of complete information and competitive markets, the strict rules approach is efficient because its virtues of clarity, simplicity, certainty, and consistency far outweigh the costs of any inequity arising out of an exceptional case. Moreover, there is empirical evidence that in a number of competitive industries, trade associations use a strict rules approach to contracts.[27] The strict rules approach is consistent with common law maxim that courts merely enforce private agreements and do not make contracts for parties.

In *Windows, Inc. v. Jordan Panel Systems Corp.*,[28] a seller of windows shipped them to buyer by freight carrier. Due to carrier's negligence, two-thirds of the shipment was damaged due to a "load shift." The issue in the case was whether the contract was a shipment contract or a delivery contract: under a *shipment* contract, once seller transfers the goods to a carrier, buyer bears the risk of loss, so buyer must sue negligent carrier; under a *delivery* contract, until the goods are delivered by carrier to buyer, seller bears risk of loss, so seller must sue negligent carrier (chapter 11). The contract provided: "All windows to be shipped properly created/packaged/boxed suitable for cross country motor freight transit and delivered to New York City." Even though the word "delivery" was used in the contract, the court said that in order to overcome the UCC's presumption favoring shipment contracts, the parties must have explicitly agreed to impose on seller the obligation of delivery at a particular destination; the court held that the language of the contract did not impose that obligation. Moreover, the court said that seller did not use any commonly recognized industry term indicating that a delivery contract was intended. The UCC stipulates that the term F. O. B. ("free on board") when used with a named place, is a destination term.[29] Had the contract said, "delivered F. O. B. New York City," the contract would have been a delivery contract. (F. O. B. city of origin would be interpreted as a shipment contract.) In effect, the court said that business persons who write commercial contracts are strictly held to know and use proper commercial terms; that is, courts do not want their time wasted construing ambiguous contracts and courts will construe ambiguous terms in contracts against parties who drafted them.

Since the strict rules, competitive market contract model assumes that parties have complete information and equal bargaining power, attempts to avoid contractual duties based on allegations of lack of good faith, fraud, or mutual mistake are narrowly construed. Parties to contracts are presumed to have read and understood contracts they have signed. Standardized contracts, especially business-to-business contracts, which may include provisions that limit or negate warranties are enforceable. In the context of complete information and competitive markets, the provisions of such standardized contracts must be efficient, or competition would cause other more preferred provisions to be offered to buyers.

Although invalidating contracts based on physical duress is not ruled out, the assumptions that markets are competitive and information is complete means that claims of economic duress at the time of contract formation will usually be unavailing. Of course, courts will be vigilant against duress during contract performance, as when one party to a contract attempts opportunistically to "renegotiate" or hold-up the other party. For example, in the well-known case of *Alaska Packers Ass'n v. Domineco*,[30] seamen, who had agreed to accept certain wages for a voyage, waited until the ship was in remote Alaskan waters to demand a raise in wages. The renegotiated wage, which was based upon a threat by seamen to destroy value by keeping the ship idle if their demands were not met, was held unenforceable.

Courts Allocate Uncontracted for Risks but Do Not Reapportion Risks

Recall that complete information does not mean certainty but that the probability distribution of all possible chance events is known at the time of contract. Given complete information, parties to a contract could, in theory, negotiate a *complete contingent contract*, that is, a contract that specified the rights and duties of both parties under all circumstances. However, a simple application of the Judge Hand test demonstrates that transacting for unlikely contingencies, especially those of small dollar magnitude, is inefficient because transaction costs of negotiating such provisions, B, exceed the expected monetary value of the losses, PL. If an unlikely contingency occurs, the policy prescription of the strict rules model, that courts enforce but to not make contracts, implies that courts will allocate risks due to changed circumstances by either excusing or not excusing performance but will not reapportion risks or share losses among parties through contract reformation. In other words, in this model, courts *fill in gaps* in contracts, if by that phrase one means that courts ask which party the risk would have been allocated if they thought about it, but they do not reapportion risks or reform contracts.

Courts implicitly use Calabresi's cheapest preventer and cheapest insurer criteria to allocate these risks (chapter 13). Courts will not excuse performance by a party who should have protected against a contingency by taking precautions, making hedge contracts, or insuring against the loss.[31] *Transatlantic Financing Corp. v. U.S.*[32] involved a contract to deliver American wheat to Iran. Before performance of the contract, a war broke out that closed the Suez Canal through which the ship had expected to transport wheat. In order to deliver the wheat, the ship changed course and went around the Cape of Good Hope. The ship owner claimed additional compensation for added expense of the voyage around the Cape. The court denied additional compensation because ship owners are the cheapest cost insurers; they can best calculate the cost of performance by alternative routes, the amount of insurance required, and are sensitive to international troubles that affect their services.

With respect to contracts for goods in the seller's possession that are destroyed without seller's fault, whether the seller or buyer is the cheapest insurer depends upon the circumstances: "The factory owner can bear the risk of fire at least cost if she can easily purchase fire insurance whose coverage includes liability for not delivering goods, whereas the customer can bear the risk at least cost if he can easily obtain substitute goods from another factory."[33]

However, in circumstances of loss where the cheapest insurer is uncertain, determining the efficient result in each case is often less important than that the law be clear as to which party, buyer or seller, bears the risk so that risks will be efficiently priced into contracts. The UCC has adopted clear default rules: seller is excused from performance if goods in seller's possession were destroyed without fault and were identified when the contract was made, or were shipped, marked or otherwise designated by seller as goods to which the contract refers. In addition, under the UCC, buyer has an insurable interest in goods once identified to the contract.[34] (By granting an insurable interest in goods in possession of the seller to buyer, the UCC greatly reduces the likelihood of over reliance on contracts, which is discussed below.) By virtue of possession, seller also has an insurable interest in goods identified to the contract. For goods not identified to the contract, seller assumes the risk of destruction.[35] Parties to a contract are allowed to change these UCC liability default rules by contract (chapter 11).

A Change in Liability Rules in Competitive Markets Is Irrelevant to Efficiency

Rottenberg's corollary, which was discussed in chapter 11, is valid in the strict rules, competitive market contract model. Recall that if the law clearly specifies whether one party (e.g., buyer, seller, employer, employee, etc.) has a liability rule specified in its favor, competitive markets will achieve the efficient use of resources; that is, the efficient outcome is independent of liability rules or property rights. With respect to a change in liability rules, the disfavored party will be compensated by a price differential equal to the benefit conferred or cost saved by the favored party; that is, the distribution of income will not be changed as a result of a change in liability rules. The UCC default rules with respect to destruction of goods in seller's possession, discussed above, are an example as are the UCC default rules discussed in chapter 11.

EXPECTATION DAMAGES

A bilateral contract for goods or services is one of mutual (simultaneous) promises. Parties enter into contracts voluntarily because each will be better off in its own estimation after performance of the agreement; that is, making a contract is a Pareto improvement. The sum of their individual gains is the joint increase in value or cooperative surplus created by the contract. For example, if a seller who values a car at $5,000 makes a contract to sell the car for $6,000 to a buyer who values the car at $7,000, each gains $1,000 and their joint surplus is $2,000. The sale of the car transfers property from one who values it less to one who values it more.

What remedy should be imposed for breach of contract? In a legal regime in which there is no remedy for breach of contract, a potential contract breaker would consider only his own gains and losses in deciding whether to fulfill a contract and would *externalize* the cost to the other party from not fulfilling his promise. If costs resulting from breach of contract to non-breaching party were not compensable, there would be little incentive to make contracts.

A remedy for breach of contract is efficient if it causes a potential contract breaker to *internalize* the other party's loss. *Expectation damages*, which has the purpose of placing the non-breaching party, whether seller or buyer, in a position as good as if the contract breaker had performed its promise, provides an incentive to perform or not perform the contract that is efficient.[36] The remedy of expectation damages assures that the non-breaching party will receive her share of joint surplus generated by the contract.

Assume that markets are competitive and buyer breaches a contract for goods. If at the time of breach, contract price, \$5,000, > market price, \$3,000, seller will incur \$2,000 in expectation damages. Seller should *cover*, that is, sell goods that buyer refuses to accept at the market price. If the breaching buyer is required to pay the amount by which the contract price exceeds the market price, seller's expectation that the contract will be performed is protected (plus incidental damages equal to the transaction costs of making an additional contract). If market price equals or exceeds contract price, seller has no expectation damages.

Conversely, assume there is a breach of contract by seller. If market price > contract price, buyer will incur expectation damages. Buyer should cover by buying substitute goods at the higher market price. If the breaching seller is required to pay the excess of the market price over the contract price, the share of the joint surplus that buyer expected to receive from performance of the contract will be protected (plus incidental damages equal to the transaction costs of making an additional contract).

EFFICIENT BREACH

Performance of the contact is socially desirable when performance yields a greater net value than nonperformance; conversely, nonperformance is socially desirable when nonperformance yields a greater net value or smaller net loss than performance; this latter case in which nonperformance is socially desirable is known as *efficient breach*.

Assume a buyer is deciding whether to breach a contract. Buyer knows in advance of breach that seller's expectation damages will be the excess of the contract price over the market price, so buyer will breach only if the net loss to it from breaching the contract and having to pay expectation damages is less than the loss from performing the contract.

As an example of efficient breach by a buyer, assume buyer has no need for the goods or services due to a fire that destroyed buyer's place of business. In this example, the loss to buyer from performing the contract exceeds the loss to seller from not performing the contract, so buyer will be better off by paying damages rather than performing the contract.

As an example of efficient breach by seller, assume seller has a contract to deliver 1,000 widgets to buyer at \$5 per widget for a total contract price of \$5,000. Just before delivery, third party who is in urgent need of widgets offers to pay \$10 per widget for 1,000 widgets for a total contract price of \$10,000. Assume further that if third party's offer is accepted, seller will not be able to complete timely delivery to buyer as required by their contract, causing buyer to lose \$3,000 in profits because of the delay in delivery. Due to complete information, seller has knowledge what buyer's losses will be before breach. If the contract is breached and buyer is paid \$3,000 in expectation damages, buyer will be as well off as if the contract were completed, seller will be better off by \$2,000, and third party will be better off because it gets the widgets. If seller agrees to pay buyer more than the \$3,000 expectation damages that buyer incurs, say $4,000, seller keeping $1,000 for itself, all three parties will be better off after the breach (a strong Pareto improvement).

SPECIFIC PERFORMANCE

Specific performance is an order by a court that prohibits a contract breaker from breaching its promise; put affirmatively, it is an order by a court that requires a party to perform its promise. In the context of strict rules, competitive market contract model, specific performance is usually an efficient remedy because the assumptions of competitive markets, complete information, and goods and services that are good substitutes for each other means that transaction costs incurred in arranging for substitute performance are roughly equal for both parties. For example, if seller breaches a contract for goods due to an inability to perform, say because a fire has destroyed its factory (before goods have been identified to the contract), in a situation in which buyer will have expectation damages because market price exceeds contract price, the complete information assumption means that seller's cover and transaction costs from having to buy substitute goods at market price are the same as buyer's cover and transaction costs from having to buy substitute goods; only one additional transaction must take place, and both parties are equally situated to find the cheapest price for substitute goods.

Conversely, if buyer breaches a contract for goods in which seller incurs expectation damages because contract price exceeds market price, buyer's transaction costs from having to accept and sell the goods on the market are the same as seller's transaction costs. In each case, only one additional transaction must take place and complete information implies that both parties are equally situated to sell goods at market price.

Specific Performance as Alternative to Efficient Breach

In competitive markets with complete information and goods that are good substitutes for each other or fungible, contracts in which breach is efficient should be specifically enforced because the windfall gain that would have accrued to the party who breached the contract will now go to the non-breaching party. This is consistent with the risk allocation function of contracts in which buyer is protected from risk of rising prices and should benefit if the market price exceeds the contract price, and seller is protected from the risk of falling prices and should benefit if market price is less than the contract price.

A continuation of the widget example will clarify this conclusion: A decree of specific performance would require seller to deliver the widgets for $5,000 to buyer who would then resell the widgets to third party for $10,000 so that buyer makes a gain of $5,000 minus its losses from not using the widgets in its factory, $3,000, for a net gain of $2,000. Alternatively, seller and buyer will bargain over the $2,000 cooperative surplus; seller could offer to split the cooperative surplus with buyer if buyer agrees to relinquish its right to specific performance.

If goods are in short supply, say due to crop blight or drought, even fungible goods may take on the uniqueness and inadequacy of damages characteristics which the law requires for specific performance. In *Campbell Soup Co. v. Wentz*,[37] plaintiff contracted with defendant for all the Chantenay red cored carrots grown on its farm. Defendant breached the contract and was selling its carrots to others at the market price which was a great deal higher than the contract price. Plaintiff used these carrots in its soups because of their uniform color, which uniformity was of "considerable commercial significance" to it. Because substitute carrots of this variety were "virtually unobtainable," the court held that specific performance was appropriate in this case.

The twentieth century saw a rise in use of specific performance in the context of cases that fit the assumptions of the competitive market model; as examples, output and requirements contracts for fungible agricultural and industrial products such as peaches, tobacco, hops, sugar beets, tomatoes, steel scrap, salvaged glass, lumber, natural gas, oil, water, electric power, and gypsum, as well as for the sale of small businesses.[38] These cases can be distinguished from cases that fit the assumptions of the imperfect competition model, where goods are not standardized, information is incomplete, and there are problems of over reliance, strategic behavior, bilateral monopoly, so that money damages to be more efficient than strict performance (chapter 20).

OVER-RELIANCE DOES NOT OCCUR DUE TO COMPLETE INFORMATION

In many breach of contract cases, there are unforeseen *consequential damages*, for example, *Hadley v. Baxendale*, in which a miller sued for lost profits for the time his mill was unable to operate due to the negligent carrier's delay in returning the mill's replacement crank shaft; the court refused to award lost profits because the carrier had not been informed of the consequences of delay. The miller's lost profits, which were unknown and unforeseen by the carrier, could be characterized as *over reliance* on the contract by the miller.

Assume buyer, a retail store, makes a contract to buy goods for sale in its store which goods are to be delivered in the future. Buyer relies on the contract by making expenditures on advertising, hiring sales personnel, or expanding its facilities, which increase the value of performance of the contract but also increase its losses if seller breaches the contract. Buyer may also increase reliance on a contract if it assumes as certain that seller will perform even though seller's factory could be destroyed by fire or costs could rise precipitously due to unforeseen events like natural catastrophes. The higher the liability for breach, the greater buyer's incentive is to rely.

Seller may rely on the contract by making expenditures on specialized materials, tools, or by hiring extra or specially trained workers, assuming as certain that buyer will perform the contract. Such expenditures by seller increase its losses if buyer breaches the contract. The higher the liability for breach, the greater seller's incentive is to rely. Seller may also increase reliance on the contract if it assumes there are few or no circumstances under which buyer will renege on the contract, such as if the buyer's place of business is destroyed by fire or the market for its product collapses.

If either party, buyer or seller, is unaware of the other's reliance expenditures, they will each underestimate the other's expectation damages from breach and will take inefficiently low expenditure against breach (chapter 20). However, in the strict rules, competitive market contract model, the assumption of complete information means that each party's reliance costs are known as is the entire probability distribution of contract risks; both reliance expenditures and risks are considered and priced into the contract if they pass the Judge Hand test, $B < PL$. Thus, due to complete information, the amount of each party's expected reliance is priced into the contract. As discussed above, by granting both buyer and seller an insurable interest in goods identified to a contract, the UCC greatly reduces the likelihood of over reliance on contracts.

Consider the case of *Metropolitan Life Insurance Co. v. RJR Nabisco, Inc.*,[39] which involved a leveraged buyout (financial leverage was defined in chapter 16), where a group of investors financed buying RJR Nabisco's stock with a great deal of bonds (debt) and little equity investment. Prior to the leveraged buyout, the plaintiff, Metropolitan Life, owned more than $340 million RJR Nabisco's existing bonds. Plaintiff claimed that the large amount of new debt increased default risk so that its existing bonds incurred a multimillion dollar loss in value; it asked the court to create an "implied" covenant not to incur the debt necessary to facilitate the leveraged buyout. The court, which ruled for defendant, noted that both parties are among the most sophisticated financial institutions, the bond indentures explicitly permitted mergers and assumption of new debt and that the market was likely to have taken into account the possibility of additional debt, including a leveraged buyout, when setting the value of Met. Life's bonds. The court said that plaintiffs seek "an additional benefit for which they did not bargain." In other words, the court believed that financial markets are likely to have priced the expected amount of reliance into the debt contract; if plaintiffs had insisted on limiting additional debt, either the interest rate on bonds would be lower or defendants would seek other lenders from whom to borrow. Plaintiffs' over reliance was unjustified.

RELIANCE DAMAGES AS MEASURED BY OPPORTUNITY COST

Reliance damages have the purpose of placing the non-breaching party in the position that she would have been if the contract had never been made, which is similar to the purpose of compensatory damages in torts. Assume that there is a breach of contract by buyer who refuses to accept and pay for goods or services because contract price exceeds market price. Since markets are competitive and goods have good substitutes or are fungible, seller had opportunities to make similar contracts with many other buyers. By making the contract that is breached, seller forgoes the opportunity of entering into myriad other contracts on identical terms, including price. Assuming that the alternative buyers would have performed these other contracts, reliance damages as measured by opportunity cost will place seller in the position it would have been in if she had not relied on this buyer but instead had contracted with an alternative buyer. In competitive markets in which goods have good substitutes or are fungible, reliance damages equal to opportunity cost are equal to expectation damages.

Conversely, in competitive markets, if there is a breach by seller and market price exceeds contract price, buyer incurs reliance damages in terms of opportunity cost, which is the same measure as expectation damages.

LIQUIDATED DAMAGES

Liquidated damages provisions in contracts are clauses in which the parties expressly provide that a specific sum shall be recoverable if contract is breached. For example, a construction contract may contain a liquidated damages provision stating that if the building contractor does not complete construction by a given day, a stipulated amount is payable in damages for every day the construction is not completed.

Since most contracts are performed, it would be inefficient to require that all contracts contain liquidated damages provisions; moreover, according to the Judge Hand test, if the transaction costs of including such provisions in contracts exceed the value of the damages and litigation costs saved from such provisions discounted by their probability of occurring, including them is inefficient.

In competitive markets, information is complete and no party has any power over any other party because there are myriad alternative parties with whom to transact; therefore liquidated damages provisions will equal ex ante predictions of expectation damages based the known probability distribution of all relevant risks. For example, a promisor may find herself in breach of a contract which has a liquidated damages provision because one of the risks that were heavily discounted as a low probability event has come to pass. However, the assumptions of complete information and competitive markets imply that all risks that would trigger the liquidated damages provision were known in advance and that the contract price was adjusted to reflect these contingencies.

Liquidated damages provisions may be based upon known and contracted-for reliance costs, such as where a contract for construction of a hotel includes a provision for $10,000 per day in liquidated damages for every day of late completion.

Thus, in this model, all liquidated damages provision are efficient and should be enforced. In the strict rules, competitive market contract model, the distinction between liquidated damages and penalties is not necessary.

CONSEQUENTIAL DAMAGES ARE FORESEEABLE DUE TO COMPLETE INFORMATION

The assumption that goods and services have good substitutes means that if there is a breach of contract by seller prior to time of performance or even at the time of performance under circumstances in which substitute goods or services are readily available on short notice, buyers will incur only expectation damages. However, if there is a material breach of contract by seller when it is too late to contract for substitute goods or services, latent defects in goods (as where the purchase of defective machinery necessitates the shutting down of an entire production assembly line), or material breach during performance of services, buyer will often incur indirect or consequential damages.

In the context of the competitive market contract model, the expected monetary value of defective performance or last minute breach is factored into the contract price; therefore a restrictive consequential damages rule would result in inefficient incentives to perform and create a moral hazard problem.[40]

Assume a pizza parlor hires a plumber to fix a broken oven, the price for his services is $500, and work is to be completed on Sunday when the pizza parlor is closed. The plumber botches the job and is dismissed. The pizza parlor owner hires a second plumber who has to wait four days for parts to repair problems created by first plumber. The second plumber charges $1,000 for her services to repair the first plumber's mistakes and fix the original problem. The pizza parlor is closed for a week. The owner is entitled to $500 in compensatory damages and a week's profits in consequential damages. Since information is complete, consequential damages are not only easily foreseeable, their probability distribution is known and their expected monetary value has been factored into the price.

In *Deitsch v. Music Company*,[41] a four-piece band agreed to perform at plaintiff's wedding reception for $295 and was paid a security deposit of $65. When the band failed to arrive at wedding reception, a friend obtained some stereo equipment to provide music. Plaintiffs contended that correct measure of damages was entire cost of the reception, $2,643.59. Defendants contended that plaintiff's damages should be measured by amount of out-of-pocket loss, the $65 deposit. The court said that where another band could not readily be obtained at last minute, plaintiffs were entitled to compensation for their distress, inconvenience, and diminution in value of their reception in the amount of $750 in addition to a refund of $65 security deposit. In this case, consequential damages are foreseeable and necessary to put plaintiffs in as good position as if band had performed as promised.

Plaintiffs' consequential damages should be measured by sum of a) lost consumer surplus (discussed in chapter 8), which is equal to the difference between the maximum amount plaintiffs would have been willing to pay for live music at their wedding and market price (presumably $295), plus b) difference in market price for live music (again $295) and value of a wedding with music supplied by stereo equipment (presumably a value less than $295).

SELLER'S LOST-VOLUME PROFIT RULE IS IRRELEVANT TO EFFICIENCY

Suppose that a buyer makes a contract to buy a good or service, but before it is delivered or performed, he reneges. Seller subsequently resells the good or service to another customer at the same price. Has the seller suffered damages (aside from incidental damages); if so, should she be compensated? Specifically, is the seller entitled to lost profits on the sale? This question is called the seller's lost-volume profit problem. If seller has a *fixed* supply, as when a private party agrees to sell his automobile to another, a breach by buyer and resale to another party at the same price would cause no loss of profit to seller. However, if the seller has a variable or unlimited supply, as when a consumer breaches a contract to by an appliance from a dealer, resale by seller raises the question whether seller is entitled to lost profit on the breach because seller could have made both sales.

Assuming seller considers buyer's decision to breach the contract (return the product) is a random event with a known probability distribution, so that the expected level of breach is independent of whether seller is entitled to lost profit, and attitudes toward the risk of breach are the same for buyers and sellers, then according to Rottenberg's corollary for a change in liability rules in competitive markets, whether seller is entitled to lost-volume profit or not, is irrelevant to efficiency: if sellers are entitled to lost profit, prices will be higher; if not, prices will be lower.[42]

In the real world, if sellers in competitive markets clearly state their return policies before buyers purchase goods and services, Rottenberg's corollary on liability rules applies. Some retailers adopt extremely liberal policies on customer returns; moreover, some department stores and mail order firms have a no-questions-asked return policy, even if the goods have been used which is an even more liberal policy than not recognizing lost-volume profits. Sellers must believe that the increase in good will from the guarantee of satisfaction provided by such a policy outweighs its costs. Of course, buyers who buy from sellers with liberal return policies pay the higher prices that cover higher costs of these sellers.

On the other hand, some retail sellers notify buyers before sale that there is a restocking fee, say of 15 percent of the price, and discount retailers often post notices of no return policies; these sellers justify their policies by stating in advance that they charge lower prices, which savings is passed on to customers.

Liberal returns policy are less likely for "big-ticket" items or items that lose value as soon as they are reclassified as used, such as automobiles. In this context, a nonrefundable deposit may be characterized as either an option to buy contract or a liquidated damages provision. A nonrefundable deposit indicates that the buyer is informed of damages in the event of breach and causes buyer to internalize the risk of nonperformance. Conversely, by stating a nonrefundable deposit before the contract is entered into, seller puts a price on the risk of return. With respect to retailing in competitive markets, Goldberg concluded that in the absence of an agreed upon nonrefundable deposit, a policy of no legal liability for lost seller's profit on retail sales is preferable because of the low enforcement costs of a policy of leaving the losses where they lie.[43]

To the extent that information is *incomplete* and the expected level of breach is *not* independent of the legal rule, the strict rules, competitive market contract model's conclusion that the legal rule with respect to seller's lost-volume profit is irrelevant to efficiency is not accurate because the information about the probability of breach will not be the same for buyers and sellers; that is, there will be problems of asymmetric information.

As to sellers, they may have superior ability to gather information about the expected value of sales based upon business experience and are in a better position to know the magnitude of lost-volume profit in the event of a breach; sellers are also in a better position to spread the risk of a breach over similar transactions than are buyers.

As to buyers, if there is no liability for breach by buyers, their superior knowledge as to their tendency to adhere to contracts will result in moral hazard and adverse selections problems. With respect to moral hazard, under a no liability rule, buyers, who, for example, intend to take only one airline flight, may make numerous contracts, each for a different time, so as to protect themselves if they are delayed. Of course, if buyers incur no penalty from delay, they are more likely to be delayed and to make more contracts due to moral hazard. With respect to adverse selections, "bad apples," those buyers who know that they are more likely to breach contracts would be more likely to make more contracts for the future. Thus, many hotels and airlines require nonrefundable deposits or fares to control moral hazard and adverse selections.

In the context of incomplete information, which will be discussed in next chapter's discretionary standards, imperfect competition contract model, the legal rule with respect to the seller's lost-volume profit is relevant to efficiency. To the extent that there is both incomplete information and limited number of buyers and sellers, the case for lost-volume profits is stronger.

NOTES

1. Grant Gilmore, *The Death of Contract* (Columbus, OH: The Ohio State University Press, 1974).
2. Louis Kaplow, "Rules Versus Standards: An Economic Analysis," *Duke Law Journal* 42 (1992): 557–629.
3. In statistical theory, Type I error is the probability that a true hypothesis will be rejected. For example, in criminal law, Type I error is the probability that a person who is presumed innocent and is innocent, will be convicted.
4. In statistical theory, Type II error is the probability that a false hypothesis will be accepted. For example, in criminal law, Type II error is the probability that a person who is presumed innocent but is guilty will go free.
5. 26 Stat. 209 (1890), as amended, 15 U.S.C.A § 1.
6. 15 U.S.C.A § 2.
7. *U.S. v. Aluminum Co. of America*, 148 F.2d 416, 429–430 (1945).
8. *Standard Oil Co. v. U.S.*, 221 U.S. 1 (1911).
9. *U.S. v. International Business Machines Corp.* (IBM), Civil Action No. 69 Civ. 200 (S.D.N.Y.).
10. 198 U.S. 45.
11. Gilmore, note 1, 42.
12. *Ibid.*, 15.
13. 9 Ex. 341; 156 Eng. Rep. 145 (1854).
14. Gilmore, note 1, 14.
15. Allan Farnsworth, "Legal Remedies for Breach of Contract," *Columbia Law Review* 70 (1970): 1145, 1216.
16. Nineteenth-century courts allowed employers to externalize the cost of industrial accidents to employees through the creation of the "unholy trinity" of defenses, which were discussed in chapter 13.
17. Gilmore, note 1, 76–77, 87–90.
18. 222 N.Y. 88 (1917).
19. §§ 5-1103–1109 N.Y. General Obligations Law (1936); Article 2, Sales, UCC § 2–205 (Firm Offers), § 2–209 (Modification, Rescission and Waiver).
20. Gilmore, note 1, 77.
21. 128 N.J. 427 (1992).

22. 350 F.2d 445 (DC Cir.1965).
23. Arthur Leff, "Economics Analysis of Law: Some Realism About Nominalism," *Virginia Law Review* 60 (1974): 451, 460–61.
24. Jane Mallor, A. James Barnes, Thomas Bowers, Michael Phillips, and Arlen Langvardt, *Business Law and the Regulatory Environment: Concepts and Cases*, 10th ed. (Boston: Irwin/McGraw-Hill, 1998), 177–178.
25. Recall that the other two models in the game theoretic framework for the law and economics of civil obligation are the unilateral contract model and the unjust enrichment—promissory estoppels and quasi-contract models. These models are not discussed.
26. Gilmore, note 1, 65.
27. Lisa Bernstein, "Symposium: Law, Economics, & Norms: Merchant Law in a Merchant Court: Rethinking the Code's Search for Immanent Business Norms," *University of Pennsylvania Law Review* 144 (1996): 1765.
28. 177 F.3d. 114 (2nd Cir. 1999).
29. UCC § 2–319(1).
30. 117 F. 99 (9th Cir. 1902).
31. Richard Posner and Andrew Rosenfeld, "Impossibility and Related Doctrines in Contract Law: An Economic Analysis," *Journal of Legal Studies* 6 (1977): 83.
32. 363 F.2d 312 (DC Cir.1966).
33. Robert Cooter and Thomas Ulen, *Law & Economics*, 6th ed. (Boston: Addison-Wesley, 2012), 351.
34. § 613.
35. § 2–501(1)(b).
36. Robert Cooter and Melvin Eisenberg, "Damages for Breach of Contract," *California Law Review* 73 (1985): 1432, 1459–1468, 1473.
37. 172 F.2d 80 (1948).
38. E. Alan Farnsworth, *Contracts*, 2nd ed. (Boston: Little, Brown, 1990), §12.6 at 862; M. T. Van Hecke, "Changing Emphases in Specific Performance," *North Carolina Law Review* 40 (1961): 1, 3–9.
39. 906 F.2d 884 (1990).
40. Melvin Eisenberg, "The Principle of Hadley v. Baxendale," *California Law Review* 80 (1992): 563, 577.
41. 6 Ohio Misc.2d 6 (1983).
42. Victor Goldberg, "Comment: An Economic Analysis of the Lost-Volume Retail Seller," *Southern California Law Review* 57 (1984): 283, 294–295.
43. *Ibid.*, 296.

Chapter 20

Discretionary Standards, Imperfect Competition Contract Model

The game theoretic framework for the discretionary standards, imperfect competition contract model assumes incomplete information and simultaneous promises. In this model, a regime of discretionary standards is more efficient than a regime of strict contract rules. Courts are concerned with incomplete and asymmetric information, good faith dealing, market power, and changed circumstances due to uncertainty.

In the context of incomplete information, there may be misunderstandings as to whether parties have made a legally enforceable agreement, that is, whether there is mutual assent. Courts, in cases involving changed circumstances, in addition to allocating risks to the cheapest insurer or preventer, may suggest risk reapportionment and contract reformation.

Sellers usually have a duty to disclose negative information that materially reduces an object's value whereas buyers usually do not have a duty to disclose positive information that materially increases an object's value. Standardized contracts used by sellers with market power, which contracts contain long paragraphs of fine print that consumers typically do not read on a take-it-or-leave-it basis, should be scrutinized by courts to determine whether their terms unduly favor sellers. (Are such contract terms Pareto inefficient?)

In imperfect competition, where goods and services are differentiated, expectation damages are more efficient than either specific performance or out-of-pocket reliance damages. If information is incomplete, over reliance may inefficiently increase expectation damages from a breach of contract. Two doctrines in law of contracts, the rule of *Hadley v. Baxendale*, and mitigation of damages, limit over reliance by buyers and sellers respectively. The liquidated damages-penalties distinction is efficient. In imperfect competition, when buyers breach contracts, sellers are entitled to lost-volume profits, plus overhead cost.

INCOMPLETE INFORMATION, MUTUAL MISTAKE, AND RISK REAPPORTIONMENT

Recall that a game has incomplete information if some or all players do not know the payoffs of other players, that is, some or all players have private information that is unknown to other players. Incomplete information may be either symmetric as in the game of poker where neither player knows the cards in the other player's hand or asymmetric where one party to a contract knows what is in the fine print but the other does not. Moreover, in addition to "risks," which are chance events whose probability distribution is known, incomplete information includes "uncertainty" which includes chance events of which actors may not be aware and whose probability cannot be calculated.

Misunderstanding and Mutual Mistake

In the context of incomplete information, there may be misunderstandings as to whether parties have made a legally enforceable agreement, that is, whether there is mutual assent. In *Frigaliment Importing Co. v. B.N.S. International Sales Corp.*,[1] the issue was, "what is a chicken?" Buyer said that "chicken" called for in the contract means a young chicken, suitable for broiling and frying. Seller argued that "chicken" means any bird of that genus, which meets contract specification on weight (2½–3 lbs.) and quality, including what it called "stewing chicken" and what buyer pejoratively termed "fowl." The court adopted the seller's interpretation because it coincided with the dictionary definition of chicken, which includes both meanings (broilers and fowl), as does the definition in the United States Department of Agriculture Regulations, to which the contract made at least oblique reference, and with at least some usage in trade.[2] However, the *Restatement (Second) of Contracts*, in an example apparently based upon this case, concluded that if neither party "had reason to know of the difference in meaning," there is no contract.[3] Where both parties have a different understanding as to the subject matter of the contract, courts often do not enforce contracts on grounds of mutual mistake.

Reapportionment of Contract Risks Due to Incomplete Information

In the context of incomplete information, there are cases in which a loss due to changed circumstances will be allocated to the cheapest insurer or cheapest preventer just as in the strict rules contract model. However, there will also be cases in which unforeseen circumstances, either due to extreme miscalculations by the parties or uncertainty, require a reapportionment of risks and reformation of contracts. These cases often exemplify the old adage: I owe the bank $60,000 and cannot pay, I have a problem; I owe the bank $600 million and cannot pay, the bank has a problem! These cases are also examples of bilateral monopoly in which a compromise, if achieved, occurs because parties see it as least evil alternative; other alternatives being bankruptcy, complete excuse on grounds of commercial impracticability, or an imposed judicial settlement.

Aluminum Co. of Amer. v. Essex Group, Inc.[4] involved a long-term contract for converting alumina into molten aluminum by a smelting process. The price was set based upon a formula, which was used as a proxy for changing production costs other than labor. The contract was profitable for the first six years, but by the tenth year losses threatened to amount to more than $75 million because the unanticipated oil crises of the 1970s drove up energy prices. Defendant alleged that there was a mutual mistake as to fundamental circumstances of the contract. The trial court reformed the price so as to yield defendant a small profit; however, the parties settled the case while the judge's decision was being appealed. Although the attempted reformation of the contract by the court in this case has been criticized on the ground that judges should not make contracts for parties,[5] this criticism, while relevant to the strict rules, competitive contract model, is misplaced in the discretionary standards, imperfect competition contract model due to incomplete information, uncertainty, strategic considerations, bilateral monopoly, and loss apportionment problems.

ASYMMETRIC INFORMATION: FRAUD, MISREPRESENTATION, AND DUTY TO DISCLOSE

Sellers have an incentive to reveal information that enhances an object's value and to conceal information that reduces its value whereas buyers have an incentive to reveal information that reduces an object's value and to conceal information that increases its value; thus, the issue of whether there is a duty to disclose asymmetric information is usually limited to whether sellers should have a duty to disclose *negative* information and whether buyers should have a duty to disclose *positive* information.

The law with respect to asymmetric information is itself asymmetric. In general, sellers have a much greater duty to disclose negative information that materially reduces an object's value than buyers have to disclose positive information that materially increases an object's value. That is, failure by sellers to disclose negative information about attributes that reduce value is often grounds for an action in fraud or misrepresentation, the remedy for which is either rescission of contract or compensatory damages at the election of buyer. Failure by buyers to disclose positive information about attributes that increase value is in general not actionable (exceptions are noted below).

There are efficiency reasons for the asymmetric legal treatment of asymmetric information possessed by buyers and sellers. For example, if the seller of the home had no duty to disclose termite infestation, every potential buyer would have to incur the cost of a termite inspection.[6] Alternatively, if seller has a duty to disclose termite infestation that it knows or should know about, then only one inspection is necessary. Thus, if sellers were not required to disclose negative information concerning latent attributes that reduce an object's value, there would be inefficient duplication of costs on the part of potential buyers.[7] If information is incomplete, courts must monitor seller's representations, half-truths, and concealments for both fraud and misrepresentation.

There are many instances in which legislatures require mandatory disclosure of information by sellers, such as fabric identification for garments, nutrition labeling for food products, and truth in lending laws that standardize the calculation of interest rates. Although social regulation can become so pervasive and costly that its costs outweigh its benefits, economic regulations that require disclosure of information, such as the laws stated in the previous sentence (as opposed to those involving complex financial or business information), are likely to enhance efficiency.

The duty of sellers to reveal negative information about latent attributes that reduce the value of a contract to buyers also provides support for legislation that requires standardized consumer contracts be written in "plain" language and judicial scrutiny of provisions in contracts on grounds of unconscionability. If each consumer (buyer) had to read all the fine print of every standardized sale, lease, service, or insurance contract (life, home, health, and auto) that he signs and had to hire a lawyer to unpack the legalese in such contracts, there would be needless duplication of effort. It is efficient for courts and legislatures to require that sellers conspicuously disclose asymmetric information they have concerning material provisions in standardized contracts that negatively affect buyers. Most consumers do not read the fine print in consumer contracts; they probably assume that courts and legislatures will protect their reasonable expectation that provisions in these contracts that negatively affect their interests will, if not held illegal, at least be subject to a conspicuous disclosure requirement. (The conditions under which the terms in standardized contracts are or are not efficient are discussed below).

On the other hand, requiring buyers to disclose positive legally acquired information[8] that enhances value is in general inefficient because it eliminates buyers' incentives to search for undervalued assets.[9] For example, after conducting extensive aerial surveys which revealed a geological anomaly indicating the presence of massive sulfide deposits, Texas Gulf Sulfur Co. purchased mineral rights options without disclosing this information to seller. When the existence of the deposit became known, seller sued on grounds that buyer intentionally misled it as to the mineral deposits. Although the suit was settled out of court,[10] in a related case, which was litigated, the court remarked that Texas Gulf Sulfur did only "[w]hat any prudent mining company would have done to acquire property in which it knew a very promising anomaly lay," when it purchased property "without causing the prospective vendors to suspect that a discovery had been made."[11] If a buyer of land had a duty to disclose to a seller the results of a secret aerial survey, buyer would have no incentive to acquire that information because seller who has a property right could refuse to sell and use the valuable information without compensating buyer or charge a price for the land which reflects the full value of the information.

Although the law is relatively clear that buyers have no duty to reveal positive information, are deliberately false statements by buyers actionable? In the famous case of *Laidlaw v. Organ*,[12] a buyer of tobacco received advance private information which was not disclosed to seller that the War of 1812 between Great Britain and the United States had ended. After the contract was transacted, public notice of the War's end was posted, and the price of tobacco soared due to the end of the British blockade of New Orleans. Seller breached the contract for the sale of tobacco and buyer sued. Chief Justice John Marshall said buyer was not bound to communicate the information because "[i]t would be difficult to circumscribe the contrary doctrine within proper limits where the means of intelligence are equally accessible to both parties." However, the Supreme Court reversed the trial court's decision in favor of the buyer and remanded the case for submission to the jury on the issue of whether "any imposition" was practiced by buyer on seller. Seller had asked buyer "'if there was any news which was calculated to enhance the price," but there was no evidence of any answer.'"[13] Since there is no subsequent record of *Organ*, it does not answer the question of whether buyers are allowed to make deliberately false statements.

If the law were that all false representations by buyers are actionable, sellers before concluding a contract would, as in *Laidlaw*, routinely ask buyers, "Do you have any private information that materially affects this transaction?" If buyers have such information, make false representations, and such representations are grounds for an action in fraud by seller, then the general rule that buyers do not have to disclose positive information would be vitiated. Thus, some false representations by buyers must be tolerated. But, to what extent?

For example, what if seller has great sentimental attachment to her home and wishes to sell the property to someone who will maintain it as a single-family home? Buyer affirmatively represents that he wants to use the property as a single-family home. After the sale is completed, seller learns that buyer intended all along to demolish the house in order to build a high-rise building. Seller sues buyer. Since this agreement involves land, which agreements are not enforceable unless in writing, both efficiency and law suggest that the decision should turn on whether the affirmative representation by buyer is in writing. Since restrictions by prior owners on current land use prevent property from being used in its highest current value, requiring land-use restrictions to be in writing tends to ensure that there was actual negotiation and consideration paid in the form of a lower price for such restrictions. The legal ground for the decision is the parol evidence rule, which is that contemporaneous oral agreements that vary or change a written contract are not admissible in evidence at trial. Although there are many exceptions to the parol evidence rule, one of which is for fraud, the rule is usually strictly enforced with respect to land contracts. In *Mitchill v. Lath*,[14] a buyer of land under a written contract attempted to testify that seller orally agreed to remove an unsightly ice house from a nearby tract as part of the deal. The oral evidence was held inadmissible. However, if the restriction on land use is bargained for, material, and in writing, it usually should be enforced. (See the discussion below of *Peevyhouse v. Garland Coal & Mining Company*.)

Are the Terms of Standardized (Adhesion) Contracts Efficient?

Since billions of transactions involving car leases, land, home, and installment sales are similar, it would be needlessly repetitive and inefficient to negotiate and write such contracts on an individual basis over and over again. The use of standardized contracts significantly reduces transaction costs below what they would be if every contract were negotiated and written individually. Standardized contracts are clearly efficient in this obvious sense. However, in the context of incomplete information and imperfect competition, sellers with market power often employ standardized contracts containing long paragraphs of fine print which consumers typically do not read on a take-it-or-leave-it basis. Query: Are these contracts which have been called "adhesion" contracts Pareto efficient; that is, are the terms in these standardized contracts the same as consumers would freely choose in competitive markets, or do their terms unduly favor the drafters, sellers? Two such contracts will be discussed.

In *Carnival Cruise Lines, Inc. v. Shute*,[15] plaintiff, who lived in Washington state, boarded a cruise ship in Los Angeles, which sailed to Mexico and back. During a guided tour of the ship's galley, she slipped on a deck mat and was injured. On the back of the ticket were twenty-five numbered paragraphs, number 8 of which contained the condition that all disputes in connection with this contract shall be litigated in a court located in Florida, the cruise line's headquarters. Plaintiff did not have an opportunity to read paragraph 8 until after the tickets were purchased. Moreover, paragraph 16(a) provided that the carrier shall not be liable to make any refund for tickets not used, so that a passenger who read the contract would probably accept the risk of having to sue in Florida rather than cancel a vacation without a refund.

The US Supreme Court upheld the forum-selection clause partly on the alleged ground that passengers benefit from reduced fares reflecting the savings that the cruise line enjoys by limiting the fora in which it may be sued. However, cost externalization always results in lower costs as, for example, polluters and toxic-waste dumpers well know. Cost externalization violates Calabresi's admonition (discussed in chapter 13), that prices that do not cover all relevant costs are inefficient and misallocate resources. (Recall the discussion in chapter 18 on products liability, in which strict liability for manufacturing defects is more efficient than negligence due to asymmetric information favoring manufacturers even though prices are higher under strict liability.)

By making it more expensive to sue the cruise line for its negligence, costs, which it should bear, are being externalized. Under *Carnival*, plaintiff from Washington was entitled to her day in court provided that she could afford to go to Florida to file suit. Query: Would the Supreme Court have upheld the forum-selection clause if it required suit in Panama, the country in which the cruise line is incorporated?

Even if such inefficient cost externalization lowers costs, whether such cost reduction will be passed on to consumers as predicted by the US Supreme Court in *Carnival* cannot be determined without an inquiry into whether cruise line industry is competitive or imperfectly competitive; that is, whether firms in these cases have market power, the power to raise prices, and thus the ability to retain the gains from inefficiently lower costs.[16]

In 1992, Congress responded to *Carnival* by amending the relevant statute, which made it unlawful for passenger vessels to employ agreements that limit liability for injury or death caused by negligence or to avoid the right "to a trial by court of competent jurisdiction." The amendment inserted the word "any" before the word "court" in the previous sentence.

If the *Carnival* decision had been allowed to stand, private parties with market power, through non-negotiated adhesion contracts, which have been called "private legislation," would have the power to overrule *World-Wide Volkswagen Corp. v. Woodson*,[17] in which the US Supreme Court held that a state may exercise personal jurisdiction over a non-resident defendant so long as there exist "minimum contacts" between defendant and the forum state; for example, Carnival's extensive advertising in plaintiff's home state.

In *AT&T Mobility LLC v. Concepcion*,[18] plaintiffs were charged sales tax on mobile phones that AT&T advertised as "free." Plaintiffs sued AT&T alleging fraud; the suit became a class action. AT&T moved to dismiss the suit on the grounds that its service contract required individual arbitration and barred class actions. The California Supreme Court held that under California consumer protection laws adhesion contracts that waive the right to sue or class actions are "unconscionable" and void.

AT&T appealed the case to the Ninth Circuit which held that since §2 of the Federal Arbitration Act (FAA) states that arbitration agreements "shall be valid, irrevocable, and enforceable, save upon grounds as exist at law or in equity for the revocation of any contract," the class action could proceed because the state's consumer protection laws were within the exception. The United States Supreme Court, in a 5–4 ruling, reversed the lower courts. Justice Scalia, writing for the court, said that the FAA which was designed to promote arbitration over more costly litigation preempted state law. In dissent, Justice Breyer said that the saving clause in the FAA allowed individual states to determine how a contract or its clauses may be revoked.

Employ the Judge Hand test to mandatory individual arbitration of consumer claims like that of *AT&T Mobility* where claim aggregation is disallowed. In other words, compare the likely burden, B, in dollars to consumer of making an arbitration claim to the expected monetary value or return, PL, in dollars to consumer from arbitration. If, in general, $B > PL$, mandatory arbitration of individual claims is an illusory right. In *AT&T Mobility,* a large corporation is allowed to externalize costs due to its misrepresentation. Query: Which is more efficient for consumers, binding arbitration for claims worth at most a few hundred dollars, where joinder with others is prohibited, or class actions? Should fine print that has not been read be allowed to undermine the meaning of negotiated terms?[19]

Moreover, since *AT&T Mobility*, mandatory arbitration clauses have proliferated in employment, credit, nursing home contracts, and so on, which contracts make the cost of redress for negligence, fraud, and malfeasance suffered by individual consumers greater than the benefit, especially when arbitrators have an incentive to favor defendants who employ them repeatedly; such clauses make fraud and negligence nearly costless for business firms with market power to exploit consumers.

EXPECTATION DAMAGES ARE MORE EFFICIENT THAN SPECIFIC PERFORMANCE

Recall from the previous chapter that a remedy for breach of contract is efficient if it causes a potential contract breaker to internalize the other party's loss; that is, the contract breaker is required to incur the cost to the other party of nonperformance as measured by the gain the other party would have realized had the contract been performed. Expectation damages provides an incentive to perform or not perform the contract that is efficient.[20] The remedy of expectation damages assures that the nonbreaching party will receive her or his share of joint surplus generated by the contract.

In the context imperfect competition, where goods and services are *differentiated*, that is, goods and services which, while not unique, do not have good substitutes, specific performance is generally not an efficient remedy because it creates bilateral monopoly and strategic hold-up problems. Consider the well-known case of *Jacob & Youngs v. Kent*,[21] in which the contract for construction of a house stated that all wrought iron pipe in the house must be standard pipe of Reading manufacture. During construction, the plumbing subcontractor used standard pipe manufactured by Reading and other pipe companies which was indistinguishable except for the brand name. After buyer had been living in the house for over six months, he learned of the switch and refused to pay the balance due on the contract. The contractor sued for the balance due and the buyer counter-claimed for damages and specific performance.

Granting specific performance would have resulted in a hold-up problem because the plumbing was then encased within the walls except in a few places where it had to be exposed, so that the cost of refitting the house with Reading pipe was greater than the market value of the house. Enforcing the buyer's entitlement with specific performance would be to enforce it with a property right. As the discussion of the Coase Theorem showed, this solution gives the party with the property right strategic (monopoly) power to hold-up or extort money from the party without the property right (recall the discussion of *Boomer v. Atlantic Cement Co.* in chapter 12). Had the court granted buyer's request for specific performance in *Jacob & Youngs*, buyer could have demanded that the construction company pay an amount up to the cost of refitting the house with Reading pipe, even though objective damages are nil.

An additional reason for not using specific performance as the usual remedy for breach of contract when there is incomplete information and imperfect competition is that it would discourage efficient breach as in *Jacob & Young*. Even if profit from redirecting resources through efficient breach of contract exceeds the buyer's loss, specific performance would discourage efficient breach. Why? Since buyer has bargaining power to get most or all of the cooperative surplus from specific performance, specific performance raises the price that buyer will demand.

A contract may be breached because the promisor is unable rather than unwilling to perform her promise, as when a strike causes disruption in specifically designed supplies. In such cases, there will often be a modification of contract with appropriate monetary damages in return for accepting alternative performance. If specific performance is the usual remedy, contract modification is less likely for non-standardized objects because specific performance raises the price that non-breaching party will demand.

With respect to unique objects, specific performance is often a more efficient remedy than monetary damages because the likelihood is high that an owner's subjective value of such property exceeds objective market value. For example, an art collector who makes a contract to purchase a Rembrandt self-portrait may not consider money damages to be an adequate substitute for seller's failure to deliver the painting. Land, because of its historically privileged position in Anglo-American law, is generally considered to be unique so that contracts involving land are also usually enforceable via actions for specific performance. However, many land contracts specifically limit the remedy for breach of contract to a forfeit of down payment.

EXPECTATION DAMAGES ARE MORE EFFICIENT THAN OUT-OF-POCKET RELIANCE DAMAGES

Reliance damages can be defined broadly as equal to *opportunity* cost or defined narrowly as equal to *out-of-pocket* costs. In the context of strict rules, competitive market contract model, products and services have good substitutes, so buyers and sellers have opportunities to make identical contracts with many other similarly situated buyers or sellers. Assuming alternative buyers or sellers would have performed, the opportunity cost measure of reliance damages is equal to expectation damages and will place non-breaching party in the position it would have been in if it had contracted with an alternative buyer.

However, in discretionary standards, imperfect competition contract model, in which goods are not standardized and competition is imperfect, calculating the opportunity cost measure of reliance damages is difficult or impossible because seller often does not have an opportunity to enter into numerous other contracts on identical terms, including price, as in the competitive market model. Thus, the measure of reliance damages to which expectation damages must be compared is either out-of-pocket costs or cost of completing performance.

Expectation damages will usually but not always be greater and more efficient than out-of-pocket cost or the cost-of-completion measure. (See *Peevyhouse v. Garland Coal & Mining Company,* discussed below, where reliance damages exceed expectation damages.) For example, a real estate agent who has agreed to find a buyer for a seller of land expects to earn a 6% brokerage fee on a selling price of $200,000, which is equal to $12,000. After the agent who spends $2,000 on advertising and selling costs finds a willing buyer for the land, seller has a change of mind and refuses to sell the land. Agent's out-of-pocket reliance damages, $2,000, is less than his expectation damages, $12,000. Since out-of-pocket reliance costs are often close to zero before performance of many contracts, if reliance costs are the usual measure of damages, parties could often breach contracts with impunity where both parties have not yet performed. Moreover, the reliance measure does not take into account loss of profit, which is both the force motivating market activity and the opportunity cost of investing capital. Thus, in imperfect competition, expectation damages is usually more efficient than out-of-pocket reliance costs because a contract breaker is required to internalize the non-breaching party's entire loss, both costs and profit.

Security Stove & Mfg. Co. v. American Ry. Express,[22] is an example where out-of-pocket reliance measure of damages was used because expectation measure of profits was impossible to calculate. The manufacturer of a gas burner informed carrier both in person and by letter that its prototype burner was to be exhibited at a convention so that deliverance before start of the convention was of utmost importance. Due to carrier's negligence, a crucial part of gas burner never arrived at the convention. Since determining whether anyone at the convention would have been interested in the gas burner prototype had it arrived on time was not possible, manufacturer sued for and won out-of-pocket reliance costs which included express charges paid to the carrier, rental cost for a booth at the convention, return freight charges on the burner parts, railroad fares, hotel expenses, and wages for the president of the company and a workman who accompanied him to the convention.

OVER RELIANCE ON CONTRACTS DUE TO INCOMPLETE INFORMATION

As discussed in the previous chapter, if information is complete and markets are competitive, strict liability on breach of contract does not cause over reliance because expected reliance costs are priced into contracts. In short, over reliance on contracts is due to incomplete information. As shown below, when information is incomplete, two doctrines in law of contracts, the rule of *Hadley v. Baxendale*, and mitigation of damages, limit over reliance by buyers and sellers respectively.

Out-of-pocket reliance expenditure, which is incurred before a contract is performed, called beneficial reliance, increases both profit to buyer if the contract is performed and risk of loss to the seller if the contract is breached. If information is incomplete, reliance expenditure may inefficiently increase expectation damages from a breach of contract. If a buyer of goods, which are to be sold in its retail store, assumes as certain that seller-manufacturer will deliver goods on time, buyer may make transaction-specific expenditures on services or assets in reliance on this contract in order to maximize the value of the goods to it without informing the other party to the contract of these expenses. For example, in anticipation of receiving goods to be sold in its store, buyer may engage in advertising without regard to the possibility of breach.

However, since there is some probability that the contract will not be performed, the efficient level of advertising expenditure in reliance on the contract is where a dollar of advertising expenditure produces a dollar of *expected* monetary profit; that is, profit to the buyer as discounted by the probability of nonperformance (another application of the Judge Hand test). If a buyer relies on the contract being performed as a certainty or discounts the probability of nonperformance at less than the true probability of nonperformance, buyer may inefficiently over rely on the contract, which over reliance inefficiently increase expectation damages.

Sellers as well as buyers may incur out-of-pocket reliance expenditures which should also be discounted by the probability of buyer's breach; for example, a seller may incur reliance costs by ordering materials or hiring workers with special skills necessary to perform the contract.

Over reliance by buyer can also occur when the cost of completing performance after a breach exceeds the market value of the completed contract. The issue in such cases is whether the cost of completing performance is the expectation measure or the reliance measure of damages. In *Jacob & Youngs*, discussed above, cost of completing performance by refitting the house with Reading pipe was implicitly characterized as an expectation measure, but it could have been described as over reliance by buyer.

Next, consider a much discussed case in which cost of completing performance after breach exceeded the value of completed contract resulting in alleged over reliance. In *Peevyhouse v. Garland Coal & Mining Company*,[23] plaintiffs leased their farm to defendant for five years in order to engage in strip mining of coal. The defendant specifically agreed in writing to perform restorative and remedial work at the end of lease period. Remedial work would have involved moving many thousands of cubic yards of dirt at a cost estimated by an expert at around $29,000. Plaintiffs sued for $25,000. Jury returned a verdict of $5,000, which was only a fraction of cost of performance, but more than total value of the farm even if remedial work was done. The court said that "where contract provision breached is merely incidental to the main purpose in view, and where the economic benefit that would result to the lessor by full performance of the work is grossly disproportionate to the cost of performance, the damages which lessor may recover are limited to the diminution in value resulting to the premises because of non-performance."[24] The court held that diminution in land value due to nonperformance of the remedial work was $300 and awarded that amount.

The dissenting judge pointed out that the contract provision which was breached was not incidental because plaintiffs had insisted that provision for remedial work be included in the contract and would not have allowed strip mining on their property unless this clause was included. In other words, the dissenter believed that plaintiff's subjective valuation or expectation measure is relevant in this case. (Query: Would plaintiff in *Jacob & Youngs* have agreed to the contract without stipulation that Reading brand pipe be used?) Moreover, the dissenter said that there were no conditions existing at the time of the suit which could not have been reasonably anticipated when the contract was negotiated and signed. In short, the majority implicitly characterized cost of completing performance as over reliance, whereas dissenter implicitly characterized cost of completing performance as expectation measure of damages. Query: What would have been the result in this case if plaintiffs had sued for specific performance? As a result of this decision, a coal company could in the future promise to restore land after strip mining regardless of cost even though it had no intention of keeping its promise. This case, decided in 1962, which ratified the despoliation of land, preceded the environmental movement of the 1970s.

A most spectacular example of over reliance on contracts occurred in the financial meltdown of 2007–2008, where buyers of over 60 trillion dollars of "credit default swaps," which are essentially "insurance" agreements that pay off if securitized mortgages default, found out that they had over relied on and underestimated the risk of default by home mortgagors. By act of Congress, financial institutions were allowed to use the term credit default swaps instead of the term insurance so that these financial instruments could be issued without reserves against potential losses.[25] Warren Buffett has called credit default swaps "financial instruments of mass destruction."[26]

THREE BILATERAL CARE CONTEXTS WHERE NO LIABILITY AND STRICT LIABILITY ARE INEFFICIENT

Consider three contexts involving bilateral care in which both strict liability and no liability are inefficient: Recall Coase's railroad sparks-crops example where high transaction costs preclude bargains; both no liability and strict liability are inefficient. No liability results in too many railroad trips and too few crops because crop damage is external to railroad's calculations; strict liability results in too few railroad trips and too many crops because lost railroad profits are external to farmers' calculations. In this context, a tax only on the railroad is efficient; the level of the tax determines the number of profitable railroad trips per day which in turn controls how close the farmers will plant their crops to the railroad tracks.

Recall automobile-pedestrian Chicken games in torts, in which there is also no bargaining between parties. Under no liability, drivers who adopt Intimidate Chicken Game strategies take less than the efficient level of care; in response, pedestrians have to walk defensively, that is, walk with more than due care. Under strict liability, pedestrians who adopt Intimidate Chicken Game strategies take less than the efficient level of care; drivers have to drive defensively, that is, drive with more than due care. In this bilateral care context (torts), liability based upon comparative negligence will cause both drivers and pedestrians to adopt the efficient levels of care.

Three, the law of contracts involves bilateral care because both sellers and buyers make promises that may be breached. Both no liability and strict liability are inefficient. No liability is inefficient because the loss to non-breaching party, whether buyer or seller, is external to calculations of contract breakers. When information is incomplete, strict liability on contract is also inefficient because a non-breaching buyer or seller may over rely on the expected performance of contract breaker, who may breach without knowledge of the non-breaching party's over reliance. Over reliance inefficiently increases the non-breaching party's expectation damages.

These three bilateral care examples have two significant differences: One, in the first two examples of bilateral care, the parties cannot bargain with each other whereas in the third example, contracts, parties do bargain. Two, in the first two examples which involve bilateral care, there is an efficient alternative to no liability or strict liability: in the railroad sparks-crops example, a tax on the railroad will achieve the efficient result; in automobile-pedestrian Chicken games, the law of comparative negligence achieves the efficient result. However, in the law of contracts, there is no alternative rule to no liability or strict liability. (As discussed in the previous chapter, if information is complete, no alternative is needed because expected reliance costs are priced into contracts.) Where information is incomplete, over reliance on contracts can occur. There are two doctrines in law of contracts that limit over reliance: the rule of *Hadley v. Baxendale* and mitigation of damages which limit over reliance by buyers and sellers respectively.[27]

HADLEY AND MITIGATION OF DAMAGES DOCTRINES LIMIT OVER RELIANCE

In the previous chapter's discussion of complete information and competitive markets, consequential damages are foreseeable in the sense that expected monetary value of all risks is known at the time of contracting and is reflected in price. However, in imperfect competition, where information is incomplete, the narrow interpretation consequential damages, known as the rule of *Hadley v. Baxendale,*[28] is efficient because it has an information forcing function to perform and is needed to prevent over reliance by buyers. Over reliance by sellers is ameliorated by the mitigation of damages doctrine.

Hadley v. Baxendale Rule Limits Over Reliance by Buyers

In *Hadley*, owners of a steam driven flour mill contracted with a common carrier to deliver the mill's broken crank shaft to engineers in Greenwich. Due to the negligence of carrier, delivery was delayed by five days. The issue was the proper measure of damages. The miller claimed that because the shaft was sent as a pattern for a new shaft and mill could not operate until it arrived, he was entitled to recover loss of profit for five days by which resumption of work was delayed due to the breach.

Defendants claimed that they were not liable for these damages because they were too remote.

Notwithstanding the plaintiffs' testimony that carrier's clerk was told that "the mill was stopped and that the shaft must be sent immediately" and "that a special entry, if required, should be made to hasten its delivery"[29] as well as the jury verdict for plaintiffs from which it can be inferred that the jury, which is the finder of fact, believed plaintiff's testimony, the Court of Exchequer Chamber held that "we find that the only circumstances here communicated by plaintiffs to defendants at the time contract was made, were, that the article to be carried was the broken shaft of a mill, and that plaintiffs were the millers of that mill."

In *Hadley*, the court went on to promulgate what has come to be known as the either-or rule: Consequential damages are such as arise *either* "according to the usual course of things," from breach of contract "*or* such as may be reasonably be supposed to have been in the contemplation of both parties, at the time they made the contract, as the probable result of the breach of it."[30]

In competitive markets, where information is complete, the *Hadley* rule has no "information forcing" function to perform because expected monetary value of consequential losses can reasonably be supposed to have been in the contemplation of both parties and factored into the price; that is, there is no over reliance. However, in imperfect competition, where information is incomplete, sellers who breach contracts often neither know specific consequential damages buyer will incur nor know the probability distribution of such damages due to uncertainty. In the incomplete information, imperfect competition context, the rule of *Hadley* is efficient because it provides a rationale for forcing buyer to communicate information concerning the consequences of breach at the time of contract rather than at time of breach so that seller can take precautions or insure against such loss. This is the information forcing argument for requiring a buyer to communicate to seller the level of reliance on the contract.

For example, assume someone sends a package through the US Post Office and does not inform or conceals from the postal clerk that the package contains a valuable diamond ring so as to avoid paying for insurance. If the package gets lost in transit, sender's failure to disclose its value forecloses damages. For Post Office to incur legal liability, sender must declare the package's value and pay an amount equal to the premium on an insurance policy for risk of loss plus transportation charge. The *Hadley* rule forces sender to reveal its level of reliance on the contract.

In other words, *Hadley* creates an efficient separating equilibrium, whereby buyers with special needs must inform sellers of their potential consequential damages in order to hold sellers liable. Since there are usually fewer buyers with special needs and risks than there are buyers with average needs and risks, efficiency requires that the law should adopt a pooling equilibrium default rule, no consequential damages unless buyers with special needs take initiative and separate themselves by informing sellers of their reliance needs.

Mitigation of Damages Doctrine Limits Over Reliance by Sellers

The mitigation of damages doctrine limits over reliance by sellers when calculating expectation damages on breached contracts. For example, assume that buyer agrees to pay $10 per shirt for 1,000 custom-made shirts which are to be sold in her retail store. Assume further that after 200 shirts are produced, accepted, and offered for sale few or none are bought. Buyer realizes belatedly that consumers abhor this shirt's style. Buyer informs seller who has yet to produce the remaining 800 shirts that it has no need of and will not accept additional shirts. Since the shirts are unsellable, resources would be wasted if seller produced shirts that are not wanted. Under the doctrine of mitigation of damages, the non-breaching party is required to use reasonable effort to minimize damages incurred after the contract breach. Mitigation of damages doctrine gives the parties efficient incentives, which in this case is not to produce the unwanted, specialized shirts; that is, seller cannot over rely on the contract.

For the shirts that should not be produced after notice of contract breach, seller is entitled to the expectation measure, which is called lost "profits" but is actually the difference between contract price and costs that were not incurred on the 800 shirts not produced (opportunity cost saved). What costs of performance were not incurred? Economists divide costs into two basic categories, fixed and variable costs. Variable costs are costs that vary with output; typical variable costs are labor, materials, and sales taxes. Fixed costs are costs that do not vary with output: typical fixed costs or overhead costs are mortgage payments on the factory and its machinery, long-term leases, and property taxes. Costs that are usually variable, like labor, can be fixed if, for example, instead of being paid an hourly wage a worker has a contract with a guaranteed annual income; costs which are usually fixed, like the cost of factory, can be variable if a factory is leased on a month-to-month basis. Moreover, all costs are variable in the long run because the question in the long run is whether to build an entirely new factory, buy a new machine, or renew a long-term contractual commitment. However, the long run is not relevant with respect to contracts that are performed over a relatively short period of time.

In the above example, what costs were avoided as a result of the 800 shirts that were not produced? Assume that the $10 price per shirt consists of $4 in variable costs (labor and materials), $5 in fixed costs (factory and equipment), and $1 in profit.[31] When seller mitigates damages by not producing 800 shirts, it saves only $4 per shirt in variable costs but still incurs fixed costs of $5 per shirt which it must pay even if no output is produced. The manufacturer reasonably expects to receive that part of the contract price that goes to defray its fixed or overhead costs. Manufacturer is entitled to $5 per shirt for fixed costs plus $1 per shirt for profit, which equals $6 per shirt multiplied by 800 shirts for expectation damages of $4,800 on the unproduced shirts, plus $2,000 for the shirts that were delivered and accepted for a total of $6,800. The UCC has adopted the measure of profit plus overhead cost as damages for non-acceptance or contract repudiation.[32] The doctrine of mitigation of damages limits over reliance on contracts by sellers.

LIQUIDATED DAMAGES-PENALTIES DISTINCTION IS EFFICIENT IN IMPERFECT COMPETITION

Since most contracts are performed, it would be inefficient to require that all contracts contain liquidated damages provisions; according to the Judge Hand test, including liquidated damages provisions in contracts is inefficient if transaction costs of including the provision, B, exceed the litigation costs saved from such provisions discounted by the risk that they will occur, PL. However, in the context of incomplete information, a liquidated provision is appropriate and even necessary when damages contemplated as a consequence of breach, should it occur, are lost profit on a new and, as yet, untried business which has no past record of profits from which damages could be estimated (there is uncertainty in addition to known risks). Absent a liquidated damages provision, such damages may be too speculative to be awarded by courts. Liquidated damages provisions also serve as efficient incentives for specific reliance when "time is of essence"; for example, a contract deadline for construction of a resort hotel that has booked its rooms in advance of construction.

In contracts that have liquidated provisions, courts make a distinction between liquidated damages provisions that are enforceable and penalties that are not enforceable. *Penalties* are damage provisions in contracts that are neither reasonably related to estimated damages at the time of contract nor to actual damages at the time of breach. There is a great deal of dispute among economists whether the distinction between liquidated damages provisions and penalties is meaningful. Economists who are in favor of enforcing all stipulated damages provisions argue that since the parties agreed to these provisions, they must be better off for having made them or they would not have agreed to them (Pareto improvement).

As discussed in the previous chapter, the liquidated damages-penalties distinction is irrelevant in the strict rules, competitive market contract model because there is complete information concerning the distribution of risks that could trigger liquidated damages provisions, and competitive markets ensure that no one has enough market power to demand liquidated damages that are greater than the risks that such provisions protect against.

However, in the discretionary rules, imperfect competition contract model where information is incomplete and where one party to a contract may have market power over another party, stipulated damages provisions that are characterized by courts as penalty clauses may be less efficient than money damages determined ex post facto.

Reasons sometimes given in support of enforcing penalty clauses in contracts are: by signing a contract with a large penalty for missing a deadline promisor conveys information as to its credibility; a penalty clause may be necessary compensation for dealing with a party who has a high risk of default; or the punitive element may be characterized as a premium on an insurance contract written in favor of the non-breaching party.

However, unequal bargaining power and extortion can always be rationalized or disguised as insurance and risk allocation (organized crime seems always to have known of this ruse). Whenever consideration for a bargain is highly disproportionate in favor of one side, perverse incentives are often created so that someone is benefited by nonperformance of the agreement. A study of high stipulated damages provisions in construction contracts concluded that such provisions not only cause contractors to raise their bids due to a fear that there may be a delay beyond their control (risk which may not be easily calculable due to the uncertainty aspect of incomplete information), such provisions also make a contractor susceptible to considerable labor pressure; a contractor may have to give in to union demands whether or not they are justified. Moreover, when labor unions make a "squeeze play" of this sort, other contractors in the area find that they must give same benefits or face considerable trouble.[33]

An alternative approach to the liquidated damages-penalty distinction would be for courts to employ a market power analysis in order to determine whether there was unequal or monopoly bargaining power on one side. If there is unequal bargaining power, courts could infer that a stipulated damages clause was probably not efficiency enhancing. However, in imperfectly competitive markets, determining whether a firm's market power is substantial and efficiency reducing or insubstantial and *de minimis* is an extremely difficult decision for courts even with extensive investigation and testimony; in federal antitrust cases in which the issue is whether a business has substantial market power, the decision often turns on whether the market is defined broadly or narrowly.[34]

Courts may have concluded implicitly that a determination of whether a stipulated damages clause is a reasonable ex ante estimate of damages which is closely related to ex post actual damages or is a penalty is less costly and more within their expertise than is an investigation into questions of market structure and monopoly power in imperfect competition. Judge Posner has noted that courts have not altered their views or rulings prohibiting penalty clauses even after being appraised of economists' arguments in favor of enforcing them.[35] In sum, in the discretionary standards-imperfect competition contract model, arguments in favor of retaining the legal distinction between liquidated damages clauses and penalties outweigh the arguments in favor of doing away with this distinction.

LOST-VOLUME PROFITS (PLUS OVERHEAD) FOR SELLERS IN IMPERFECT COMPETITION

In the previous chapter, it was shown that in competitive markets, if sellers inform buyers ex ante whether or not they require lost-volume profit for breach of contract, there will be a compensating price differential such that if sellers have liberal return policies, prices for goods will be higher, and if they do not allow returns or there is a restocking fee, prices for goods will be lower. In competitive markets, the lost-volume seller rule is an example of Rottenberg's corollary for liability rules in competitive markets, where a change in legal liability rule has no effect on either efficiency or income distribution.

In competitive markets, all sellers are price takers not price makers; for example, all farmers receive the same market price, over which they have no control, for grain. In competitive markets, the challenge for sellers is to produce output cheaply enough to make a profit at the market price. A seller's output is limited by the law of diminishing returns. In competitive markets, sellers essentially have a production problem, not a selling problem.

However, in imperfect competition, sellers usually sell differentiated products (not all products have good substitutes) and face downward sloping demand curves; they set their own price; they are price makers not price takers as in competitive markets. They set price so as to cover all costs including a reasonable profit. Sellers are often uncertain where buyers are located and so must incur considerable time and expense to find them. These sellers cannot avail themselves of organized stock or commodities markets as can sellers of grain, metals, and common stock. In imperfect competition where information is incomplete, the challenge for sellers is to find enough buyers to make a profit. In these markets, sellers have a selling or marketing problem.[36] When demand declines in imperfectly competitive markets, especially for expensive or "big-ticket" items, sellers usually adjust by reducing the quantity produced rather than by reducing price so as not to spoil or cannibalize their markets which they perceive as limited; in imperfect competition, demand must be depressed for a considerable time before price reductions occur. When contracts are breached in imperfectly competitive markets, sellers are entitled to lost-volume profits and contribution to fixed costs.

In *R.E. Davis Chemical Corp. v. Diasonics, Inc.*,[37] the manufacturer agreed to sell medical diagnostic equipment, magnetic resonance imaging (MRI), for $1,500,000 to buyer, who breached the contract. After reselling the equipment to a third party for the same price, seller sued for lost-volume profit. The court said that in addition to having the capacity to make an additional sale, seller must also show that it would have been profitable for the seller to produce the additional unit and that it has a predictable and finite number of customers.[38] Seller was awarded damages for reasonable overhead cost and lost profits amounting to $453,050 ($453,050/$1,500,000 = 30% of contract price).[39] The court noted that seller possessed the capacity to manufacture one more MRI and also demonstrated that it was "beating the bushes for all possible sales."[40] The court's hunting metaphor captures the spirit of uncertain demand in imperfect competition.

A more difficult case is *Copenhaver v. Berryman*,[41] which involved a five-year contract to provide and operate laundry equipment within an apartment complex. With forty-seven months remaining on the contract, the owner of apartment complex terminated the contract, removed plaintiff's laundry machines, and replaced them with his own equipment. Within six months after breach, plaintiff's equipment, eight washing machines and four driers, was placed in use in other locations. Plaintiff claimed damages for lost profit on laundry operations for forty-seven months amounting to $13,886.58. According to plaintiff, equipment was comparatively easy to obtain; the major factor in expanding business was obtaining desirable locations. Plaintiff contended that business expansion was limited by the breach because he was forced to place the equipment in question into a location that he would have acquired anyway and that defendant should not receive benefit of plaintiff's diligent business practices. However, between the time of breach and time of trial, plaintiff's business expanded by adding 350 machines to fourteen to fifteen new locations. The court limited plaintiff's lost profits to the six-month period on the ground that income generated after the breach by the same equipment should be considered in mitigation of damages; it awarded damages of $3,525.84.

Can the two lost-volume profit cases be distinguished by characterizing the market in *R.E. Davis* as imperfectly competitive, whereas the market in *Copenhaver* has elements of competitive and imperfectly competitive markets? Although plaintiff in *Copenhaver* said demand was limited by the difficulty in finding desirable locations, commercial laundry facilities provide a standardized or fungible service; moreover, plaintiff substantially expanded his business by the time of trial. The court's limitation of damages to lost profit for the six-month period before the equipment was redeployed is consistent with view that the market is closer to the competitive model than imperfect competition model.

NOTES

1. 190 F.Supp. 116 (S.D.N.Y. 1960).
2. In *Shrum v. Zeltwanger*, 559 P.2d 1384 (Wyo. 1977), the issue was, "what is a cow?" Did the term include a "heifer" which had never had a calf? The court held that the word "cow" had "within the corral of this case, no plain and ordinary meaning," so that there must be a trial to determine which parties' interpretation was correct.
3. (St. Paul, MN: American Law Institute, 1981), § 201(3), illustration four.
4. 499 F.Supp. 53 (W.D.Pa.1980).
5. John Dawson, "Judicial Revision of Frustrated Contracts: The United States," *Boston University Law Review* 64 (1984): 1, 35–36.
6. *Obde v. Schlemeyer*, 56 Wash.2d. 672 (1960).
7. Jack Hirshleifer, "The Private and Social Value of Information and the Reward to Inventive Activity," *American Economic Review* 61 (1971): 561–574.
8. Buyers who have trespassed on seller's land in order to acquire information are legally liable to sellers for valuable information obtained illegally as have those who acquire information illegally, such as from "insider" trading or violation of a fiduciary duty.
9. Steven Shavell, "Acquisition and Disclosure of Information Prior to Sale," *Rand Journal of Economics* 25 (1994): 20–36.
10. Anthony Kronman, "Mistake, Disclosure, Information and the Law of Contracts," *Journal of Legal Studies* 7 (1978): 1, 20–21. [1–34].
11. *Leitch Gold Mines, Ltd. V. Texas Gulf Sulphur*, 1 Ontario Reports 469, 492–493 (1969).
12. 15 U.S. (2 Wheat.) 178 (1815).
13. E. Allan Farnsworth, *Contracts*, 2nd ed. (Boston: Little, Brown, 1990), 256.
14. 247 N.Y. 377 (1928).
15. 499 U.S. 585 (1991).
16. Victor Goldberg, "Institutional Change and the Quasi-Invisible Hand," *Journal of Law & Economics* 17 (1974): 461, 462, 483–490.
17. 444 U.S. 286 (1980).
18. 563 U.S. 333 (2011).
19. E. Allan Farnsworth and William Young, *Contracts: Cases and Materials*, 5th ed. (New York: Foundation Press, 1995), 416.
20. Robert Cooter and Melvin Eisenberg, "Damages for Breach of Contract," *California Law Review* 73 (1985): 1432, 1459–1468, 1473.
21. 230 N.Y. 239 (1921).
22. 227 Mo.App. 175 (1932).
23. 382 P.2d 109 (Okla. 1962).
24. 382 P.2d 114.
25. Commodity Futures Modernization Act of 2000.
26. Berkshire Hathaway Annual Report 2002.
27. Thomas Miceli, *The Economic Approach to Law*, 2nd ed. (Palo Alto, CA: Stanford University Press, 2009), 114–117.
28. 9 Ex. 341, 156 Eng. Rep. 145 (1854).
29. 156 Eng.Rep. 147.

30. 156 Eng.Rep. 151 (emphasis added).
31. In economic theory, a "normal" profit is equal to the opportunity cost on the capital invested in business.
32. § 2–708(2).
33. A. L. Elliott, *A Study of Liquidated Damages on Highway Contracts* (Sacramento, CA: Division of Highways, 1956), 21–22.
34. *U.S. v. DuPont*, 351 U.S. 377 (1956); *FTC v. Staples, Inc.*, 970 F.Supp. 1066 (DDC 1997).
35. *Lake River Corp. v. Carborundrun Co.*, 769 F.2d 1284 (7th Cir.1985)(Posner, J.).
36. Cooter and Eisenberg, note 20, 1456.
37. 826 F.2d 678 (1987).
38. 826 F.2d 683–684. Marginal cost is an increasing function of output due to the law of diminishing returns; therefore, seller should have to prove that both units could be produced for less than the contract price. Charles Goetz and Robert Scott, "Measuring Sellers' Damages: The Lost-Profits Puzzle," *Stanford Law Review* 31 (1979): 323, 332–33, 346–47.
39. *R.E. Davis Chemical Corp. v. Diasonics, Inc.*, 924 F.2d 710 (1991).
40. 924 F.2d 711.
41. 602, S.W.2d 540 (Tex. App. 1980).

Chapter 21

Decision Theory, Suit, and Settlement

A potential plaintiff's willingness to sue depends on the probability of prevailing at trial, magnitude of monetary judgment, legal costs of trial, and method by which legal costs are allocated between parties. The defendant's willingness to go to trial or settle depends on the same factors.

This chapter's analysis of the risks of litigation uses *decision theory*, which is a combination of decision trees from sequential game theory and expected monetary value, two techniques which have been emphasized in this book.

First, decision theory is used to evaluate plaintiff's willingness to sue under the American rule versus British rule for allocating litigation costs. Then, the optimism suit and settlement model, which attributes settlement of a lawsuit to a convergence in plaintiffs and defendants' estimates as to how likely plaintiff is to win the suit, is discussed. Finally, the effect of a contingent fee arrangement on a lawyer's willingness to sue and lawyer-client conflict of interest are discussed.

DECISION ANALYSIS

When a dispute arises, the initial decision is whether to file a lawsuit. If a party decides to file suit, the next decision is whether to go to trial or attempt to settle the dispute. If the parties settle, the suit ends; if the plaintiff decides to go to trial, whether the plaintiff wins at trial depends upon a chance event. If plaintiff loses, plaintiff must decide whether to file an appeal or end the suit. If plaintiff decides not to appeal, the suit ends; if plaintiff decides to appeal, whether plaintiff wins or loses the appeal depends upon a chance event. If plaintiff loses the appeal, the suit ends. If plaintiff wins the appeal, the appellate court remands the suit back to the trial court with instructions on how to proceed. For example, the appellate court may say that the jury's decision in favor of the defendant is being reversed because the trial court gave erroneous and prejudicial instructions to the jury. The appellate court remands the case back to the trial court for a new trial, ordering it to give instructions to the jury in conformity with its opinion. Alternatively, if defendant loses at trial, it has similar decisions to make.

American Rule Versus British Rule for Allocation of Costs of a Lawsuit

Figure 21.1 depicts plaintiff's decision whether or not to file a lawsuit in an automobile collision case and possible outcome of trial should it occur. At the intersection of *tree branches*, which are called *nodes*, there is either a square which indicates that an actor must make a decision among branches or a circle which indicates that the branch followed is determined by chance; the probability that a branch following a circle will be chosen is listed alongside each branch. The payoffs, whether due to chance or decision, are listed at end of each branch. Each chance node of a decision tree is evaluated by its expected monetary value via backward induction from *right* to *left* as described below. (Recall the backward induction solution for a Repeated Prisoners' Dilemma in chapter 7.) When all branches of a decision tree have been evaluated, the branch with the highest value is chosen; two lines are used to strike off branches that are eliminated.

In Figure 21.1, starting from the right, the chance node is evaluated. If plaintiff decides to sue, there are two possible outcomes from trial: plaintiff is awarded \$100,000 with a probability of 60% or loses the case and gets \$0 with a probability of 40%. The plaintiff's litigation costs are \$25,000. Under the American rule for allocation of litigation costs, each litigant pays its own costs irrespective of win or loss, so litigation costs are a fixed cost for plaintiff that she incurs as a result of going to trial.

A risk neutral plaintiff's expected monetary judgment of the lawsuit is: $E(J_P)$ = (\$100,000)(.6) + \$0(.4) = \$60,000, which is placed under the chance node. Next, the decision node (square) on the left is evaluated. If plaintiff's chooses Sue, the net expected payoff is \$60,000 minus litigation costs of \$25,000, which equals \$35,000; if

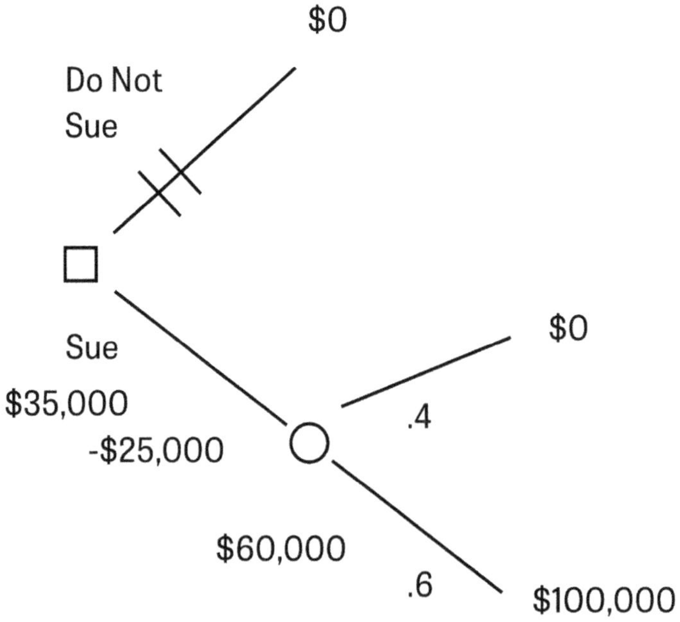

Figure 21.1 American Rule for Litigation Costs. *Source*: Created by the author.

plaintiff chooses Do Not Sue, the payoff is \$0. Since \$35,000 > \$0, the branch with \$0 payoff is struck off with two strokes through it, indicating that this branch will not be chosen.[1] The result of this example on the willingness to sue under the American rule for the allocation of litigation costs is easily generalized: under the American rule, a risk neutral plaintiff will file suit only if expected monetary judgment from litigation, $E(J_P)$, is greater than the cost of litigation, C_P:

$$E(J_P) > C_P$$

Assume same facts as above but that allocation of litigation costs is determined by the British rule, where loser pays both parties' litigation costs, which in this suit are \$25,000 for each party. Under the British rule, whether plaintiff will incur litigation costs is dependent upon whether plaintiff wins or loses the lawsuit so litigation costs are part of the expected monetary value calculation, not a fixed cost. The decision analysis based upon these facts is shown in Figure 21.2.

In Figure 21.2, starting with the chance node on the right, the expected monetary judgment is: $E(J_P)$ = (\$100,000)(.6) + (–\$25,000 – \$25,00)(.4) = \$60,000 – \$20,000 = \$40,00, which is placed under the chance node. Next, evaluate the decision node on the left; since \$40,000 > \$0, the Do Not Sue branch with the \$0 at the end of the branch is struck off with two strokes through it, indicating that this branch is not chosen. Note that because the probability plaintiff will win the suit is greater than 50%, the expected judgment of the lawsuit to plaintiff is greater under the British rule (\$40,000), where loser pays litigation costs than under the American rule (\$35,000) where each party pays its own litigation costs; conversely, if plaintiff's probability of winning the suit is less than 50%, plaintiff's expected judgment of the lawsuit is greater for plaintiff under the American rule.

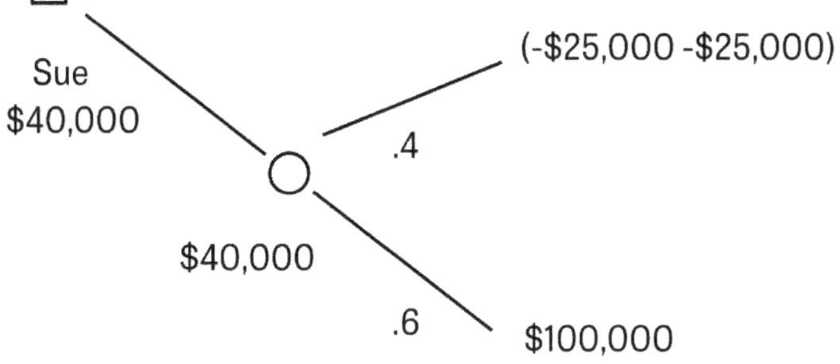

Figure 21.2 British Rule for Litigation Costs. *Source*: Created by the author.

The result of this example on the willingness to sue under the British rule for the allocation of litigation costs is easily generalized: plaintiff's probability of winning the case is p, plaintiff's probability of losing the case is (1−p), and parties' litigation costs are C_P and C_D; the necessary condition for a risk neutral plaintiff to sue is that plaintiff's probability of winning the suit multiplied by monetary judgment (winning plaintiff does not have to pay litigation costs) is greater than the plaintiff's probability of losing the suit multiplied by the sum of the litigation costs.

$$(p)J > (1-p)\left(C_P + C_D\right)$$

If there is a high probability of winning a large judgment, plaintiffs will sue under both American and British rules for allocating litigation costs; if there is a low probability of winning a small judgment, plaintiffs will not sue under both rules for allocating litigation costs.

The rules have different incentives if there is a low probability of winning a large judgment or a high probability of winning a small judgment: if there is a low probability of winning a large judgment, a plaintiff is more likely to sue under the American rule because plaintiffs who lose do not have to pay defendant's litigation costs. And, if there is a high probability of winning a small judgment, plaintiffs are more likely to sue under the British rule because a winning plaintiff's litigation costs are paid by defendant.

Many lawyers believe that the American rule tends to encourage "lottery" lawsuits (low probability-high judgment) by plaintiffs who are risk preferrers rather than risk neutral and who think of such suits as a chance either to win the big prize in a lottery or to "blackmail" defendants who are risk adverse and would prefer to settle such suits rather than "bet the farm" or firm on one lawsuit. Others believe that the British rule overly discourages lawsuits by risk adverse plaintiffs who fear large legal costs should they lose the suit. One promising reform proposal would adopt the British rule but cap litigation costs at no more than that spent by the losing side. This modified British rule does not allow defendants to adopt a strategy of discouraging lawsuits by threatening a plaintiff with crushing legal costs.

The British rule's advantage over the American rule in encouraging meritorious small claims where plaintiffs have a high probability of winning is somewhat offset by American small claims courts, where suits under $5,000 are litigated in an informal setting (neither trial record nor lawyers), or class actions, where small legal claims that are similar can be combined in one suit. The class action has the added attraction of economies of scale over multiple single suits. However, the class action has two major shortcomings: first, the costs of identifying the members of a class and giving each member his pro rata share of the judgment may result in very small awards; second, the lawyers for the class, who bear the legal costs and are often the real parties suing, have an incentive to settle with the defendant for a small judgment and a large legal fee. However, if deterrence (confronting violators with the costs of their violations) is more important than compensation of victims, these two shortcomings of the class action are not serious.[2]

Suit or Settlement?

Assume that plaintiff has filed a lawsuit in an automobile collision case and defendant has offered to settle for $200,000. Assume further that there are three possible outcomes from trial: plaintiff is awarded $1,000,000 with a probability of 30%, awarded $200,000 with a probability of 40%, or loses the case and gets $0 with a probability of 30%; the cost of trial and attorney fees are $50,000. Plaintiff must decide whether to sue or settle. The decision analysis based upon these facts is shown in Figure 21.3.

Begin by evaluating the chance node on the right: the expected monetary judgment from litigation is: $E(J_P) = (\$1,000,000)(.2) + (\$200,000)(.4) + (\$0)(.5) = \$280,000$; this number is placed under the chance node. Next, evaluate the decision node (square) on the left where the decision between Sue or Settle must be made. If Settle is chosen, plaintiff receives $200,000; if Sue is chosen, the net payoff from litigation is $280,000, minus the cost of litigation under the American rule, –$50,000, which equals $230,000. Comparing the results of the two possible decisions, $230,000 > $200,000, so a risk neutral plaintiff should sue. The branch with the $200,000 Settle label is struck off with two strokes through it, which indicates that this branch will not be chosen.

Antitrust Suit

Consider an antitrust suit in the context of decision theory. Assume that plaintiff, widget manufacturer, alleges that there is a price-fixing conspiracy among defendants who sell a necessary input in the widget manufacturing process. Defendants have offered to pay plaintiff $2 million in order to settle the antitrust suit. Alternatively, if the suit goes to trial, the probability of winning the suit is 70% and the probability of losing is 30%. If plaintiff wins the suit, the cost of the suit is $0 because the relevant statute authorizes successful plaintiffs in antitrust suits to recover costs of suit, including a reasonable attorney's fee. If plaintiff losses the suit, it pays court costs plus attorney's fees totaling $0.5 million.

The damage award will depend upon proof of how much plaintiff was overcharged by the price-fixing conspiracy multiplied by three because the relevant statute

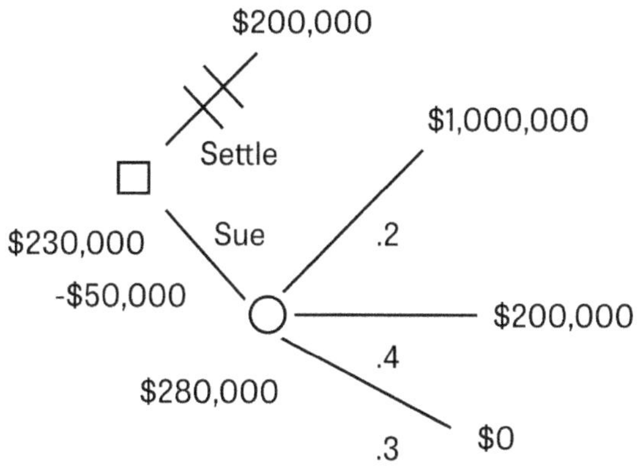

Figure 21.3 Sue or Settle. *Source*: Created by the author.

authorizes triple damages in order to promote deterrence. Assume there is an 80% probability that plaintiff will be able to prove a low total overcharge of \$0.5 million multiplied by 3 for a total damage award of \$1.5 million, and a 20% chance that plaintiff will be able to prove a high total overcharge of \$3 million multiplied by 3 for a total damage award of \$9 million. A decision tree analysis is constructed below in Figure 21.4.

Evaluate the decision tree from right to left; at the chance node on the right, expected monetary damages is: $E(D_P)$ = (\$1.5 million.)(.8) + (\$9 million)(.2) = \$3 million, which is placed below the right chance node. Next, at the chance node on the left, expected monetary judgment of the lawsuit is: $E(J_P)$ = (\$3 million)(.7) + (−\$0.5 million)(.3) = \$1.95 million.

In Figure 21.4, on the left, there is a square indicating that a decision between settle or sue must be made. Plaintiff should settle rather than sue because \$2 Mil. > \$1.95 Mil; the Sue branch is struck off with two strokes.

Sensitivity Analysis

An experienced lawyer will usually estimate probabilities and monetary values of potential litigation based upon past experience with similar cases. Recall from chapter 15, if values of variables are normally distributed, the further an estimate is from the mean of damages in prior cases, the less likely such an estimate is. This suggests that if estimates are normally distributed, past experience is a good guide to the future. However, if values are not normally distributed, extreme values are more likely. An analyst may implicitly assume that values are normally distributed without having enough data to test this assumption statistically. Moreover, even if estimates are normally distributed, an analyst must ask, how sensitive is my decision to changes in probabilities and monetary values?

Sensitivity analysis is concerned with the question of whether the conclusions of decision analysis are robust or stable when probabilities or monetary values in the analysis are changed. For example, in the antitrust lawsuit discussed above, the difference in decision between settling for \$2 million and the expected judgment from suing, \$1.95 million, is only \$50,000. Even a slight change in assumptions, such as

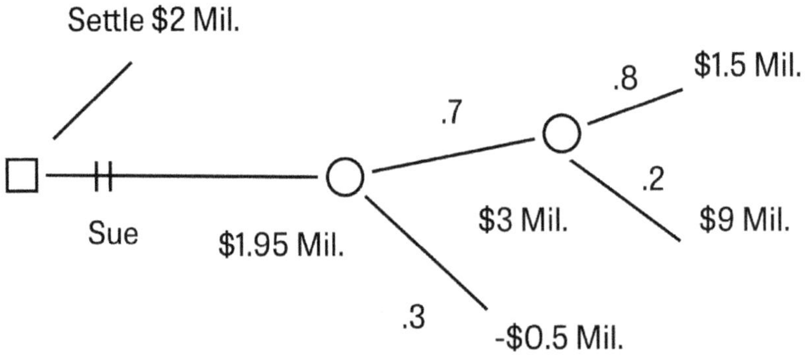

Figure 21.4 Antitrust Suit or Settlement. *Source*: Created by the author.

increasing the probability of winning the suit by 1.5%, would cause the decision to change from settling to sue. This means that the decision whether to settle or sue in this case is extremely sensitive to a change in assumptions as to the probability of winning the suit or the amount of damages awarded. Given that the decision whether to sue is extremely sensitivity to a small changes in assumptions, the argument for settling for a sum certain, $2 million, is strengthened.

OPTIMISM MODEL OF SETTLEMENT

The optimism model which is based upon expected monetary value attributes settlement of a lawsuit to a convergence in plaintiffs and defendants' estimates as to how likely the plaintiff is to win a lawsuit.

Optimism Model Theorem

If both parties are risk neutral and have equal estimates of the probability that the plaintiff will win, that is, $p_P = p_D$, *then* there is always a positive range of settlement, which is equal to the sum of the litigation costs saved.

If the parties' litigation costs are equal, then the settlement range is equal to twice one party's litigation costs. If parties settle, they save litigation costs, which is the cooperative surplus over whose distribution they can bargain. These conclusions are described formally as follows:

The minimum settlement plaintiff will accept under the American rule is equal to the expected monetary judgment to plaintiff, $E(J_P)$, *minus* litigation costs, C_P:

$$E(J_P) - C_P = \text{minimum settlement plaintiff will accept}$$

The maximum settlement defendant will offer under the American rule is equal to the expected monetary judgment defendant will pay based upon its estimate of the probability plaintiff will win, $E(J_D)$, *plus* litigation costs, C_D:

$$E(J_D) + C_D = \text{maximum settlement defendant will offer}$$

The necessary condition for settlement is that the minimum settlement plaintiff will accept must be less than the maximum settlement defendant will offer; that is, the settlement range must be positive.

$$E(J_P) - C_P < E(J_D) + C_D$$

For example, if $p_P = p_D = 0.7$, J = $100,000, and C = $10,000 for each party, the fundamental theorem of optimism model holds because minimum settlement plaintiff will accept, which is $E(J_P) - C_P$ = [.7($100,000) + .3($0)] – $10,000 = $60,000, is less than maximum settlement defendant will pay, which is $E(J_D) + C_D$ = [.7($100,000)

+ .3($0)] + $10,000 = $80,000; therefore the settlement range is $60,000 – $80,00 = $20,000, or twice litigation costs.

Eight Corollaries

A number of corollaries follow from the fundamental theorem of the optimism model:

Corollary I: If plaintiff's estimate of the probability of winning suit is optimistic relative to defendant's estimate plaintiff will win, $p_P > p_D$, settlement range is reduced or eliminated.

For example, if $p_P = 0.8$, $p_D = 0.5$, J = $100,000, and C = $10,000 for each party, minimum settlement plaintiff will accept is [.8($100,000) + .2($0)] – $10,000 = $70,000; maximum settlement defendant will pay is [.5($100,000) + .5($0)] + $10,000 = $60,000; therefore settlement is not possible.

Corollary II: If plaintiff's estimate of the probability of winning suit is optimistic relative to defendant's estimate plaintiff will win, the larger the amount in controversy, the less likely is a positive settlement range.[3]

However, since money received in future is of less value than money received in present, delay reduces the real value of the amount in controversy and thereby tends to promote settlement. Adding prejudgment interest to eventual judgment tends to offset defendant's advantage from delay.

Corollary III: If either or both parties is risk adverse, settlement range increases; if either or both parties is a risk preferrer, settlement range decreases.

Corollary IV: The existence of a positive settlement range is a necessary but not sufficient condition of settlement. (If Coase Theorem were true, existence of a positive settlement range would be a sufficient condition of settlement.)

Settlement negotiation between plaintiff and defendant is a strategic bargaining situation; negotiations may not be successful if either party demands more than half of the cooperative surplus from settlement and other party refuses to accede to this demand.

Corollary V: The effect of an increase in the settlement range is ambiguous.

As settlement range increases, the cooperative surplus increases which increases possibility of settlement; on the other hand, as settlement range increases, a party who adopts an uncooperative or hard bargaining strategy has more to gain, so an increasing settlement range exacerbates the conflict over the distribution of the cooperative surplus from settling (chicken game) and therefore decreases probability of settlement.

Effect of prejudgment interest on settlement is ambiguous. By increasing the amount in controversy, prejudgment interest tends to reduce the probability of settlement (Corollary II); however, since prejudgment interest also increases settlement range, its effect is ambiguous.

Corollary VI: Pretrial discovery should increase the probability of settlement by causing a convergence of $E(V_P)$ and $E(V_D)$.

As a result of pretrial discovery, parties to a suit will become better informed about the strength of their opponent's case so that estimates of the probability plaintiff will win and monetary judgment should plaintiff win can be expected to converge; that is, the necessary condition for settlement that plaintiff's minimum settlement is less than defendant's maximum settlement is more likely to be true.

However, to the extent that discovery is expensive relative to trial costs, settlement becomes less likely once substantial discovery costs have been incurred. In other words, when a substantial portion of litigation costs are incurred before trial, it may be worth going to trial in order to win the judgment, J, rather than settle for the significantly lower expected judgment, E(J).

The side with more financial resources, the "deep pocket," can often use the discovery process to inflict a high cost on other side in terms of money and time required to respond to discovery demands in order to force a settlement on terms favorable to it. Although judges may deter most blatant examples of this strategy, this is another example where individual self-interested behavior may not increase social welfare.

Corollary VII: "Repeat players," those who are involved in a succession of similar lawsuits, are less likely to settle.

Repeat player is a party who is concerned with precedential or strategic value of a case, as for instance, when a manufacturer is contesting many pending product liability suits or an insurance company is litigating whether an expensive and rare treatment is covered by its health insurance policy. When a defendant in a suit is a repeat player, she is less likely to settle a case because settlement will embolden other plaintiffs and potential plaintiffs.

Corollary VIII: A "nuisance" suit will be settled if defendant has higher litigation costs than plaintiff.

A "nuisance" suit is one that both parties believe will be decided in favor of defendant if it goes to trial; that is, $E(J_P) = E(J_D) = 0$. If the defendant has higher litigation costs than plaintiff (C_P = \$1,000, C_D = \$3,000) because defendant is a real estate developer for whom construction delays are expensive or because defense lawyers who work for large law firms or insurance companies have higher opportunity costs than do plaintiff's lawyers, the settlement range is:

$$E(J_P) - C_P < E(J_D) + C;$$
$$\$0 - \$1{,}000 < \$0 + \$3{,}000$$
$$-\$1{,}000 < \$3{,}000.$$

If parties bargain cooperatively they will share the cooperative surplus ($4,000) equally, $2,000 each, and settle suit for $1,000 which is half way between –$1,000 and +$3,000. If both parties have equal litigation costs, a nuisance suit will fail because plaintiff's minimum settlement is equal to but not less than defendant's maximum settlement; that is, since the cooperative surplus is $0, plaintiff's threat of a trial is not credible because he stands to lose as much as defendant in terms of litigation costs.

CONTINGENT FEE CONTRACTS AND LAWYER-CLIENT CONFLICT

It used to be commonly believed by the public that the contingent fee system was barely ethical; lawyers who took cases on a contingent fee basis were derogatorily called "ambulance chasers." However, economic analysis of contingent fees emphasizes two positive aspects: one, the contingent fee system is an efficient risk distribution system; risk neutral lawyers accept and pool many cases with similar risks from clients who may not be willing to sue on their own because they are risk adverse and cannot evaluate the probability of winning the suit.

Two, the information asymmetry between a knowledgeable lawyer and an uninformed client is at the root of the lawyer-client conflict of interest, which is an example of the more general principal-agent problem (chapter 16). When a client (principal) hires a lawyer (agent) to represent him in a lawsuit or other legal matter and pays the lawyer an hourly rate or a flat fee for her services, the client who generally does not know the law has little or no way of monitoring and evaluating the lawyer's performance. Hourly billing gives lawyers an incentive to devote a great deal of time on a case because the cost of spending extra, inefficient hours on the case is externalized to the client. A flat fee-for-service gives lawyers an incentive to devote little time to each case and to take on many cases because the reduced benefit from spending less than the efficient amount of time on a case is externalized to the client. The contingent fee contract reduces the lawyer-client conflict of interest by making their interests more compatible. (Recall the discussion of moral hazard and incentive compatibility in chapter 16.)

However, in terms of willingness to sue, contingent fee contracts require a higher judgment relative to hourly billing and fee-for-service contracts. Since the lawyer is making the decision whether or not to sue under a contingent fee contract and the lawyer gets, say, 1/3 of the settlement, then, under the American rule for allocation of litigation costs, the necessary condition for suit is:

$$(1/3)E(J_P) > C_P = E(J_P) > 3C_P$$

In other words, in order for a suit to be initiated under a contingent fee contract, the expected judgment from the litigation must be greater than three times litigation costs. To the extent that a lawyer is risk adverse, is unable to estimate the probability of winning a case accurately due to the randomness in the jury system, or is expected to pay for litigation expenses in anticipation of a settlement or judgment, contingent fee contracts will reduce but not eliminate the lawyer-client conflict.

If contingent fee contracts were structured so that lawyers received a fixed share of the proceeds from the suit, they would have a powerful incentive to settle rather than try cases. In order to reduce this aspect of the lawyer-client conflict of interest, contingent fee contracts are often structured so that the lawyer's share is based on a sliding scale which depends on whether the case is settled, goes to trial, or is appealed. A common practice is for the lawyer to take 33 percent of the plaintiff's award if the case is settled without trial; 40 percent if the plaintiff wins at trail; and 50 percent if a judgment for the plaintiff is affirmed on appeal. Given such an incentive structure, the client has less need to monitor and evaluate a lawyer's performance. But, how much incentive should lawyers be given to litigate rather than settle cases in a legal system in which filing fees and court costs which are paid by private parties come nowhere close to covering the full costs of litigation? Although some public subsidy of litigation costs is justified because litigation that establishes and clarifies legal rules provides benefits to society, that is, positive externalities, it is possible to provide too great a subsidy for litigation. For example, if litigation were totally subsidized so that it became a free good, there would be an inefficiently large number of lawsuits. Thus, totally eliminating the lawyer-client conflict of interest with respect to the lawyer's incentive to settle rather than try cases may not be efficient. Moreover, since well over 95 percent of civil suits are settled out of court, would our heavily subsidized legal system be overwhelmed with litigation if a significantly lower percentage of suits were settled?

NOTES

1. Louis Kaplow and Steven Shavell, *Decision Analysis, Game Theory, and Information* (New York: Foundation Press, 2004).
2. Steven Shavell, "Suit, Settlement, and Trial: A Theoretical Analysis under Alternative Methods for Allocation of Legal Costs," *Journal of Legal Studies* 11 (1982): 55–81.
3. Even if, as seems reasonable, litigation costs, C, are a function of the amount in controversy, J, Corollary II is true as long as litigation costs grow at a slower rate than the amount in controversy. Thus, it may not be possible to settle a lawsuit with a plaintiff who is optimistic about winning a very large judgment (no positive settlement range).

Bibliography

Ackerman, Bruce. *Economic Foundations of Property Law*. Boston: Little, Brown, 1975.
Akerlof, George. "The Market for Lemons: Uncertainty and the Market Mechanism." *Quarterly Journal of Economics* 84 (1970): 488–500.
Ashenfelter, Orley, and Michael Greenstone. *National Bureau of Economic Research, Working Paper No. 9094*. Cambridge, MA: NBER, 2002.
Axelrod, Robert. *The Evolution of Cooperation*. New York: Basic Books, 1984.
Baird, Douglas, Robert Gertner, and Randall Picker. *Game Theory and the Law*. Cambridge, MA: Harvard University Press, 1994.
Baumol, William. *Economic Theory and Operations Analysis*, 3rd edition. Englewood Cliffs, NJ: Prentice Hall, 1972.
———. "On Taxation and Control of Externalities." *American Economic Review* 62 (1972): 307–322.
Bernstein, Lisa. "Symposium: Law, Economics, & Norms: Merchant Law in a Merchant Court: Rethinking the Code's Search for Immanent Business Norms." *University of Pennsylvania Law Review* 144 (1996): 1765–1821.
Brown, John P. "Toward an Economic Theory of Liability." *Journal of Legal Studies* 2 (1973): 323–349.
Buchanan, James, Robert Tollison, and Gordon Tullock, eds. *Toward a Theory of the Rent-Seeking Society*. College Station: Texas A&M University Press, 1980.
Burton, Steven. *An Introduction to Legal Reasoning*. New York: Aspen, 1985.
Calabresi, Guido. "First Party, Third Party, and Product Liability Systems: Can Economic Analysis of Law Tell Us Anything about Them." *Iowa Law Review* 60 (1984): 833–851.
———. *The Cost of Accidents: A Legal and Economic Analysis*. New Haven, CT: Yale University Press, 1970.
———. "Transaction Costs, Resource Allocation, and Liability Rules—A Comment." *Journal of Law and Economics* 11 (1969): 67–73.
———. "The Decision for Accidents: An Approach to Non Fault Allocation of Costs." *Harvard Law Review* 78 (1965): 713–745.
———. "Some Thoughts on Risk Distribution and the Law of Torts." *Yale Law Journal* 70 (1961): 499–553.
Calabresi, Guido, and Jon Hirschoff. "Toward a Test for Strict Liability in Torts." *Yale Law Journal* 81 (1972): 1055–1085.
Calabresi, Guido, and Douglas Melamed. "Property Rules, Liability Rules, and Inalienability: One View of the Cathedral." *Harvard Law Review* 85 (1972): 1089–1128.

Caldwell, Bruce J. "Clarifying Popper." *Journal of Economic Literature* 29 (1991): 1–33.
Camerer, Colin. *Behavioral Game Theory: Experiments in Strategic Interaction*. Princeton, NJ: Princeton University Press, 2011.
Camerer, Colin F., and Richard Thaler. "Anomalies: Ultimatums, Dictators and Manners." *Journal of Economic Perspectives* 9 (1995): 210–219.
Cho, In-Koo, and David Kreps. "Signaling Games and Stable Equilibria." *The Quarterly Journal of Economics* 102 (1987): 179–222.
Cirace, John. "When are Law and Economics Isomorphic." *Golden Gate University Law Review* 39 (2009): 183–220.
———. "When Judges Balance Interests Through Trade-Offs, They Implicitly Use Economic Theory to Order Cases." *University of Detroit Mercy Law Review* 74 (1996): 27–62.
———. "A Theory of Negligence and Products Liability." *St. John's Law Review* 66 (1992): 1–71.
Coase, Ronald. "The Problem of Social Cost." *Journal of Law and Economics* 3 (1960): 1–44.
———. "The Nature of the Firm." *Economica, New Series* 4 (1937): 386–405.
Cole, David. "Symposium: Assessing the Rehnquist Court's Parting Words on Criminal Justice: The Liberal Legacy of *Bush v. Gore*." *Georgetown Law Journal* 94 (2006): 1427–1474.
Cooter, Robert. "The Cost of Coase." *Journal of Legal Studies* 11 (1982): 1–33.
Cooter, Robert, and Melvin Eisenberg. "Damages for Breach of Contract." *California Law Review* 73 (1985): 1432–1481.
Cooter, Robert, and Thomas Ulen. *Law and Economics*, 6th edition. New York: Addison-Wesley, 2014.
Dawson, John. "Judicial Revision of Frustrated Contracts: The United States." *Boston University Law Review* 64 (1984): 1–38.
Dewey, Donald. *The Theory of Imperfect Competition: A Radical Reconstruction*. New York: Columbia, 1969.
Dixit, Avinash, and Barry Nalebuff. *Thinking Strategically: The Competitive Edge in Business, Politics, and Every Day Life*. New York: W. W. Norton, 1991.
Dworkin, Ronald. "Is Wealth a Value?" *A Matter of Principle*. Cambridge, MA: Harvard University Press, 1985.
———. *Law's Empire*. Cambridge, MA: Belknap, 1986.
Eisenberg, Melvin. "The Principle of Hadley v. Baxendale." *California Law Review* 80 (1992): 563–613.
Elliott, A. L. *A Study of Liquidated Damages on Highway Contracts*. Sacramento, CA: Division of Highways, 1956.
Epple, Dennis, and Richard Romano. "Competition between Private and Public Schools, Vouchers, and Peer-Group Effects." *American Economic Review* 88 (1998): 33–62.
Esterbrook, Frank. "Ways of Criticizing the Court." *Harvard Law Review* 95 (1982): 802–832.
Farnsworth, E. Alan. *Contracts*, 2nd edition. Boston: Little, Brown, 1990.
———. "Legal Remedies for Breach of Contract." *Columbia Law Review* 70 (1970): 1145–1216.
Farnsworth, E. Allan, and William Young. *Contracts: Cases and Materials*, 5th edition. New York: Foundation Press, 1995.
Fuller, Lon. "The Forms and Limits of Adjudication." *Harvard Law Review* 92 (1978): 353–409.
Gilmore, Grant. *The Death of Contract*. Columbus, OH: The Ohio State University Press, 1974.
Ginsburg, Ruth Bader. "Speaking in a Judicial Voice." *New York University Law Review* 67 (1992): 1185–1209.
———. "Some Thoughts on Autonomy and Equality in Relation to *Roe v. Wade*." *North Carolina Law Review* 63 (1985): 375–386.

Goetz, Charles, and Robert Scott. "Measuring Sellers' Damages: The Lost-Profits Puzzle." *Stanford Law Review* 31 (1979): 323–373.
Goldberg, Victor. "Comment: An Economic Analysis of the Lost-Volume Retail Seller." *Southern California Law Review* 57 (1984): 283–298.
———. "Institutional Change and the Quasi-Invisible Hand." *Journal of Law & Economics* 17 (1974): 461–492.
Goodman, John. "An Economic Theory of the Evolution of the Common Law." *Journal of Legal Studies* 7 (1978): 393–406.
Greer, Douglas F. *Business, Government, and Society*, 3rd edition. New York: Macmillan, 1993.
Hardin, Garrett. "Dr. Pangloss Meets Cassandra." *The New Republic* (October 28, 1981), 31–34.
———. "The Tragedy of the Commons." *Science* 162 (1968): 1243–1248.
Himmelstein, David, and Miraya Jun, et al., "A Comparison of Hospital Administrative Costs in Eight Nations: US Costs Exceed All Others by Far." *Health Affairs* 33 (2014): 1586–1594.
Hirschfelder, Rosemary, and John Hirschfelder. *Introduction to Discrete Mathematics*. Pacific Grove, CA: Brooks/Cole, 1991.
Hirschleifer, Jack. "The Private and Social Value of Information and the Reward to Innovative Activity." *American Economic Review* 61 (1971): 561–574.
Hirschleifer, Jack, and John Riley. *The Analytics of Uncertainty and Information*. Cambridge: Cambridge University Press, 2003.
Holmes, Jr., Oliver Wendell. *The Common Law*. Cambridge, MA: Harvard University Press, 1963.
Hotelling, Harold. "Stability in Competition." *The Economic Journal* 39 (1929): 41–57.
Hungerford, Thomas, and Gary Solon. "Sheepskin Effects in the Returns to Education." *Review of Economics and Statistics* 69 (1987): 175–177.
Kaplow, Louis. "Rules versus Standards: An Economic Analysis." *Duke Law Journal* 42 (1992): 557–629.
Kaplow, Louis, and Steven Shavell. *Decision Analysis, Game Theory, and Information*. New York: Foundation Press, 2004.
Keynes, John M. *The General Theory of Employment, Interest, and Money*. New York: Palgrave Macmillan, 2007.
———. *A Tract on Monetary Reform*. Cambridge: Cambridge University Press, 1923.
Klein, Benjamin. "The Selection of Disputes for Litigation." *Journal of Legal Studies* 13 (1984): 1–54.
Kronman, Anthony. "Mistake, Disclosure, Information and the Law of Contracts." *Journal of Legal Studies* 7 (1978): 1–34.
Kuper, Simon, and Stefan Szymanski. *Soccernomics*. New York: HarperCollins, 2016.
LaFave, Wayne, and Jerold Israel. *Criminal Procedure*, 2nd edition. St. Paul, MN: Thompson West, 1992.
Laibson, David, and John List. "Principles of (Behavioral) Economics." *American Economic Review: Papers and Proceedings* 105 (2015): 385–390.
Landes, Elizabeth, and Richard Posner. "The Economics of the Baby Shortage." *Journal of Legal Studies* 7 (1978): 323–348.
Leff, Arthur. "Economics Analysis of Law: Some Realism about Nominalism." *Virginia Law Review* 60 (1974): 451–482.
Levi, Edward. *An Introduction to Legal Reasoning*. Chicago: The University of Chicago Press, 1948.
Lewis, Michael. *Moneyball*. New York: W. W. Norton, 2013.
Llavador, Humberto, John E. Roemer, and Joaquim Silvestre. *Sustainability for a Warming Planet*. Cambridge, MA: Harvard University Press, 2015.

Llewellyn, K. N. *The Bramble Bush*. New York: Oceana Publications, 1960.
Mailath, George. "Do People Play Nash Equilibrium? Lessons from Evolutionary Game Theory." *Journal of Economic Literature* 36 (1998): 1347–1374.
Mallor, Jane, A. James Barnes, Thomas Bowers, Michael Phillips, and Arlen Langvardt. *Business Law and the Regulatory Environment: Concepts and Cases*, 10th edition. Boston: Irwin/McGraw-Hill, 1998.
Malthus, Thomas. *An Essay on the Principle of Population as It Affects the Future Improvement of Society*. Cambridge: Cambridge University Press, 2016.
Mandelbrot, Benoit, and Richard Hudson. *The Misbehavior of Markets: A Fractal View of Risk, Ruin, and Reward*. New York: Basic Books, 2006.
Mansfield, Edwin. *Microeconomics: Theory and Applications*, 6th edition. New York: W. W. Norton, 1988.
Marshall, Alfred. *Principles of Economics*, 8th edition. New York: Palgrave Macmillan, 1920.
Miceli, Thomas. *The Economic Approach to Law*, 2nd edition. Palo Alto, CA: Stanford University, 2009.
Mill, John S. *On Liberty*. Harmondsworth, UK: Penguin Books, 1974 (1859).
Mueller, Dennis. *Public Choice II*, revised edition. Cambridge: Cambridge University Press, 1989.
Nader, Ralph. *Unsafe at Any Speed: The Designed-in Dangers of the American Automobile*. New York: Pocket, 1966.
Nassim, Taleb. *The Black Swan*. New York: Random House, 2007.
National Association of Attorneys General. *Horizontal Merger Guidelines of the National Association of Attorneys General*. Washington, DC: Bureau of National Affairs, 1987.
Nordhaus, William D. *A Question of Balance: Weighing the Options on Global Warming Policies*. New Haven, CT: Yale University Press, 2008.
Note. "Automobile Design Liability: Larsen v. General Motors and its Aftermath." *University of Pennsylvania Law Re*view 118 (1969): 299–312.
Nowak, John, and Ronald Rotunda. *Constitutional Law*, 4th edition. St. Paul, MN: Thomson West, 1991.
Nowak, Martin, with Roger Highfield. *Super Cooperators: Altruism, Evolution, and Why We Need Each Other to Succeed*. New York: Free Press, 2011.
Nozick, Robert. *The Nature of Rationality*. Princeton, NJ: Princeton University Press, 1993.
Olson, Mancur. *The Logic of Collective Action: Public Goods and the Theory of Groups*. Cambridge, MA: Harvard Economic Studies, 1971.
Petersen, H. Craig. *Business and Government*, 4th edition. New York: HarperCollins, 1993.
Piketty, Thomas. *Capital in the Twenty-First Century*. Cambridge, MA: Belknap Press, 2014.
Polinsky, A. Mitchell. *An Introduction to Law and Economics*, 4th edition. New York: Wolters Kluwer, 2011.
Posner, Richard. *Economic Analysis of Law*, 9th edition. New York: Wolters Kluwer, 2014.
———. *Not A Suicide Pact: The Constitution in a Time of National Emergency*. New York: Oxford, 2006.
Posner, Richard, and Andrew Rosenfeld. "Impossibility and Related Doctrines in Contract Law: An Economic Analysis." *Journal of Legal Studies* 6 (1977): 83–118.
Priest, George. "The Current Insurance Crisis and Modern Tort Law." *Yale Law Journal* 96 (1987): 1521–1590.
———. "The Common Law Process and the Selection of Efficient Rules." *Journal of Legal Studies* 6 (1977): 65–82.
Rabin, Robert. "Indeterminate Risk and Tort Reform." *Journal of Legal Studies* 14 (1985): 633–643.

Rasmussen, Eric. *Games and Information: An Introduction to Game Theory*, 4th edition. Cambridge, MA: Blackwell, 2006.
Rawls, John. *A Theory of Justice*. Cambridge, MA: Belknap Press, 1971.
Rottenberg, Simon. "The Baseball Players' Labor Market." *Journal of Political Economy* 64 (1956): 242–258.
Rubin, Paul. "Why Is the Common Law Efficient?" *Journal of Legal Studies* 6 (1977): 51–63.
Rudman, Warren. U.S. Senator. "60 Minutes," *CBS Television Network*, program broadcast Sunday, December 6, 1993, 7–8 PM.
Schelling, Thomas. *The Strategy of Conflict*. Cambridge, MA: Harvard University Press, 1980.
Schumpeter, Joseph. *Capitalism, Socialism, and Democracy*. New York: Harper, 1942.
Shavell, Steven. "Acquisition and Disclosure of Information Prior to Sale." *Rand Journal of Economics* 25 (1994): 20–36.
———. "Torts in Which Victim and Injurer Act Sequentially." *Journal of Law and Economics* 26 (1986): 589–612.
———. "Suit, Settlement, and Trial: A Theoretical Analysis under Alternative Methods for Allocation of Legal Costs." *Journal of Legal Studies* 11 (1982): 55–81.
Smith, Adam. *An Inquiry into the Nature and Causes of the Wealth of Nations*. Modern Library edition. New York: Random House, 1937.
Smith, Jessica, and Carla Medalla. *Health Insurance Coverage in the United States* 2013. Washington, DC: US Census Bureau, 2014.
Solow, Robert M. "Intergenerational Equity and Exhaustible Resources." *Review of Economic Studies* 41 (1974): 29–45.
Spence, A. Michael. "Job Market Signaling." *Quarterly Journal of Economics* 87 (1973): 355–374.
Stiglitz, Joseph. "The Private Uses of Public Interests: Incentives and Institutions." *Journal of Economic Perspectives* 12 (1998): 3–22.
———. *Economics*. New York: W. W. Norton, 1997.
———. "Incentives and Risk Sharing in Sharecropping." *Review of Economic Studies* 41 (1974): 219–255.
Sunstein, Cass R. *One Case at a Time: Judicial Minimalism in the Supreme Court*. Cambridge, MA: Harvard University Press, 2001.
Tussman, Joseph, and Jacobus tenBroek. "The Equal Protection of the Laws." *California Law Review* 37 (1949): 341–381.
US Department of Transportation. *Guidance on Treatment of the Economic Value of a Statistical Life in U.S. Department of Transportation Analyses*. Washington, DC: Department of Transportation, 2015.
Van Hecke, M. T. "Changing Emphases in Specific Performance." *North Carolina Law Review* 40 (1961): 1–22.
Varian, Hal. *Intermediate Microeconomics: A Modern Approach*, 7th edition. New York: W. W. Norton, 2006.
Vickrey, William. "Automobile Accidents, Tort Law, Externalities, and Insurance: An Economist's Critique." *Law & Contemporary Problems* 33 (1968): 464–487.
Viscusi, W. Kip. *Risk by Choice: Regulating Health and Safety in the Workplace*. Cambridge, MA: Harvard University Press, 1983.
Viscusi, W. Kip, John Vernon, and Joseph Harrington Jr. *Economics of Regulation and Antitrust*, 3rd edition. Cambridge, MA: MIT Press, 2000.
Waldman, Don, and Elizabeth Jensen. *Industrial Organization: Theory and Practice*. New York: Routledge, 1997.
World Bank Data. *Health Expenditure, Total (% of GDP)*. Washington, DC: World Bank Group, 2015.

CASES

Alaska Packers Ass'n v. Domineco, 117 F. 99 (9th Cir. 1902).
Aluminum Co. of Amer. v. Essex Group, Inc., 499 F.Supp. 53 (W.D.Pa.1980).
AT&T Mobility LLC v. Concepcion, 563 U.S. 333 (2011).
Baxter v. Ford Motor Co., 168 Wash. 456 (1932).
Beshada v. Johns-Manville Products Corp., 90 N.J. 191 (1982).
Blyth v. Proprietors of Birmingham Waterworks Co., 156 Eng. Rep. 1047 (1856).
Boomer v. Atlantic Cement Co., 26 N.Y.2d 219 (1970).
Bowers v. Hardwick, 478 U.S. 186 (1986).
Brown v. Board of Education, 347 U.S. 483 (1954).
Bryant v. Lefever, (1879) 4 CPD 172.
Butterfield v. Forrester, 103 Eng. Rep. 926 (K.B. 1809).
California v. ARC America Corp., 490 U.S. 93 (1989).
Carnival Cruise Lines, Inc. v. Shute, 499 U.S. 585 (1991).
Carpenter v. Double R Cattle Co., 105 Idaho 320 (1983), rev'd. 108 Idaho 602 (1985).
Copenhaver v. Berryman, 602, S.W.2d 540 (Tex. App. 1980).
Davies v. Mann, 152 Eng. Rep. 588 (Ex. 1842).
Davis v. Consolidated Rail Corp., 788 F.2d 1260 (7th Cir. 1986).
Deitsch v. Music Company, 6 Ohio Misc.2d 6 (1983).
Devlin v. Smith, 89 N.Y. 470 (1882).
Dreisonstock v. Volkswagenwerk, A.G., 489 F.2d 1066 (4th Cir. 1974).
Dyson v. General Motors Corp., 298 F.Supp. 1064 (1969).
Ellithorpe v. Ford Motor Co., 503 S.W.2d 516 (Tenn. 1973).
Evans v. General Motors Corp., 359 F.2d 822 (7th Cir.), cert. denied, 385 U.S. 836 (1966).
Fisher v. Johnson Milk Co., 174 N.W.2d 752 (Mich. 1970).
Frericks v. General Motors Corp., 317 A.2d 494 (Md. Ct. Spec. App. 1974).
Frigaliment Importing Co. v. B.N.S. International Sales Corp., 190 F.Supp. 116 (S.D.N.Y. 1960).
FTC v. Staples, Inc., 970 F.Supp. 1066 (DDC 1997).
Greenman v. Yuba Power Products, 59 Cal.2d 57 (1963).
Griswold v. Connecticut, 381 U.S. 479 (1965).
Hadley v. Baxendale, 9 Ex. 341; 156 Eng. Rep. 145 (1854).
Henningsen v. Bloomfield Motors, Inc., 32 N.J. 358 (1960).
Hirabayashi v. United States, 320 U.S. 81 (1943).
Illinois Brick Co. v. Illinois, 431 U.S. 720 (1977).
Islamic Republic of Iran v. Boeing Co., 771 F.2d. 1279 (9th Cir.1985).
Jackson v. Leach, 152 A. 813 (Md. 1931).
Jacob & Youngs v. Kent, 230 N.Y. 239 (1921).
Korematsu v. United States, 323 U.S. 214 (1944).
Laidlaw v. Organ, 15 U.S. (2 Wheat.) 178 (1815).
Lake River Corp. v. Carborundrun Co., 769 F.2d 1284 (7th Cir.1985).
Larsen v. General Motors Corp., 391 F.2d 495 (8th Cir. 1968).
Lawrence v. Texas, 539 U.S. 558 (2003).
Leitch Gold Mines, Ltd. V. Texas Gulf Sulphur, 1 Ontario Reports 469, 492–493 (1969).
LeRoy Fibre Co. v. Chicago, Milwaukee & St. Paul Ry., 232 U.S. 340 (1914).
Levi v. Southwest Louisiana Electric Membership Coop., 542 So.2d 1081 (La. 1989).
Lochner v. New York, 198 U.S. 45.

Malewicz v. City of Amsterdam, 362 F.Supp.2d 298 (2005).
Mapp v. Ohio, 367 U.S. 643 (1961).
Mazetti v. Armour Co., 75 Wash. 622 (1913).
Menzel v. List, 24 N.Y.2d 91 (1969).
Metropolitan Life Insurance Co. v. RJR Nabisco, Inc., 906 F.2d 884 (1990).
Mickle v. Blackmon, 252 S.C. 202 (1969).
Mitchill v. Lath, 247 N.Y. 377 (1928).
Nada v. Ford Motor Co., 509 F.2d 213 (7th Cir. 1974).
Obde v. Schlemeyer, 56 Wash.2d. 672 (1960).
O'Brien v. Muskin Corp., 94 N.J. 169 (1983).
Peevyhouse v. Garland Coal & Mining Company, 382 P.2d 109 (Okla. 1962).
Phillips v. Kimwood Machine Co., 269 Or. 48 (1974).
Pittsley v. Houser, 875 P.2d 232 (Idaho Ct. App. 1994).
Planned Parenthood v. Casey, 505 U.S. 833 (1992).
Plessy v. Ferguson, 163 U.S. 537 (1896).
Posey v. Clark Equip. Co., 409 F.2d 560 (7th Cir. 1969).
Powell v. State, 510 S.E.2d 18 (Ga. 1998).
Prah v. Maretti, 108 Wis.2d 223 (1982).
Prentis v. Yale Mfg. Co., 421 Mich. 670 (1984).
R.E. Davis Chemical Corp. v. Diasonics, Inc., 826 F.2d 678 (1987); 924 F.2d 710 (1991).
Roe v. Wade, 410 U.S., 113 (1973).
Saint Barnabas Hospital Center v. County of Essex, 111 N.J. 916 (1988).
Schneider v. Chrysler Motor Corp., 401 F.2d 549 (8th Cir. 1968).
Security Stove & Mfg. Co. v. American Ry. Express, 227 Mo.App. 175 (1932).
Shrum v. Zeltwanger, 559 P.2d 1384 (Wyo. 1977).
Sony Corporation of America v. Universal City Studios, Inc., 464 U.S. 417 (1984).
Spur Industries, Inc. v. Del Webb Development Co., 108 Ariz. 178 (1972).
Standard Oil Co. v. U.S., 221 U.S. 1 (1911).
Statler v. Ray, 195 N.Y. 478 (1909).
Sturges v. Bridgman, (1879) 11 C.D. 852.
Tennessee Trailways, Inc. v. Ervin, 438 S.W. 733 (1969).
Thomas v. Winchester, 6 N.Y. 397 (1852).
Transatlantic Financing Corp. v. U.S., 363 F.2d 312 (D.C. Cir.1966).
Turner v. General Motors Corp., 514 S.W.2d 497 (Tex. 1974); 584 S.W.2d 844 (Tex. 1979).
U.S. v. Aluminum Co. of America, 148 F.2d 416 (1945).
U.S. v. DuPont, 351 U.S. 377 (1956).
U.S. v. International Business Machines Corp. (IBM), Civil Action No. 69 Civ. 200 (S.D.N.Y.).
U.S. v. Leon, 468 U.S. 897 (1984).
Utah v. Strieff, 579 U. S. ____ (2016).
Vincent v. Lake Erie Transp. Co., 109 Minn. 456 (1910).
Weeks v. United States, 232 U.S. 383 (1914).
Weichert Co. Realtors v. Ryan, 128 N.J. 427 (1992).
Williams v. Walker-Thomas Furniture Co., 350 F.2d 445 (D.C. Cir.1965).
Willis v. Chrysler Corp., 264 F.Supp. 1010 (D.C. Tex. 196).
Windows, Inc. v. Jordan Panel Systems Corp., 177 F.3d. 114 (2nd Cir. 1999).
Wood v. Lucy, Lady Duff-Gordon, 222 N.Y. 88 (1917).
World-Wide Volkswagen Corp. v. Woodson, 444 U.S. 286 (1980).

US CONSTITUTION, STATUTES, AND RESTATEMENT OF LAW

Commodity Futures Modernization Act of 2000.
Family and Medical Leave Act of 1993, P.L. 103–03 (February 5, 1993).
Genetic Information Nondiscrimination Act (GINA) of 2008.
National Traffic and Motor Vehicle Safety Act of 1966.
N.Y. General Obligations Law §§ 5–1103–1109 (1936).
Restatement (Second) of Contracts. St. Paul, MN: American Law Institute, 1981.
Restatement (Third) of Torts: Products Liability. St. Paul, MN: American Law Institute, 1998.
Sherman Act, 15 U.S.C.A § 1, § 2.
Uniform Commercial Code, Article 2 (Sales).
US Constitution, Amendment XIV.

Index

About the Author

John Cirace is professor emeritus of economics at Lehman College, The City University of New York, and, for twenty-five years, was adjunct professor of law at Brooklyn Law School where he taught law and economics. His published articles are in the fields of industrial organization and regulation and law and economics.

Lightning Source UK Ltd.
Milton Keynes UK
UKHW050411200720
366568UK00024B/464